The Closed World

Inside Technology
edited by Wiebe E. Bijker, W. Bernard Carlson, and Trevor Pinch

Wiebe E. Bijker, *Of Bicycles, Bakelites, and Bulbs: Toward a Theory of Sociotechnical Change*

Wiebe E. Bijker and John Law, editors, *Shaping Technology/Building Society: Studies in Sociotechnical Change*

Stuart S. Blume, *Insight and Industry: On the Dynamics of Technological Change in Medicine*

Louis L. Bucciarelli, *Designing Engineers*

Geoffrey C. Bowker, *Science on the Run: Information Management and Industrial Geophysics at Schlumberger, 1920–1940*

H. M. Collins, *Artificial Experts: Social Knowledge and Intelligent Machines*

Paul N. Edwards, *The Closed World: Computers and the Politics of Discourse in Cold War America*

Pamela E. Mack, *Viewing the Earth: The Social Construction of the Landsat Satellite System*

Donald MacKenzie, *Inventing Accuracy: A Historical Sociology of Nuclear Missile Guidance*

Donald MacKenzie, *Knowing Machines: Essays on Technical Change*

The Closed World
Computers and the Politics of Discourse in Cold War America

Paul N. Edwards

The MIT Press
Cambridge, Massachusetts
London, England

©1996 Massachusetts Institute of Technology

All rights reserved. No part of this book may be reproduced in any form or by any electronic or mechanical means (including photocopying, recording, or information storage and retrieval) without permission in writing from the publisher.

This book was set in Baskerville by Pine Tree Composition, Inc. and printed and bound in the United States of America.

Library of Congress Cataloging-in-Publication Data

Edwards, Paul N.
  The closed world : computers and the politics of discourse in Cold
War America / Paul N. Edwards
    p. cm.—(Inside technology)
  Includes bibliographical references and index.
  ISBN 0-262-05051-X (alk. paper)
  1. Computers—History. 2. Military art and science—Data
processing—History. I. Title. II. Series.
QA76. 17.E34 1996
306.2—dc20   95-24674
                                                             CIP

For Gabrielle

Contents

Preface ix
Acknowledgments xvii

1
"We Defend Every Place": Building the Cold War World 1

2
Why Build Computers?: The Military Role in Computer Research 43

3
SAGE 75

4
From Operations Research to the Electronic Battlefield 113

5
Interlude: Metaphor and the Politics of Subjectivity 147

6
The Machine in the Middle: Cybernetic Psychology and World War II 175

7
Noise, Communication, and Cognition 209

8
Constructing Artificial Intelligence   239

9
Computers and Politics in Cold War II   275

10
Minds, Machines, and Subjectivity in the Closed World   303

Epilogue: Cyborgs in the World Wide Web   353

Notes   367
Index   429

Preface

The primary weapons of the Cold War were ideologies, alliances, advisors, foreign aid, national prestige—and above and behind them all, the juggernaut of high technology. Nuclear warheads, intercontinental jet bombers, ICBMs, and satellites formed the war's strategic backbone. Even this close to the Cold War's end, we can begin to see how the existence of these technologies shaped patterns of thought in both military and civilian life. The strange but compelling logic of nuclear deterrence required ever-expanding, ever-improving, ruinously expensive arsenals of high-technology weapons; these weapons played a double role as arms and as symbols of power, prowess, and prestige. Without their intercontinental reach and prodigious destructive force, the Cold War could never have become a truly global contest. Without their power as symbols—both of apocalypse and of the success or failure of entire social orders—it could never have become so truly total, such a titanic battle for hearts and minds as well as hands.

Of all the technologies built to fight the Cold War, digital computers have become its most ubiquitous, and perhaps its most important, legacy. Yet few have realized the degree to which computers created the technological possibility of Cold War and shaped its political atmosphere, and virtually no one has recognized how profoundly the Cold War shaped computer technology. Its politics became embedded in the machines—even, at times, in their technical design—while the machines helped make possible its politics. This book argues that we can make sense of the history of computers as tools only when we simultaneously grasp their history as metaphors in Cold War science, politics, and culture.

Historiography is always guided by specific tropes, genres, and plot structures. These are what enable any narrative to achieve its sense of coherence—its sense of "being a story" and therefore of embodying a possible truth. The favored tropes and plots of any given history reflect the perspective of the historian and influence how material is deemed relevant.[1]

This book is built around a history of computers as a central technology of Cold War military forces, on the one hand, and as an axial metaphor in psychological theory, on the other. It thus responds to a large existing literature on computers and cognitive science, one that is dominated by two major genres of historiography. As Michael Mahoney has observed, these genres correspond to the two major historical sources of digital computer technology: machine logic, on the one hand, and machine calculation, on the other.[2]

The first genre is an intellectual history in which computers function primarily as the embodiment of ideas about information, symbols, and logic. This is the version of history that is usually told by computer scientists,[3] cognitive scientists, and historians concerned with the evolution of concepts. It focuses on the logical power and psychological insights made possible by the computer. According to the canonical story, recounted in most textbooks as well as in practically every major study of cognitive psychology and artificial intelligence, computer technology opened the intellectual door to the solution of certain essentially philosophical problems about knowledge, perception, and communication first posed in a primitive fashion by the ancient Greeks. The invention of programmable machines capable of complex symbolic processing made it possible to pose such questions in ways more tractable to the theoretical and experimental methods of scientific inquiry.

The standard lineage here runs from Plato's investigations of the foundations of knowledge and belief, through Leibniz's rationalism, to Lady Lovelace's notes on Charles Babbage's Analytical Engine and Boole's *Laws of Thought* in the nineteenth century. In the twentieth century, Alan Turing's invention of the digital logic machine in 1936 (and later his test for machine intelligence), Norbert Wiener's cybernetic theory, the McCulloch-Pitts theory of neuron nets as Turing machines, and John von Neumann's comparisons of the computer and the brain are taken as founding moments. This tale reaches its climax in the 1956 consolidation of artificial intelligence research by John McCarthy, Marvin Minsky, Allen Newell, and Herbert Simon and the movement of information theory into psychology through the work of George Miller, J. C. R. Licklider, and others in the late 1950s. In the late 1980s, the rise of parallel distributed processing and neural networks marks ever-closer convergence in the study of human and machine intelligence. Most of the characters in this version of the story are philosophers, scientists, or mathematicians concerned with logic. Very late in the game, some are psychologists using concepts of information processing to extend or overturn behaviorist models in favor of new cognitive theories. All are intellectuals and theorists.[4]

Some pose this story as one in which the confusion of philosophy is gradually replaced by the precision and clarity of science. Others tell it as one of convergence and complementarity between scientific progress and philosophical ideas, with computers finally opening the way to experimental modeling of otherwise untestable theories. But every story in this genre makes the history of computers and cognitive science primarily a history of ideas. The genre's major tropes are progress—knowledge expanding through a steady stream of new discoveries—and revolution. The "computer revolution" of the 1950s linked calculation to communication, control, and simulation through new theories of information and symbolic programming. The "cognitive revolution" of the late 1950s carried these ideas to a wide range of previously divergent disciplines, such as psychology, linguistics, and neuroscience, bringing them new direction, linkage, and coherence.[5]

The second genre is an engineering/economic history focusing on computers as devices for processing information. This version, naturally enough, is the one told by computer engineers, business historians, and others with a primary interest in the technology and its social impacts. It focuses on the insights and technical capabilities needed to create and develop the computer as a digital calculating machine. This version typically begins (after bows to the abacus and the early calculating machines of Pascal, Leibniz, and others) in the 1820s with Charles Babbage's Difference Engine and his later, more sophisticated Analytical Engine.[6] It then runs from Herman Hollerith, the 1890s inventor of punched-card tabulation, to Vannevar Bush and his 1930s analog differential analyzers. George Stibitz, John Atanasoff, and Howard Aiken and their World War II–era electromechanical calculators are next, followed by Presper Eckert, John Mauchly, Turing, and von Neumann and their stored-program electronic computers, completed just after the war's end. The story moves on to UNIVAC, IBM, and the many other corporate developers of commercial computer equipment, as computers spread from military and university installations through factories and offices and into the home. In the 1980s, personal computers and computer networks cap the triumphant tale of a rising information economy, the vigorous backbone of an equally potent information society.

While some of the characters are the same, this story is mainly about engineers, their inventions, and the companies that developed and sold them. Usually these histories focus on technical and economic facts about computers: how they came to be invented and commercialized, the jobs they automated or transformed, and the technological achievements that build on their abilities. The computer's effects, in this story, are practical. The computer is a business machine, a tool of science, and a complex, ingeniously designed artifact. Information is a commodity rather than a concept. Calculation, control, and communication are significant because they allow the economic exploitation of information. Like its intellectual-history counterpart, the major tropes of this engineering/economic history are progress (of technology rather than ideas) and revolution. The inventors and engineers behind the "computer revolution," here driven by inevitable market forces, are cast as towering visionaries whose technological achievements have led or will lead to major social change.[7]

These parallel stories exist for a reason. Computers display, Janus-like, a double aspect. They consist simultaneously of hardware, whose heritage lies within the history of technology, and software, whose ancestry lies in mathematics and formal logic. With few exceptions, these realms remained largely separate until World War II, when the machine calculation (hardware) and machine logic (software) traditions converged in the stored-program electronic digital computer. This partly explains the divided character of computer historiography, which to date has consisted largely of accounts written by authors descended from one of these two traditions.

What is important here is that the tropes and plot lines of both genres impose requirements that lead authors to ignore or downplay phenomena outside the laboratory and the mind of the scientist or engineer. Both versions of the story explain developments in a given field solely from the perspective of actors within it. As Mahoney puts it, the authors of this "insider history... take as givens... what a more critical, outside viewer might see as choices."[8] There is little place in such accounts for the influence of ideologies, intersections with popular culture, or political power. Stories based on the tropes of progress and revolution are often incompatible with these more contingent forms of history.

This book is built on an implicit critique of existing computer historiography. Instead of progress and revolution, the plot structure I shall use emphasizes contingency and multiple determination. I shall cast technological change as technological choice, tying it to political choices and socially constituted values at every level, rendering technology as a product of complex interactions among scientists and engineers, funding agencies, government policies, ideologies, and cultural frames.

The Closed World emphasizes linkages and problems ignored or downplayed by most of the extant texts. My ultimate goal is to prove the utility of a more integrated historical approach by showing how ideas and devices are linked through politics and culture. To this end, I follow both concepts and technologies through many layers of relationships: of individual scientists and engineers with military technical problems; of computer development projects with military agencies and their problems; of large-scale political trends with the direction and character of research; of computer metaphors with scientific research programs; and of cultural productions with scientific and technological changes.

As for my own tropes, I will employ the concepts of *metaphor* and *discourse*. This latter term, not an entity but an analytical construct, refers to an ensemble of heterogeneous elements loosely linked around material "supports," in this case the computer. Discourses, in my usage, include techniques, technologies, metaphors, and experiences as well as language. *Closed-world discourse* articulated geopolitical strategies and metaphors (such as "containing" Communism) in and through military systems for centralized command and control. *Cyborg discourse* articulated metaphors of minds as computers in and through integrated human-machine systems and technologies of artificial intelligence. Though scientists, engineers, and politicians tried to hold them separate, Cold War popular culture grasped the intimate connection between the closed world and the cyborg. Closed-world drama, in film and fiction, repeatedly dramatized them together, articulating the simultaneous construction of the material realm of technology, the abstract realm of strategy and theory, and the subjective realm of experience.

One criticism that may be leveled against my account is that I understate or ignore the roles of the computer industry and of economic incentives in the development of computer technology. Thus, some will conclude, I inevitably overstate the influence of military agencies and their priorities. In reply, I would emphasize two points. First, during the period I discuss, the role of industry was often closely coordinated with military plans.[9] Second, I am writing against the backdrop of a field almost entirely dominated by engineering/economic studies. This book attempts a kind of "counterhistory," a corrective to perspectives that create the impression of an inevitable progress driven by impersonal market forces and technical logics. In rendering technological change as a matter of politically significant choices, and technological metaphor as a fundamental element of culture and politics, I aim to set the history of computers on a new course.

Some will also object that to treat Cold War politics as an American theater is to ignore its leading actor, namely the Soviet Union. On this view, Soviet actions and policies tightly constrained the grand strategy I will describe. The decisions taken by U.S. leaders were often the only politically possible responses to the Soviet military buildup and its declared hostility to Western powers. The closure of Soviet society forced American decisionmakers to act on the basis of small amounts of very uncertain information. Lacking firm intelligence and facing awesome weapons against which no real defense was possible, prudence dictated only one possible course: a massive military buildup on every conceivable axis.[10] Therefore, picturing the Cold War as a drama constructed by and within the United States reverses the causal arrow, granting American leaders a wider range of choices than they actually had.

I believe that this view apologizes too easily for the vigorous American role in history's most expensive and dangerous arms race. My own concept of that role will become clear enough in what follows. Yet even the most benign possible view of American goals and tactics in the Cold War is compatible with the argument I make in this book. It was precisely the political closure of Soviet society that allowed it to function in American political discourse as an enigmatic and terrifying Other. This Other led, incontestably, an ideological as well as a politico-military life, and it was in part this double nature that produced the strange political culture of the closed world. American leaders probably did have more choices than they realized. But in my view, the *reasons* they did not realize this had everything to do with the discourse they applied to the interpretation of their situation. My picture of politics as discourse in no way denies the reality of events or the important influence of Soviet actions. Instead, it frames them in a way that allows their strategic dimensions to be grasped simultaneously as technological forms and subjective spaces—as design imperatives and prescriptions for political identities as well as descriptions of objective situations.

This book views three apparently disparate histories—the history of American global power, the history of computing machines, and the history of mind and subjectivity as reflected in science and culture—through the lens of the American political imagination. The links, I will argue, are partial and sometimes contradictory, and for this reason my presentation is kaleidoscopic, often more collage than linear narrative. But the links are real, and without some attempt to trace their patterns any effort to understand the place of computers in American culture is incomplete. Not only as tools but also as models and metaphors, computers connect cognitive psychology and artificial intelligence to high-technology warfare and to the institutional structure of the modern state.

A very different kind of history lies hidden in the interstices of the received intellectual and engineering/economic histories. It is a story neither of ideas alone nor of machines and their effects, but of ideas, experiences, and metaphors in their interaction with machines and material change. At its heart are political trends and cultural transitions ignored by the canonical tale. By exploring not *the* history of computers and computer metaphors, but *a* history, constructed as a counterpoint to those the field itself has generated, this book attempts to shift the focus of historical inquiry from the power of computer technology and the truth of cognitive science to their meaning for political and cultural identity.

In writing *The Closed World* I have tried to consider the needs of a variety of audiences. Historians of technology and science will find a revisionist account of the coevolution of computers and theories of mind. Political and cultural historians will find an extended exploration of the interplay between technology, politics, and culture in the Cold War. People interested in cultural studies and science and technology studies will, I hope, find the concepts of discourse, political identity, subjectivity, and closed-world vs. green-world drama useful as analytical tools. Computer professionals will find a different perspective on the larger history surrounding computer science and engineering, as well as detailed accounts of major computer projects of the 1950s. Finally, for the growing numbers of scientifically and technologically literate generalists, the book tenders an extended exploration of the subtle and intricate relations of science and engineering with the evolution of modern society.

To speak plainly and intelligibly to such diverse groups, I have attempted wherever possible to avoid technical language or analytical jargon. Where necessary, I have provided concise definitions of significant terms, but technical details are rarely crucial to my argument. Readers completely unfamiliar with the principles of computer technology may, however, wish to keep a book like Joseph Weizenbaum's *Computer Power and Human Reason* by their side for ready reference. For readers interested in a more extended historical survey of the intellectual problems and conceptual frameworks of cognitive psychology, I suggest Howard Gardner's *The Mind's New Science*. Notes throughout the text provide pointers to the existing historical literature, but readers entirely unacquainted with work in this field might wish to peruse Stan Augarten's highly readable and comprehensive (though sometimes inaccurate) *Bit by Bit: An Illustrated History of Computers* or Kenneth Flamm's definitive two-volume economic history, *Targeting the Computer* and *Creating the Computer*.

Finally, I want to mention two terminological points. I will be discussing several kinds of computers, including mechanical calculators and electromechanical analog computers. My convention throughout will be to use the term "computer" in its ordinary modern sense as referring to the electronic stored-program digital computer.[11] When speaking of manual, analog, and pre-electronic computing technologies, I will specify them as such.

In general I have avoided the use of gendered pronouns, except in direct quotations from other authors and when discussing groups, such as Air Force pilots, historically made up exclusively or overwhelmingly of men. References to "him" or "his" should thus be taken literally. Although gender ideology is only a tertiary theme in this book, I have tried in this way to gesture toward historical links (which I have discussed elsewhere) between the traditionally extreme masculinity of the military world, the political culture of democracies, and the rationalistic masculinity of many computer cultures.[12]

Acknowledgments

"We Bokononists," writes the narrator of Kurt Vonnegut's novel *Cat's Cradle*, "believe that humanity is organized into teams, teams that do God's Will without ever discovering what they are doing. Such a team is called a *karass*." Each person's *karass* consists of all the others who, in one way or another, have touched that person's life in significant ways. One's *karass* is one's life team, the group of people through whom the meanings of one's life unfold.

Across the ten years since I first conceived this book, I have accumulated debts so vast they can barely be recognized, let alone repaid. Many of those who brought me ideas, information, and personal renewal were people I scarcely knew; many of those who mattered most will probably never read these pages. Nevertheless, it is with great delight that I attempt here to thank the friends, colleagues, and acquaintances who gave some part of their minds, their hearts, and their lives to the making of this book—the members of my *karass*.

The Closed World began its life as a doctoral dissertation in the Board of Studies in History of Consciousness of the University of California at Santa Cruz, under the incomparable Donna Haraway. The many references to her work scattered across the pages of this book merely hint at how much she shaped its contours. As a scholar, colleague, and friend, Donna contributed not only her vast knowledge and insight but her example of boldness and integrity in thought, work, and life. Robert Meister, another superlative teacher, taught me how to understand the connections between subjectivity and political systems. His political theory and his commentaries on my work lie behind my words here, on nearly every page. To Donna and Bob, writers and thinkers *sui generis*, this book owes much of whatever uniqueness it may possess.

Richard Gordon provided crucial guidance and material support, as well as an intellectual home, through his Silicon Valley Research Group, where I carried out much of the original dissertation research. At Stanford University, Terry Winograd—an external advisor on my Ph.D. committee—offered extensive comments on the dissertation and vital, ongoing support. Other History of Consciousness faculty, especially Barbara Epstein and Hayden White, contributed significantly to the broader education this book reflects. Finally, I would never have survived graduate school without the graceful, dignified, and loving assistance of Billie Harris, administrative assistant to the History of Conciousness Board (a pathetically impoverished title when compared with her actual role).

Another key intellectual resource was Sherman Hawkins of Wesleyan University, whose course in Shakespeare turned me back toward the humanities as an undergraduate. My title phrase "the closed world" and its concept belongs to him (though I have taken it in a much different direction). Sherman's Prospero in Amy Seham's 1980 production of *The Tempest* will always be, for me, the standard by which all others are judged. His inspiration and example as scholar, teacher, and poet live within me.

The price of a relatively jargon-free text has been a relative paucity of direct references to certain literatures and traditions that deeply inform my theoretical perspective. I am particularly indebted, first, to the sociology and history of science and technology represented by such figures as Bruno Latour, Michel Callon, Steve Woolgar, Trevor Pinch, Thomas Parke Hughes, Wiebe Bijker, John Law, Donald MacKenzie, Steven Shapin, and Simon Schaffer. My analytic frame was shaped by concepts deriving from this tradition, such as actor-network theory, inscription devices, heterogeneous engineering, technoscience, technological frames, entrepreneurial engineers, and the social construction of science and technology. Second, my work owes much to poststructuralist critical theorists such as Roland Barthes, Michel Foucault, Jacques Derrida, James Clifford, Fredric Jameson, Louis Althusser, and Hayden White. The ideas of discourse, narrative, and subjectivity so central to this book grew largely from reflection on these thinkers' work. Third, philosophical studies of artificial intelligence by Hubert Dreyfus, John Searle, Joseph Weizenbaum, Douglas Hofstadter, Harry Collins, and Terry Winograd form part of this book's deeper background. Last but not least, the interpretive sociology of computer communities represented by the work of Sherry Turkle, Rob Kling, Jon Jacky, Lucy Suchman, and Leigh Star played a major part in my thinking about the relations among technology, metaphors, science, and subjectivity.

As the dissertation slowly became a book, many people contributed comments on its various pieces. I want to thank especially Trevor Pinch, Lynn Eden, Susan Douglas, Lily Kay, Larry Cohen, Sherman Hawkins, and my anonymous referees for the MIT Press. For other intellectual contributions in many forms, I am indebted especially to Sherry Turkle, Peter Taylor, Langdon Winner, Chris Hables Gray, Marc Berg, Kenneth Flamm, Sheila Jasanoff, Diana Forsythe, Eric Griffiths, Judith Reppy, Timothy Lenoir, Allucquère Rosanne Stone, Rick Zinman, Zoë Sofoulis, Gregg Herken, Howard Gardner, Mitchell Ash, Marc Berg, Geof Bowker, Leigh Star, Jon Jacky, Gene Rochlin, Michael Smith, Hubert Dreyfus, Phil Agre, Merritt Roe Smith, Lucy Suchman, Judy Yang, Adele Clark, Bruno Latour, Eric Roberts, Paul Forman, Dom Massaro, and Steve Heims.

Kären Wieckert generously donated some of the research for chapter 3, and also helped shape the whole direction of this book. Cynthia Enloe and the late Sally Hacker read early papers and helped me publish them; their enormously generous encouragement was responsible both for public successes and for private renewals of confidence. James Capshew offered conversation, friendship, and professional contacts at an important early stage of my career.

The debt I owe my students, who kept me honest and forced me to turn vague gesticulations into clear concepts, is immeasurable. If the truth be told, many of the ideas in this book really belong to them. Above all I want to thank the students of UC Santa Cruz and Cornell University who joined me for courses on "Minds, Machines, and Subjectivity" between 1988 and 1992. The interpretations of science fiction in chapter 10, especially, owe much to their fresh, sharp eyes.

Completing this work would not have been possible without generous material support. For their key roles in my professional development, I would like to thank especially Peter Taylor, Sheila Jasanoff, Terry Winograd, Tim Lenoir, and Brian C. Smith. Judith Reppy aided in securing a research grant from the MacArthur Foundation through Cornell's Peace Studies Program. The Institute on Global Conflict and Cooperation of UC San Diego provided a grant in support of the final year of dissertation writing, as well as the opportunity to attend summer schools on international security and disarmament in Europe and the United States in 1988 and 1989. The UCSC

xx   Acknowledgments

Humanities Division, the History of Consciousness Board, and the Silicon Valley Research Group all granted modest research funding at various points along the way.

For the love and friendship that is the *sine qua non* of any project on this scale, I reserve my deepest thanks for Todd Jones, Rick Zinman, Anne Martin, Chris Austill, Alison Cook Sather, Jennifer Homans, Sarah Rabkin, Hasanna Fletcher, Susan Bodnar, Alaine Perry, and Robin White, extraordinary friends and exceptional human beings. The memory of my housemate Walter "Butterfly" Blumoff, who died of AIDS in 1991, comes to me often, as much an inspiration in death as he was in life. Others whose importance in my life and work far exceeds my ability to thank them include Dori Schack, Eli Zinman, Linda Holiday, Ruth Hille, Randy Austill, Joelle Hochman, Peter Taylor, Ann Blum, Vann Taylor, Joel and Marilyn Ray, Laurie Ray, Holly Hirzel, Sally Miller, Yukiko Katagiri, Archie Mackenzie, David Salesin, Judy Yang, Carol Fitzgerald, Anne Cushman, Vic Liptak, Eric Griffiths, and Randi Chazan. Robert Field, Mary Coville, and Pam Roby helped me with writing blocks. Finally, the rock-solid love of my family, Carl, Janet, Craig, and Michael Edwards, lies at the very foundation of my life. From their unwavering strength I gathered the courage and resolve to overcome severe setbacks. This is my *karass*, my life team. I could not have imagined a better one.

Finally, for my wife, Gabrielle Hecht, words cannot express the depth of my gratitude and my love. Her editorial suggestions shaped the book's basic form, while her keen critical eye helped render it readable, coherent, and intellectually honest. Without her help, love, and unwavering encouragement this book might never have been finished.

The Closed World

Strategic Air Command Headquarters, Offut Air Force Base, 1962. A Ballistic Missile Early Warning System computer monitor (r) displays a polar map of North America. Courtesy Library of Congress, *US News & World Report* collection.

1
"We Defend Every Place": Building the Cold War World

This book is about computers, as machines and as metaphors, in the politics and culture of Cold War America.

As machines, computers controlled vast systems of military technology central to the globalist aims and apocalyptic terms of Cold War foreign policy. First air defenses, then strategic early warning and nuclear response, and later the sophisticated tactical systems of the electronic battlefield grew from the control and communications capacities of information machines. As metaphors, such systems constituted a dome of global technological oversight, a *closed world*, within which every event was interpreted as part of a titanic struggle between the superpowers. Inaugurated in the Truman Doctrine of "containment," elaborated in Rand Corporation theories of nuclear strategy, tested under fire in the jungles of Vietnam, and resurrected in the impenetrable "peace shield" of Ronald Reagan's Strategic Defense Initiative, the key theme of closed-world discourse was global surveillance and control through high-technology military power. Computers made the closed world work simultaneously as technology, as political system, and as ideological mirage.

Both the engineering and the politics of closed-world discourse centered around problems of *human-machine integration*: building weapons, systems, and strategies whose human and machine components could function as a seamless web, even on the global scales and in the vastly compressed time frames of superpower nuclear war. As symbol-manipulating logic machines, computers would automate or assist tasks of perception, reasoning, and control in integrated systems. Such goals, first accomplished in World War II–era anti-aircraft weapons, helped form both cybernetics, the grand theory of information and control in biological and mechanical systems, and artificial intelligence (AI), software that simulated complex symbolic thought.

At the same time, computers inspired new psychological theories built around concepts of "information processing." Cybernetics, AI, and cognitive psychology relied crucially upon computers as metaphors and models for minds conceived as problem-solving, self-controlling, symbol-processing systems. The word "cyborg," or cybernetic organism, captures the strategic blurring of boundaries inherent in these metaphors.[1] *Cyborg discourse*, by constructing both human minds and artificial intelligences as information machines, helped to integrate people into complex technological systems.

The cyborg figure defined not only a practical problem and a psychological theory but a set of *subject positions*. Cyborg minds—understood as machines subject to disassembly, engineering, and reconstruction—generated a variety of new perspectives, self-interpretations, and social roles. These identities were most thoroughly explored in science fiction, where cyborg figures ranged from the disembodied, panoptic AIs of *Colossus: The Forbin Project* and *2001: A Space Odyssey* to the mechanical robots of *Star Wars* and the engineered biological androids of *Blade Runner*. But in a world increasingly structured by and theorized in terms of information processing devices, cyborg subjectivity was not only fictional but real.[2] Cyborgs were subjective devices nested inside the larger technological systems of the closed world. Hence this book also probes subjectivity and political identity in a real world of cyborgs by exploring the dramatic worlds of science fiction in books and film.

In exploring these ideas, I will develop three major theses. First, I will argue that the historical trajectory of computer development cannot be separated from the elaboration of American grand strategy[3] in the Cold War. Computers made much of that strategy possible, but strategic issues also shaped computer technology—even at the level of design. Second, I will link the rise of cognitivism, in both psychology and artificial intelligence, to social networks and computer projects formed for World War II and the Cold War. Here again the grand strategy of the postwar era influenced the form and content of major research programs, culminating in an abstract theory of intelligence as heuristic information processing and in a new interdiscipline, "cognitive science." Finally, I will suggest that cyborg discourse functioned as the psychological/subjective counterpart of closed-world politics. Cyborg discourse yielded up new possibilities for experience, identity, and political action within a total Cold War operated by global information and control systems. Where closed-world discourse defined the architectures of a political narrative and a technological system, cyborg discourse molded culture and subjectivity for the information age. Cyborgs, with minds and selves reconstituted as information processors, found flexibility, freedom, and even love inside the closed virtual spaces of the information society.

This chapter sets the stage for the book's argument with three short scenes from the closed world. The scenes are drawn from across the Cold War's historical span; each illustrates one of the book's major divisions. My goal is to enact the book's themes and their interplay here, at the outset, before proceeding to more detailed analysis. Like most dramas and all history, our first scene begins *in medias res*—in the middle of the story—in the night skies over Southeast Asia, riven by the sounds and furies of a terrible war.

Scene 1: Operation Igloo White

In 1968 the largest building in Southeast Asia was the Infiltration Surveillance Center (ISC) at Nakhom Phanom in Thailand, the command center of U.S. Air Force Operation Igloo White. Inside the ISC vigilant technicians pored over banks of video displays, controlled by IBM 360/65 computers and connected to thousands of sensors strewn across the Ho Chi Minh Trail in southern Laos.

The sensors—shaped like twigs, jungle plants, and animal droppings—were designed to detect all kinds of human activity, such as the noises of truck engines, body heat, motion, even the scent of human urine. When they picked up a signal, it appeared on the ISC's display terminals hundreds of miles away as a moving white "worm" superimposed on a map grid. As soon as the ISC computers could calculate the worm's direction and rate of motion, coordinates were radioed to Phantom F-4 jets patrolling the night sky. The planes' navigation systems and computers automatically guided them to the "box," or map grid square, to be attacked. The ISC central computers were also able to control the release of bombs: the pilot might do no more than sit and watch as the invisible jungle below suddenly exploded into flames. In most cases no American ever actually saw the target at all.

The "worm" would then disappear from the screen at the ISC. This entire process normally took no more than five minutes.

Operation Igloo White ran from 1967 to 1972 at a cost ranging near $1 billion a year. Visiting reporters were dazzled by the high-tech, white-gloves-only scene inside the windowless center, where young soldiers sat at their displays in air-conditioned comfort, faces lit weirdly by the dim electric glow, directing the destruction of men and equipment as if playing a video game. As one technician put it: "We wired the Ho Chi Minh Trail like a drugstore pinball machine, and we plug it in every night."

Official claims for Igloo White's success were extraordinary: the destruction of over 35,000 North Vietnamese and Pathet Lao trucks, each carrying some 10,000 pounds of supplies destined for the communist insurgency in South Vietnam. Had these figures been accurate, a conservative estimate would still have put the cost of interdiction in the neighborhood of $100,000 for each truck destroyed—the truck and the supplies inside it usually being worth a maximum of a few thousand dollars.

But the official estimates, like so many other official versions of the Vietnam War, existed mainly in the never-never land of military public relations. In 1971 a Senate subcommittee report pointed out that the figure for "truck kills claimed by the Air Force [in Igloo White] last year greatly exceeds the number of trucks believed by the Embassy to be in all of North Vietnam." Daytime reconnaissance flights rarely located the supposedly destroyed vehicles. The Vietcong were supposed to have "dragged" the trucks' carcasses into the jungle during the night, but in many cases this idea was pure delusion. The guerrillas had simply learned to confuse the American sensors with tape-recorded truck noises, bags of urine, and other decoys, provoking the release of countless tons of bombs onto empty jungle corridors which they then traversed at their leisure. Traffic over the Ho Chi Minh Trail continued, harassed but far from "interdicted."

The antiseptic efficiency of the ISC was belied by the 13,000 civilian refugees created by American operations along the Ho Chi Minh Trail[4]—as well as by the loss of an estimated 300–400 American aircraft involved in Igloo White operations. In the end, despite more than four years of intensive computer-controlled bombardment of their heavy-equipment supply lines, the communists were able to field a major tank and artillery offensive *inside* South Vietnam in 1972. Nevertheless, Igloo White was counted, officially, as an important success that had managed to destroy up to 90 percent of the equipment sent down the Ho Chi Minh Trail.[5]

Operation Igloo White's centralized, computerized, automated method of "interdiction" resembled a microcosmic version of the whole United States approach to the Vietnam War. Under Robert McNamara, the Department of Defense completed a process of centralization begun by President Truman, making the service secretaries responsible to the Secretary of Defense in practice as well as in principle. McNamara achieved this goal by seizing control of the military budget. Wielding financial power like a bludgeon, he forced the services to coordinate their purchasing and therefore to coordinate their planning as well.

To control the budget, McNamara introduced a cost-accounting technique known as the Planning Programming Budgeting System (PPBS), which was built on the highly quantitative tools of systems analysis. The PPBS was therefore a natural application for the computer, at the time still a very expensive, fascinating novelty that could generate authoritative-sounding simulations and ream after ream of cost-benefit calculations. Gregory Palmer notes that while it often served more as a heuristic or ideal, "in its pristine form, PPBS was a closed system, rationally ordered to produce carefully defined outputs."[6] Lyndon Johnson regarded the PPBS as so successful that in 1965 he ordered all federal agencies to adopt it.

As the United States became more and more deeply involved in Vietnam, the McNamara Defense Department's administrative centralization and rationalization was extended to a strategic and sometimes even a tactical centralization within the White House and the Office of the Secretary of Defense (OSD). After President Johnson ordered U.S. bombing of North Vietnam in 1965, McNamara and his assistants ran the air war in Southeast Asia from the Pentagon, integrating information and target lists prepared by military agencies all over the world. The OSD literally micromanaged the bombing campaign, specifying the exact targets to be attacked, weather conditions under which missions must be canceled or flown, and even the precise qualifications of individual pilots.[7] Even Johnson himself sometimes took part in targeting decisions.

As Martin van Creveld points out in his masterful study of command in war, the availability of new technologies and techniques of management was a large part of the reason for this entirely novel situation.

During the two decades after 1945, several factors ... caused the American armed forces to undergo an unprecedented process of centralization. In the first place, there was the revolutionary explosion of electronic communication and automatic data processing equipment, which made effective worldwide command and control from Washington a practical technological proposition. Second, there was the preoccupation during the 1950s with the need for failproof positive control systems to prevent an accidental outbreak of nuclear war, a preoccupation that led first to the establishment of the Worldwide Military Command and Control System (WWMCCS) in 1962 and then to its progressive extension from the Strategic Air Command, for which it had originally been designed, down to the conventional forces. New administrative techniques, such as cost-benefit analysis with its inherent emphasis on the pooling of resources and the careful meshing of each part with every other, further contributed to the trend toward central management, as did the appearance on the market of the data processing hardware needed to make it possible.[8]

The elements of this list of factors are worth close attention. High-technology communications and computing equipment, nuclear weapons and Cold War nuclear anxiety, quantitatively oriented "scientific" administrative techniques, and the global objectives of U.S. military power combined to drive forward the centralization of command and control at the highest levels. At the same time, this drive created serious—and in the case of Vietnam, finally fatal—impediments both to effective action and to accurate understanding of what was going on in the field. Van Creveld calls these disruptions the "information pathologies" of that war.

In Operation Igloo White we see how these techno-strategic developments were played out on a regional scale: centralized, remote-controlled operations based on advanced computing and communications gear; an abstract representation of events (sensors, maps, grids, "worms") justified in terms of statistics; and a wide gap between an official discourse of overwhelming success and the pessimistic assessments of independent observers, including American soldiers on the ground. Like McNamara's PPBS, Igloo White was a "a closed system, rationally ordered to produce carefully defined outputs." These "outputs" were not only military but rhetorical in character.

From start to finish the Cold War was constructed around the "outputs" of closed systems like Igloo White and the PPBS. Its major strategic metaphor, "containment," postulated an American-led policing of closed Communist borders. Its major military weapon, the atomic bomb, became the cultural representative of apocalypse, an all-or-nothing, world-consuming flame whose ultimate horizon encircled all conflict and restructured its meaning. Cold War military forces took on the character of systems, increasingly integrated through centralized control as the speed and scale of nuclear war erased the space of human decision-making and forced reliance on automated, preprogrammed plans. The official language of the Cold War, produced by think tanks such as the Rand Corporation, framed global politics in the terms of game-theoretic calculation and cost-benefit analysis.

None of this—metaphors, weapons, strategy, systems, languages—sprang into being fully formed. We must therefore ask: How and why did global military control come to seem a "practical technological proposition," as van Creveld puts it? How did tradition-bound military services, oriented toward leadership based on battlefield experience, become transformed into managers of automated systems embodying preprogrammed plans based on abstract strategies? What held the official strategic discourse of the Vietnam era together, in the face of what could have been construed as glaring evidence of failure? What enabled the fantasy of global control through high-technology armed forces to persist throughout the Cold War, at the highest levels of government, as exemplified in President Reagan's Strategic Defense Initiative?

This book locates a key part of the answer to these questions at the intersections of politics, culture, and computer technology, in the ways computers and the political imagination reciprocally extended, restricted, and otherwise transformed each other. Like other elements of the post-World War II high-technology arsenal, such as the atomic bomb, the long-range jet bomber, and the intercontinental ballistic missile, computers served not only as military devices and tools of policy analysis but as icons and metaphors in the cultural construction of the Cold War. As H. Bruce Franklin has put it, "American weapons and American culture cannot be understood in isolation from each other. Just as the weapons have emerged from the culture, so too have the weapons caused profound metamorphoses in the culture."9

I use the phrase "closed-world discourse" to describe the language, technologies, and practices that together supported the visions of centrally controlled, automated global power at the heart of American Cold War politics. Computers helped create and sustain this discourse in two ways. First, they allowed the practical construction of central real-time military control systems on a gigantic scale. Second, they facilitated the metaphorical understanding of world politics as a sort of system subject to technological management. Closed-world discourse, through metaphors, techniques, and fictions as well as equipment and salient experiences, linked the globalist, hegemonic aims of post-World War II American foreign policy with a high-technology military strategy, an ideology of apocalyptic struggle, and a language of integrated systems.

The Postwar World as a Closed System

In early 1947, because of a fiscal crisis, Great Britain withdrew its support from anticommunist forces in Greece and Turkey. To take up the slack, President Truman drove through a military aid package by "scaring hell out of the American people" in a major speech before Congress.[10] In it he pictured communism as global terrorism with implacable expansionist tendencies. The speech implied that the United States would henceforth support anticommunist forces anywhere in the world.

In June, the administration announced the European Recovery Plan or "Marshall Plan," an aid package of unprecedented proportions designed to help reconstruct European industrial capitalism as well as to correct a huge U.S. export surplus, to create a common market within Europe, and to integrate Germany into the European economy. That same year the term "Cold War" came into common use to describe the overt, but militarily restrained, conflict between East and West. East-West relations remained essentially frozen for the next six years—and the Truman Doctrine of "containment" became the essential U.S. policy toward communism for more than four decades.

Containment, with its image of an enclosed space surrounded and sealed by American power, was the central metaphor of closed-world discourse. Though multifaceted and frequently paradoxical, the many articulations of this metaphor usually involved (a) globalism, (b) a many-dimensional program with ideological, political, religious, and economic dimensions, and (c) far-reaching military commitments that entailed equally far-reaching domestic policies. The rhetoric of American moral leadership that underlay the idea of containment can be traced back to the colonial vision of a City on a Hill, while the idea of an American sphere of influence dates to the Monroe Doctrine.[11] Closed-world political discourse differed from its predecessors, however, in its genuinely global character, in the systematic, deliberate restructuring of *American* civil society that it entailed, and in its focus on the development of technological means to project military force across the globe.

The language of global closure emerged early in the Truman administration as a reflection of perceived Soviet globalist intentions.

Truman's young special counsel Clark Clifford, in an influential secret 1946 report, wrote that the Soviets saw conflict with the West as inevitable and sought "wherever possible to weaken the military position and influence of the United States abroad." "A direct threat to American security," Clifford concluded, "is implicit in Soviet foreign policy."[12] With the Monroe Doctrine in the background, American policy soon progressed to a globalism of its own.

Then Undersecretary of State Dean Acheson was one of the principal architects of containment. In pushing the aid package for Greece and Turkey that became the occasion for the Truman Doctrine, Acheson used the analogy of "rotten apples in a barrel" whose "infection" would spread throughout the world if unchecked. The ambiguity of Acheson's container metaphor is instructive. Was the United States the lone active agent in the scene, reaching in from outside the barrel to remove the bad apple? Or was the United States inside the barrel as well, one of the apples to whom "infection" might spread if nothing were done? Such ambiguities ruled the political culture of the Cold War era. That culture saw communism both as an external enemy to be contained or destroyed by overt economic manipulation, covert political intervention, and military force, and as an internal danger to be contained by government and civil surveillance, infiltration, public denunciations, and blacklisting.

The military dimension of closed-world discourse followed from the United States' role as the new hegemonic power within what historians such as Fernand Braudel and Immanuel Wallerstein have called the "capitalist world-system."[13] World-systems theory holds that the intrinsic logic of capitalism drives it to seek international economic integration: the elimination of trade barriers of all sorts (economic, political, social, and military) to foster free-market exchange. Capitalism, as a purely economic force, knows no geography. (Nation-states, on the other hand, tend to pursue policies of economic autarky, seeking to maximize their own well-being within a geographical territory or trading bloc by establishing a balance of power.) Those politico-economic units that succeed in remaining outside the capitalist world-system are either part of the "external world" (self-sufficient empires), or are unconnected to large-scale economic systems (subsistence communities). According to this theory, when a single hegemonic power emerges within the world-system, its structural position leads it to attempt to force other nations to abandon autarky in favor of free trade and free capital flows.[14]

The United States, as the only combatant nation to emerge unscathed from World War II, became the hegemon of the postwar period. The USSR became the predominant organizing force of the "external world" outside capitalist markets. Thus the world-system formed one kind of closed world, while the Soviet Union and its satellites formed another. The Cold War struggle occurred at the margins of the two, and that struggle constituted the third closed world: the system formed from the always-interlocking traffic of their actions.

Both the military and the economic logic of containment had an ambiguous character. American goals were simultaneously

- to enclose the Soviet Union (seeing it as a closed society, an empire),

- to enclose the capitalist nations (seeing capitalism as a closed system, shielding it from the supposedly penetrating force of communist politico-economic doctrines), and ultimately

- to extend the capitalist world-system to enclose the entire world by penetrating and exploding the closed Soviet sphere.

In an ideologically laden metaphor, this last goal was normally spoken of as "opening up" the world to the free market.

So the world of the Truman Doctrine and McCarthyism was closed in a triple sense. On one reading the closed world was the repressive, secretive communist society, surrounded by (contained within) the open space of capitalism and democracy. This was the direct intent of the containment metaphor. But on another reading, the closed world was the capitalist world-system, threatened with invasion. It required defenses, a kind of *self*-containment, to maintain its integrity. In the third and largest sense, the global stage as a whole was a closed world, within which the struggle between freedom and slavery, light and darkness, good and evil was being constantly joined in every location—within the American government, its society, and its armed forces as well as abroad. Each side of the struggle had, in effect, a national headquarters, but the struggle as a whole went on everywhere and perpetually.

Under the Truman Doctrine and the Marshall Plan, the world had become a system to be both protected and manipulated by the United States. Within the quasi-religious American mythos, no ideological space remained for other conflicts. Truman's construction of the bilateral world in his 1947 speech presented Congress with a simple binary decision: democracy or Stalinist communism, freedom or slavery, good or evil.[15] Bilateralism created a systematic vision of the world by making all third-world conflicts parts of a coherent whole, surrogates for the real life-or-death struggle between the Free World and its communist enemies.

The Truman administration gradually articulated a "defensive perimeter" that ran southward along the Iron Curtain, then eastward across Greece, Turkey, Israel, and Iran to southern Asia. From America's Pacific coast the line stretched along the Aleutian chain to Japan, Taiwan, South Korea, the Philippines, and Vietnam. This "perimeter" essentially enclosed the Soviet Union within a circle of forward air bases and politico-military alliances. By 1950, with the U.S. entry into the Korean War, the administration had defined American interests in totally global terms. In a memo to Congress defending the President's right to commit troops on his own authority, Acheson argued that "the basic interest of the United States is international peace and security. The United States has, throughout its history, . . . acted to prevent violent and unlawful acts in other states from depriving the United States and its nationals of the benefits of such peace and security." The North Korean invasion of South Korea constituted exactly such a disruption. Summing the terms of his equation, Acheson concluded that the North Korean action represented a "threat to the peace and security of the United States and to the security of United States forces in the Pacific."[16]

Under such a definition of national security, the U.S. umbrella covered the globe. When the incorrigible General Douglas MacArthur wanted to roll back the Chinese as well as the North Koreans, Truman was forced to relieve him of the Korean command. But in the ensuing Senate investigation, MacArthur became a national hero for declaring the struggle against communism a "global proposition." "You can't let one-half of the world slide into slavery and just confine yourself to defending the other," the general told the senators. "What I advocate is that *we defend every place*, and I say we have the capacity to do it. If you say we haven't, you admit defeat."[17]

Truman repudiated MacArthur, exposing the difference between the rhetoric of Cold War and the limits of political will. But MacArthur's strong words merely carried Acheson's doctrine to its logical conclusion. Acheson, in the MacArthur hearings, explained that Korea itself mattered very little. Rather, American security now depended not only upon strategic might but also upon ideological power. To demonstrate the free world's strength, the United States must now actively repel communist aggression anywhere in the world.[18] But John Foster Dulles, Eisenhower's secretary of state, would prefer MacArthur's language to Acheson's: "a policy which aims only at containing Russia is, in itself, an unsound policy.... It is only by keeping alive the hope of liberation, by taking advantage of whatever opportunity arises, that we will end this terrible peril which dominates the world."[19] Dulles threatened "massive retaliation"—implying nuclear force—in response to communist aggression anywhere in the world.

National Security Council Resolution 68 (NSC-68), probably the most important document of the Cold War, was also the most forthright expression of what James Chace and Caleb Carr have called the "universalization of threats to American security."[20] It held that

[t]he implacable purpose of the slave state to eliminate the challenge of freedom has placed the two great powers at opposite poles.... The assault on free institutions is world-wide now, and in the context of the present polarization of power a defeat of free institutions anywhere is a defeat everywhere.... [It is no longer] an adequate objective merely to seek to check the Kremlin design, for the absence of order among nations is becoming less and less tolerable. This fact imposes on us, in our own interests, the responsibility of world leadership.... [T]he cold war is in fact a real war in which the survival of the free world is at stake.[21]

In these and similar words the architects of closed-world discourse articulated a new language along with their political strategy and military posture, intimately linking metaphors, beliefs, and ideologies to practices, policies, and technologies of the Cold War in the dark and all-encompassing theater of apocalypse.

Characterizing the Closed World

A "closed world" is a radically bounded scene of conflict, an inescapably self-referential space where every thought, word, and action is ultimately directed back toward a central struggle. It is a world radically divided against itself. Turned inexorably inward, without frontiers or escape, a closed world threatens to annihilate itself, to implode.[22]

The term descends from the literary criticism of Sherman Hawkins, who uses it to define one of the major dramatic spaces in Shakespearean plays. Closed-world plays are marked by a unity of place, such as a walled city or the interior of a castle or house. Action within this space centers around attempts to invade and/or escape its boundaries. Its archetype is the siege, with the *Iliad* as originary model; war, either literal or figurative, is its driving force. Notably, the closed world includes not just the sealed, claustrophobic spaces metaphorically marking its closure, but the entire surrounding field in which the drama takes place. The dividing conflict which drives social action in the closed world finds parallels in the inward psychological division of characters, such as Hamlet, torn between the power and the impotence of rationality and between the necessity and the choking restriction of social convention. In tragedy this leads to self-destruction (e.g., Hamlet or Romeo) and in comedy to exorcism of these forces (e.g., the punishment of Malvolio).[23]

The alternative to the closed world is not an open world but what Northrop Frye called the "green world."[24] The green world is an unbounded natural setting such as a forest, meadow, or glade. Action moves in an uninhibited flow between natural, urban, and other locations and centers around magical, natural forces—mystical powers, animals, or natural cataclysms (e.g., *A Midsummer Night's Dream*). Green-world drama thematizes the restoration of community and cosmic order through the transcendence of rationality, authority, convention, and technology. Its archetypal form is the quest, in which characters struggle to integrate (rather than overcome) the world's complexity and multiplicity. The green world is indeed an "open" space where the limits of law and rationality are surpassed, but that does not mean that it is anarchical. Rather, the opposition is between a human-centered, inner, psychological logic and a magical, natural, transcendent one.[25]

The "closed world" of this book is political and ideological rather than literary. But since historiography always involves a dramatic reconstitution of a disorderly past, it has much in common with its literary cousins.[26] Postwar American politics, as well as those of divided Europe, were in fact dominated by the same unity of place that characterizes closed-world drama. The stage was the globe as a whole, truly a world divided against itself as never before. The action was one of attempts to contain, invade, or explode a closed communist world symbolized by phrases like "the Iron Curtain" and physically instantiated by the Berlin Wall. At the same time the globe itself was seen as a closed whole, a single scene in which the capitalist/communist struggle was the only activity and from which the only escape was the technological utopia of space travel. The United States reconceived itself—building upon the political heritage of Manifest Destiny and the religious iconography of the City on a Hill[27]—as the manager, either directly or by proxy, of the entire global political, economic, and military scene.

In the closed world of the Cold War, all military conflict took place beneath the black shadow of nuclear arms. It was war in a military world where mutual and total annihilation, even the end of all human life, was the overarching possibility within which all other conflicts were articulated. Paradoxically, ultimate weapons also produced ultimate limits to military power. After 1949, nuclear weapons could deliver only the hollowest and most Pyrrhic of "victories." Against the contradictions and the terror of nuclear arms, war itself became as much an imaginary field as a practical reality.

Inside the closed horizon of nuclear politics, simulations became more real than the reality itself, as the nuclear standoff evolved into an entirely abstract war of position. Simulations—computer models, war games, statistical analyses, discourses of nuclear strategy—had, in an important sense, more political significance and more cultural impact than the weapons that could not be used. In the absence of direct experience, nuclear weapons in effect forced military planners to adopt simulation techniques based on assumptions, calculations, and hypothetical "rules of engagement." The object for each nuclear power was to maintain a winning *scenario*—a theatrical or simulated win, a psychological and political effect—rather than actually to fight such a war. *Actual* outcomes no longer mattered, since the consequences had become too enormous to be comprehended and too dangerous to be tested. The world of nuclear arms became by its very grossness and scale a closed world, a lens through which every other political struggle must be seen. For those who contemplated its strategy, nuclear war could only be understood as a many-leveled game.

The Cold War's portent as an economic and material fact could not be grasped apart from its metaphorical and cultural dimensions. Weapons of war were also understood to be focal elements of the economy, of national politics, and of scientific research. Computers were a primary example of this inseparability of weapon from tool, tool from metaphor, and metaphor from political action. They were a key factor in the massive increases in the speed and scale of warfare through their implementations in systems designed for air defense, military command and control, data analysis, and satellite surveillance and, from the early 1960s, as components of self-guided and "smart" weapons such as guided missiles, cruise missiles, and advanced jet aircraft. But computers were also of immense symbolic and practical importance in the ideological worlds of the Cold War and the Vietnam War, for which they represented a potential for total oversight, exacting standards of control, and technical-rational solutions to a myriad of complex problems.

"Closed-world discourse" thus names a language, a worldview, and a set of practices characterized in a general way by the following features and elements.

- *Techniques* drawn from engineering and mathematics for modeling aspects of the world as closed systems.
- *Technologies*, especially the computer, that make systems analysis and central control practical on a very large scale.
- *Practices of mathematical and computer simulation* of systems, such as manufacturing processes and nuclear strategy, in business, government, and the military.
- *Experiences* of grand-scale politics as rule-governed and manipulable, for example by means of the power of nuclear weapons or of Keynesian economic intervention.
- *Fictions, fantasies, and ideologies*, including such visions as global mastery through air power and nuclear weapons, global danger from an expansionist "evil empire," and centralized, instantaneous, automated command and control.
- A *language* of systems, gaming, and abstract communication and information that relied on formalisms to the detriment of experiential and situated knowledge. This language involved a number of key *metaphors*, for example that war is a game and that command is control.

In the last part of this chapter we will examine in detail the concept of discourse, which links these heterogeneous elements. Then, in chapters 2–4 (and again in chapter 9), we will explore how computers were pressed into service as material and metaphorical supports for closed-world discourse, and thus for America's role in post-World War II geopolitics. First, however, let us scan another scene from the closed world: the roots of cyborg discourse in Alan Turing's universal machines.

Scene 2: Turing's Machines

In 1950 Alan Turing, the British mathematician who invented the theory of universal digital computation, devised an "imitation game" in which a computer is programmed to simulate human thought processes. A person attempts to discern the difference between the computer and a "real" person by interrogating them both through a terminal. This game became known as the Turing test for machine intelligence. The questions of whether it is the right test, whether a computer will ever pass it, and exactly what it would mean for one to do so have since become foci of long and intense debates in artificial intelligence and philosophy of mind.[28]

The possibility that machines could carry out mental operations had occurred to Turing from the moment of his first major mathematical discovery, in 1935–36 (published as "On Computable Numbers, with an application to the *Entscheidungsproblem*" in 1937), if not before. Turing had considered the relationship between the infinite set of "configurations" of a simple imaginary computing machine—known today as the "universal Turing machine," of which all possible digital computers are more or less incomplete instances—and the mental states of human beings. A human "computer" performing the operations of a Turing machine by hand would necessarily, on Turing's view, proceed through a sequence of discrete mental states directly analogous to the states of the machine. "The operation actually performed is determined ... by the state of mind of the [human] computer and the observed symbols. In particular, they determine the state of mind of the computer after the operation is carried out. ... We may now construct a machine to do the work of this computer."[29]

Elsewhere in his 1937 paper Turing made clear that the essential move in this analogy was to reduce each "state of mind" of the (human) computer to a single unit. This could be done by translating any complex operation into a series of definite steps. This, of course, is the basic principle of operation of the digital computer. Any mechanical computer would necessarily perform each step in the course of performing the operation; ergo, the steps would functionally define discrete mental states. The mechanical computer might then be said to have a kind of mind, or alternatively, the human computer could be defined as a machine. In 1937 Turing left this point hanging. His central, and successful, aim was to construct a mathematical proof that almost *any* problem that could be precisely formulated could be solved by a sufficiently powerful Turing machine.[30]

A less grand objective, namely automatic calculation, led George Stibitz of Bell Laboratories, Howard Aiken of Harvard University, John Atanasoff of Iowa State University, and others to start developing prototypes of electronic and electro-mechanical automatic digital calculators independently in the late 1930s. Their work, however, was generally ignored, like Charles Babbage's prescient nineteenth-century design for an Analytical Engine, an enormous symbol-manipulating machine with many of the features essential to a true digital computer, including a memory, programmability, conditional loop capability, and a central processing unit.[31] It was not until the war, with its urgent demands for advanced technology, that the Turing machine's revolutionary implications were carried into practice.

In 1939 Turing began working with a team of scientists at the Government Code and Cypher School (GCCS) in Bletchley Park, near London. Early in the war, British intelligence had received working copies of sophisticated German cipher machines called "Enigma" and "Fish" used to encode secret messages. The machines themselves were only part of the cryptological problem, however, since their codes required keys that were frequently changed. Manual methods could not uncover the keys fast enough to make intercepted messages useful. Turing's group at Bletchley Park was charged with developing computational devices to automate and speed up the decrypting process. One of these machines, the 1943 "Colossus," was a true electronic digital computer. One version had 2400 vacuum tubes, was programmable (though it could not store programs internally, the critical advance that created modern computers), and was in some ways more advanced than the far larger American ENIAC. Because the Colossus remained a military secret after the war, the ENIAC has often mistakenly been designated the first electronic digital computer, even though it was not fully operational until 1946 and even though the ENIAC research team was aware, at least in general terms, of Turing's wartime work.[32]

The Colossus and other devices Turing helped invent successfully decoded many thousands of German command messages. German confidence in the Enigma and British secrecy about Turing's "Ultra" project were so high that the Germans never traced the source of their security leaks to the Allies' code-breaking activities. Without this intelligence, Allied forces might have suffered even greater defeats in the first years of the war. The protection of trans-Atlantic shipping from the dreaded U-boats, for example, relied heavily on the work of Turing's decryption group. Churchill placed the GCCS work among his top priorities and personally ordered that the group's requests for personnel and equipment be instantly and fully satisfied.[33] "I won't say that what Turing did made us win the war, but I daresay we might have lost it without him," Turing's wartime statistical clerk I. J. Good said afterwards.[34]

By 1950 Turing had worked for a decade at designing, building, and operating digital computers. As his research progressed, Turing elaborated his belief in the possibility of machine intelligence. In a famous prediction, he wrote that within fifty years it would be possible "to program computers... to play the imitation game so well that an average interrogator will not have more than 70 percent chance of making the right identification after five minutes of questioning." In 1991, forty-one years after Turing's prediction, computers fooled five of ten judges in a limited version of the Turing test restricted to a single area of knowledge such as wine-tasting or romantic love.[35]

But another of Turing's 1950 predictions has received far less attention, though it was in many ways more important and more profound. He wrote:

The original question, "Can machines think?" I believe to be too meaningless to deserve discussion. Nevertheless I believe that at the end of the century the use of words and general educated opinion will have altered so much that one will be able to speak of machines thinking without expecting to be contradicted.[36]

This prediction—not those that herald the actual existence of thinking machines—is the second major theme of this book. For Turing was clearly right on this score, and far sooner than he thought. Even in his own day computers we would now think of as almost pathetically primitive were known in the popular press as "giant brains."[37] By the late 1980s phrases like "expert systems," "artificial intelligence," and "smart" and even "brilliant weapons" were part of the everyday vernacular of the business and defense communities and the popular press. Prominent philosophers argued for the naturalness of the "intentional stance" (the attribution of purposes, goals, and reasoning processes) in describing some of the actions of computers.[38] Within certain subcultures, such as computer hackers and child programmers, highly articulated descriptions of the computer as a self with thoughts, desires, and goals, and of the human self as a kind of computer or program, were commonplace.[39]

Cyborgs

In tandem with closed-world politics, new conceptions of psychological processes—"cognitive" psychology and artificial intelligence—began their rise to scientific ascendancy during the Cold War. Wartime work on integrating humans into combat machines helped produce "cybernetic" theories of information and communication that applied equally to the machines and their human components. New theories of brain function were tightly linked with concepts of digital logic stemming from Turing's ideas. By 1956 the concept of "artificial intelligence" had been invented and laboratory research on computerized minds begun.

In psychology the new view, then still unnamed, opposed behaviorism's emphasis on external observables and simple conditioning with complex internal-process models based on metaphors of computers and information processing. It reached maturity in the middle 1960s with the publication of Ulric Neisser's *Cognitive Psychology*.[40] By the late 1970s cognitive psychology had been integrated with artificial intelligence, linguistics, and neuropsychology to form a new interdiscipline known as "cognitive science." Successful inheritor of the failed ambitions of cybernetics, cognitive science views problems of thinking, reasoning, and perception as general issues of symbolic processing, transcending distinctions between humans and computers, humans and animals, and living and nonliving systems.

This new and powerful conception of psychology evolved in a reciprocal relationship with a changing culture of subjectivity for which computers became, in Sherry Turkle's words, a "second self." As she has shown, the analogy between computers and minds can simultaneously decenter, fragment, and reunify the self by reformulating self-understanding around concepts of information processing and modular mental programs, or by constituting an ideal form for thinking toward which people should strive. Interactive relationships with information machines provided an experiential grounding for this reconstituted self and its values. At the same time they helped establish the sense of a vast and complex world inside the machine. Mid-1980s cyberpunk science fiction named the world within the computer "cyberspace."[41] With the emergence of global computer networks and "virtual reality" technologies for creating and inhabiting elaborate simulated spaces, by the 1990s cyberspace became a reality. It held, and holds, an irresistible attraction for many of the millions who spend much of their daily lives "logged in."[42]

World War II-era weapons systems in which humans served as fully integrated technological components were a major source of the ideas and equipment from which cognitivism and AI arose. These were the first exemplars of a new type of device able to mediate or augment human sensory or communications processes and perform some decision or calculation functions on their own, almost always with electronics and computers. American military forces began to integrate their human and technological components on a gigantic scale through their C^3I (command, control, communications, and intelligence) systems. The smooth functioning of such machines, their tightly constrained time scales, and the requirement of continuous, 24-hour preparedness demanded that all components react predictably, that they follow orders and transmit information exactly as specified. In such highly integrated systems, the limited, slow, error-prone characteristics of human perception and decision-making had to be taken into account. This required a theory of human psychology commensurable with the theory of machines.

Contemporary high-technology armed forces employ a second generation of computerized weapon systems that take computer-assisted control to its logical conclusion in fully automatic and, potentially, autonomous weapons. Automatic weapons are self-controlled devices that use internal sensory capacities to track their targets, usually under microprocessor or other computer control. Examples include cruise missiles, torpedoes, and "killer satellites." Autonomous weapons, by contrast, would be self-controlled not only in tracking targets but in identifying them and making the decision to attack. These would include any launch-on-warning nuclear defense system, the autonomous tanks funded by the Defense Advanced Research Projects Agency's 1983 Strategic Computing Initiative, and the space-based nuclear defense system envisioned by the Strategic Defense Initiative.

The word I will use to describe these and similar technologies, ranging from artificially augmented human bodies and human-machine systems to artificial intelligences, both real and hypothetical, is "cyborg." Cyborg figures—blends of organism and machine—pervade modern culture, from the person with a pacemaker or artificial hip to AI-controlled automated factories to fictional robots and androids. Though multiply determined, these figures received their first and fullest articulation on the high-technology battlefield.

Turing thus predicted the emergence of a language of intelligent machines that I will call "cyborg discourse." This discourse is primarily concerned with the psychological and cultural changes in self-imagining brought on by the computer metaphor. Typically, cyborg discourse focuses on the psychological, metaphorical, and philosophical aspects of computer use, rather than on their political, social, and material dimension. It is both an account and an expression of the view that the computer is an "object to think with," in Turkle's phrase. Research in artificial intelligence, parallel distributed processing, cognitive psychology, and philosophy of mind forms a part of this discourse. So do social phenomena such as hacker communities and cultural expressions such as cyberpunk science fiction. While closed-world discourse is built around the computer's capacities as a tool of analysis and control, cyborg discourse focuses on the computer's mind-like character, its generation of self-understanding through metaphor.

Cyborg discourse is the field of techniques, language, and practice in which minds are constructed as natural-technical objects (in Donna Haraway's phrase) through the metaphor of computing.[43] It includes the following elements.

- *Techniques* of automation and integration of humans into mechanical and electronic systems, especially computerized systems.
- The computer as a *technology* with linguistic, interactive, and heuristic problem-solving capacities.
- *Practices* of computer use. Cyborg discourse became increasingly prominent as computers spread out of scientific and military centers into business, industry, and, in the 1980s, the home.
- *Experiences* of intimacy with computers and of connection to other people through computers, particularly in coherent communities focused on computers, such as hackers. Turkle's phrase "second self" captures the subjective depth of such experiences.
- *Fictions* and *fantasies* about cyborgs, robots, and intelligent machines, increasingly prominent in science fiction and popular culture. Scientific theories of artificial intelligence and cognitive psychology also formed cores for *ideologies* of human minds as manipulable machines, projecting their future integration with computers.

- *Languages* of formal representation of thought processes, such as computer languages, formal semantics, and theories of human information processing.
- *Metaphors* building on the computer's formal and mechanical features: the brain as a set of digital switches, the mind as a set of programs.

Like closed-world discourse, cyborg discourse as an analytical construct offers a vantage point that cuts across the divisions between the intellectual history of cognitive science and the engineering-economic history of computers. Cyborg discourse is also political, though the politics in question are more often socio-cultural than governmental.

The nature of this political structure is revealed most tellingly when the two discourses are articulated simultaneously. This happens almost anywhere that artificial intelligence experts, Defense Department planners, or communities of computer users discuss their visions of the future. But it occurs most explicitly and directly in near-future science-fiction novels and films.

Scene 3: Cyborgs in the Closed World

The closed world of computer-controlled global hegemony and the image of the computer as a cyborg, a mind-like artifact, come together powerfully in *The Terminator* (1984), a relatively low-budget science-fiction/horror film directed and co-written by James Cameron.

The Terminator opens in Los Angeles in the year 2029 A.D. amid the rubble and smoke of a nightmarish post-holocaust world. We later learn that an all-out nuclear exchange has been initiated by the "Skynet computer built for SAC-NORAD by Cyberdyne Systems. New, powerful, hooked into everything, trusted to run it all. They say it got smart. A new order of intelligence. Then it saw all people as a threat, not just the ones on the other side. It decided our fate in a microsecond: extermination." The few remaining human beings eke out a miserable existence in grimy underground bunkers, crawling out at night to do battle with the robot killing machines that are now the masters of the planet. Their one major asset in this battle is a savvy leader who seems to have special insight into the enemy, a man named John Connor.

To finish off the human resistance, the machines send a cyborg (a combination of machine and organism) back in time to the pre-holocaust present (not insignificantly, 1984). The Terminator's mission is to find and kill Sarah Connor, mother-to-be of John Connor. But the resistance learns of this gambit and is also able to send a soldier, Kyle Reese, back in time to warn and protect Ms. Connor.

The relevant Sarah Connor turns out to be the third person of the same name listed in the Los Angeles telephone directory. While the Terminator mechanically seeks out and murders the first two, Reese has a chance to find the real target and starts following her. When the cyborg attacks, he blasts it repeatedly with a shotgun at close range, but this only stops it for the few seconds Reese and Connor need to escape.

The basic structure of the plot from this point on is standard horror-movie fare about a helpless woman pursued by an unstoppable monster/man and rescued by a (male) good guy in an ever-escalating orgy of violence. After many narrow escapes and Kyle's eventual death, Sarah finally destroys the Terminator (now reduced to a robotic skeleton) by crushing it in a metal press inside a deserted automated factory.

Arnold Schwarzenegger plays the Terminator with a terrifying mechanical grace. Completely devoid of emotion, within seconds of his appearance on the screen he kills two young men just to take their clothes. His mechanical nature is repeatedly emphasized through a number of devices. He has a seemingly symbiotic relationship with all kinds of machines: for example, he starts cars by merely sticking his fingers into the wiring. When shot, he sometimes falls, but immediately stands up again and keeps on lumbering forward. We see him dissect his own wounded arm and eye with an X-Acto knife, revealing the electro-mechanical substrate beneath his human skin. Perhaps most frightening of all, he is able to perfectly mimic any human voice, enabling him to impersonate a police officer and even Sarah's own mother.

What makes the Terminator so alien is not only his mechanical body but his computerized, programmed mind. At times we see the world through his eyes: the picture becomes graphic and filtered, like a bit-mapped image viewed through infrared goggles. Displays of numbers, flashing diagrams, and command menus appear superimposed on his field of vision. The Terminator speaks and understands human language, and his reasoning abilities, especially with respect to other machines (and weapons), are clearly formidable. But he is also a totally single-minded, mechanical being. Kyle warns Sarah that the

Terminator "can't be bargained with. It can't be reasoned with. It doesn't feel pity, or remorse, or fear. And it absolutely will not stop—*ever*—until you are dead." The Terminator thus blends images of a perverse, exaggerated masculine ideal—the ultimate unblinking soldier, the body-builder who treats his body as a machine—with images of computer control and robotic single-mindedness, complete with an alien subjective reality provided by the Terminator's-eye sequences.

The film is built around the idea of a final, apocalyptic struggle to save humanity from its own creations, first from computer-initiated nuclear holocaust and second from the threat of self-aware, autonomous machines grown beyond the limits of human control. But a strong subtheme provides an unusual and very contemporary twist. Sarah Connor begins the film as a waitress whose major problem in life seems to be trying to get a Friday night date. Resentfully, sometimes angrily ("Come on. Do I look like the mother of the future? I mean, am I tough? Organized? I can't even balance my checkbook"), under the relentless pressure of the Terminator's pursuit, she is educated about the threats the future holds and her role as progenitor of the future savior. She learns to make plastic explosive, bandages Kyle's gunshot wound, and listens carefully as he instructs her in the importance of resistance, strength, and fighting spirit. She proves how far she has come in the film's final moments, when the wounded Reese flags as the cyborg approaches. Sarah, hardened and strong, drags him from the Terminator's path, shouting *"On your feet, soldier!"* in a voice that rings with determination. She, not Kyle, is the one who finally destroys the Terminator, in one of the film's most powerful moments. In the end she is transformed into a tough, purposeful mother-to-be, pregnant by Kyle, packing a gun, driving a jeep, and heading off into the sunset and the oncoming storm as heroically as any cowboy.

The Terminator thus offers a new kind of heroine: a single mother who will be both source and model for a race of soldiers fighting for humanity against machines. When Sarah asks Kyle what the women of the future are like, he replies tersely, "Good fighters," and in a dream-memory we see him and a female partner on a combat mission against the machines. In this portrait women take up arms and emerge as men's allies and equals in an increasingly dangerous, alien, and militarized world. The subplot of *The Terminator* is about arming women for a new role as soldiers, outside the more traditional contexts of marriage and male protectorship. The message is also that women are the final defense against the apotheosis of high-technology, militaristic masculinity represented by the Terminator—not only because they harbor connections to emotion and love, as in more traditional imagery, but because they are a source of strength, toughness, and endurance: "good soldiers."

The social reality of 1984 held extraordinary resonances with *The Terminator*'s themes. Public anxiety about nuclear weapons, revelations of epidemic computer failures in NORAD early warning systems, and the Strategic Defense Initiative created a highly charged context for the theme of computer-initiated nuclear holocaust. News stories about "survivalist" movements abounded. Meanwhile a rising tide of robot-based automation in industry, a new wave of computerization in workplaces based on new personal-computer technology, and the Strategic Computing Initiative's controversial proposals for autonomous weapons matched the film's theme of domination by intelligent machines. (Indeed, one of the film's more effective devices is the constant visual reference to the ubiquity of machines and computers: robots, cars, toy trucks, televisions, telephones, answering machines, Walkmans, personal computers.)

With respect to gender issues, the film took its cue from two social developments. First, the highest rates of divorce and single motherhood in history grounded the film's elevation of a single woman to heroine status. Second, starting in the mid-1970s women had become increasingly important as soldiers. Indeed, women filled 10 to 13 percent of all U.S. military jobs by 1985, and there were serious proposals to increase the ratio to 50 percent in the Air Force (since physical strength is not a factor in high-tech jobs like flying jet fighters, and women are apparently able to handle high G-stresses better than men and thus to stay conscious longer during power turns). So it was not much of a stretch for the film to find a model for women of the future in the armed forces.

The iconography of closed-world discourse is reflected in almost every element of this film. The Terminator's terrifying, mechanical single-mindedness and the references to the Skynet "defense network computers" are archetypal closed-world images. The Terminator's mind is inflexible but within its limits extremely clever; the Skynet system is "hooked into everything." The ambiance throughout *The Terminator* is that of closed-world drama. The setting is a grim, usually dark, urban landscape. Almost all of the action occurs in enclosed spaces, and much of it takes place at night. Virtually no natural objects or landscapes appear in the film. Scenes from the world of 2029 A.D. take this imagery to an extreme, with nothing remaining above ground but the rubble and twisted girders of blasted buildings and the charred remains of dead machines. Human dwellings are underground, dirty, furnished with weaponry, canned goods, and the burned-out hulks of television sets, now used as fireplaces. Only two scenes occur in a natural setting: the few hours Sarah and Kyle spend resting in a wooded area (though even here they huddle in a semienclosed space under a bridge), and the final scene in which Sarah drives off toward the mountains of Mexico in a jeep. Thus, in a pattern we will see repeatedly in closed-world discourse, the green world is the final refuge—when there is one—from apocalypse.

Cyborg imagery is also prominent in the film. The Terminator is a liminal figure: a computerized machine that can pass as a man; a living organism whose core is a metallic, manufactured robot; a thinking, reasoning entity with only one purpose.[44] He seems to be alive, but he cannot be killed. He talks, but has no feelings. He can be wounded, but feels no pain. In a flashback (to the future), we learn that the Terminators were created to infiltrate the bunkers of the resistance. Dogs, however, can sense them. Dogs, of course, are liminal figures of another sort, connecting humans with the animal, the natural, and the wild, with the green world.

The Terminator is a military unit, like Kyle Reese, but he is a caricature of the military ideal. He follows his built-in orders unquestioningly, perfectly, and he has no other reason for living. But Kyle, too, has an intense single-mindedness about him, likewise born of military discipline. He dismisses his gunshot wound with a disdainful "Pain can be controlled." He speaks of an emotionless life in a future world where humans, like the machines they fight, live a permanent garrison lifestyle. The Terminator is the enemy, but he is also the self, the military killing machine that Kyle, too, has become—and that Sarah herself must become if humanity is to survive. Humans have built subjective, intelligent military machines but are reduced to a militaristic, mechanical, emotionless subjectivity in order to fend off their own products.

The fictional world of *The Terminator* draws our attention to the historical and conceptual ways in which closed-world and cyborg discourses are linked. Just as facts—about military computing, artificial intelligence, nuclear weapons, and powerful machines—give credibility to fiction, so do fictions—visions of centralized remote control, automated war, global oversight, and thinking machines—give credibility and coherence to the disparate elements that comprise these discourses. We cannot understand their significance without understanding these linkages.

Closed-world discourse helped guide U.S. military policy into an extreme reliance on computers and other high-technology weapons.[45] It also supported many U.S. attempts to manipulate world politics. Cyborg discourse collaborated with closed-world discourse both materially, when artificial intelligence technologies and human/machine integration techniques were used for military purposes, and metaphorically, by creating an interpretation of the inner world of human psychology as a closed and technically manipulable system. Cyborg discourse is the discourse of human automata: of cybernetic organisms for whom the human/machine boundary has been erased. Closed-world discourse represents the form of politics for such beings: a politics of the theorization and control of systems.[46] Thus the third theme of this book is the interactive construction of facts and fictions through the creation of iconographies and political subject positions—maps of meaning, possible subjectivities, narrative frames—within the dramatic spaces of the closed world.

Tools, Metaphors, and Discourse

I will argue throughout this book that tools and metaphors are linked through discourse. But what is "discourse"? How does it work? How does it connect technology with theory, ideology, and subjectivity? Before proceeding with more historical investigations, I want to step back in order to develop some conceptual apparatus. This section explores the nature of computers, the relation between tools and metaphors, and the theory of discourse upon which this book relies. (Readers whose eyes glaze over at the word "theory" should feel free to skip this section, referring back to it only as necessary.)

What Are Computers?

Computers are clearly *tools* or machines, technical levers usefully interposed between practical problems and their solutions. But two essential features distinguish computers from all other machines. These are (a) their ability to store and execute programs that carry out conditional branching (that is, programs that are controlled by their own results to an arbitrary level of complexity), and (b) their ability to manipulate any kind of symbolic information at all, including numbers, characters, and images. These features allow the same computer to "be" many different machines: a calculator, a word processor, a control system, or a communication device.

Unlike classical Aristotelian machines, computers do not perform physical work. They can only control other machines that do, such as lathes, printers, or industrial robots. To do this, they transform information—programs, specifications, input from sensors—into control signals. Computers have little in common with hammers, cooking utensils, power drills, and the other devices that come to mind most readily in connection with the word "tool." They resemble more closely things like rulers and blueprints, tools whose main function is to connect ideas and concepts to the material world. For the most part, computers are tools for *organizing* rather than performing physical work: tools for the mind. Computers are language machines, *information machines*; they are—to pun on modern jargon—hyper texts: active, interactive, hyperactive, self-activating language and code.[47]

The computer's extraordinary flexibility and its special nature as an information machine make it attractive as an analogy for other complex processes less well understood. Thus the computer has also become a culturally central *metaphor* for control, for scientific analysis, and for the mind.[48] Sherry Turkle has described MIT students' use of computer jargon to talk about their human relationships: one student said she needed to "debug" herself through psychotherapy and referred to her "default solutions" for dealing with men.[49] The distinguished artificial intelligence researcher Marvin Minsky has described minds as miniature societies in which "dumb agents" analogous to small programs compete for resources, develop coalitions and enmities, and behave in sometimes unpredictable ways, in an aggregation producing intelligence as a kind of by-product.[50] Because of the computer's abilities and its complexity, this metaphorical dimension can reach beyond descriptive convenience. The computer can become a simulated world, an electronic landscape within which new experiences and relationships are possible. For heavy users, the computer can become a kind of virtual reality—a domain of experience and a way of life.

Tools as Metaphors
What is the relation between computers as tools and computers as metaphors?

In *Computer Power and Human Reason*, MIT computer scientist Joseph Weizenbaum compares computers to clocks. Like computers, clocks are machines that do no physical work. Weizenbaum calls clocks and computers "autonomous machines," as opposed to the "prosthetic machines" that extend the human physical ability to alter or move about within the material world. An autonomous machine, "once started, runs by itself on the basis of an internalized model of some aspect of the real world." Weizenbaum points out that autonomous machines and the internalized models they embody have had profound effects on human experience. His meditation on the clock (following Mumford) is worth quoting at length:

> Where the clock was used to reckon time, man's regulation of his daily life was no longer based exclusively on, say, the sun's position over certain rocks or the crowing of a cock, but was now based on the state of an autonomously behaving model of a phenomenon of nature. The various states of this model were given names and thus reified. And the whole collection of them superimposed itself on the existing world and changed it.... The clock had created literally a new reality.... Mumford [makes] the crucial observation that the clock "disassociated time from human events and helped create the belief in an independent world of mathematically measurable sequences: the special world of science." The importance of that effect of the clock on man's perception of the world can hardly be exaggerated.[51]

The clock was a machine whose primary function was metaphorical. The operation of clocks came to stand for and to structure both the physical process and the personal experience of the passage of time, drawing all aspects of time together under the aegis of a universal symbol. The name of the machine remains embedded in our contemporary concept of time, visible whenever someone responds to the question "What time is it?" with the answer "It's three o'clock." The example demonstrates the possibility of a machine's having subtle, profound, and *material* effects *solely through its function within a system of ideas*.

All tools, including clocks and computers, have both practical and metaphorical or symbolic dimensions. This is true for reasons also noted by Weizenbaum: "tools, whatever their primary practical function, are necessarily also pedagogical instruments. They are pregnant symbols in themselves. They symbolize the activities they enable, i.e., their own use.... A tool is also a model for its own reproduction and a script for the reenactment of the skill it symbolizes."[52] The experience of using any tool changes the user's awareness of the structure of reality and alters his or her sense of the human possibilities within it. Weizenbaum mentions the tool's effect on an individual's "imaginative reconstruction" of the world. In a technological culture, that effect extends beyond the phenomenology of individual experience to large elements of the society as a whole. In cases such as the clock or the automobile it can help create wholesale changes in culture.

Language is a prominent element in this "imaginative reconstruction." Complex tools like computers and cars evolve complex languages for talking about their functioning, their repair, and their design. Beyond the demands of practical interaction, linguistic metaphors drawn from tools and machines are extremely commonplace. One may speak of "hammering home" a point in an argument, "cutting through" bureaucratic "red tape," "measuring" one's words, having a "magnetic" personality, "steering" someone in the right direction, an argument's being "derailed" or "on track," and so on. Tools and their uses thus form an integral part of human discourse and, through discourse, not only shape material reality directly but also mold the mental models, concepts, and theories that guide that shaping.

Tools shape discourse, but discourse also shapes tools. In fact, I will argue that tools like the computer must be considered elements of discourse, along with language and social practices. Metaphors can be not merely linguistic but experiential and material as well. This is what makes metaphors such as the computer political entities.

Concepts of Discourse

But what is "discourse," and how does it integrate tools, metaphors, politics, and culture? To understand both the meaning of this term and my reasons for adopting it, it will help to consider some related concepts that I might have used in its place: *ideology, paradigm, worldview*, and *social construction*.

Raymond Williams defines ideology as "the set of ideas which arise from a given set of material interests."[53] Historically, especially in the Marxist tradition, this concept has been important in focusing attention on the relationships between the material conditions of existence (natural resources, human abilities and needs, the current state of technology, etc.) and social systems, shared beliefs, legal codes, and state structures. Analysis of ideology has concentrated on how political and social power emerges from those relationships.

Unfortunately, "ideology" also carries with it a strong secondary sense, that of "illusion, abstract and false thought." This is connected with the frequent intent of analysts who use the term to expose "the ways in which meaning (or signification) serves to sustain relations of domination."[54] This common connotation tends to identify "ideology" with those beliefs and cultural constructions that suppress dissent or revolt by obscuring the true sources of oppression and redirecting the energy of social unrest into channels controlled by a dominant class. Everyday usage often makes "ideology" a pejorative term distinguishing distorted, false, or socially retrograde ideas from true knowledge. Also, there is a long Marxist tradition (e.g., theories of base/superstructure relations) that regards ideology as a pure *product* of material conditions and the acceptance of ideological beliefs as "false consciousness."[55]

Terry Eagleton has recently attempted to rehabilitate the term by rendering it as the subset of discourse that deals with "those power struggles which are somehow central to a whole form of social life."[56] This is an important clarification and a sense I wish to carry forward into my own usage of "discourse." But my purpose is to identify not only politically central struggles but also contests *and collaborations* over issues that have more to do with knowledge and subjectivity than with state politics. Therefore I prefer to reserve the term "ideology" for its narrower sense, with its implications of distortion and false consciousness inherent in beliefs that emerge from particular material conditions. I intend "discourse," by contrast, to be both broader and more neutral with respect to the truth or falsity of belief, emphasizing the constructive and productive elements of the interaction of material conditions with knowledge, politics, and society.

A second alternative would be the concept of a "paradigm," as used by Thomas Kuhn and his followers. Kuhn's notion of a scientific paradigm emphasized the development of coherent structures of thought and practice centered around exemplars, or foundational experiments, and basic theoretical concepts. The exemplar(s) implicitly define a set of rules governing the choice and construction of research problems, theories, methods, and instrumentation. Once a paradigm is established, normal scientific practice consists essentially of puzzle solving, or elaborating the paradigmatic theory and working out its experimental consequences. Anomalous results, though almost always present, are simply disregarded until their weight builds to a crisis point, or until a new fundamental theory appears to challenge the established one.

At this point a revolutionary transition occurs, often quite quickly. A new foundational experiment and/or theory redefines or replaces basic terms, and the scientific community re-forms around the new paradigm. The new paradigm is said to be "incommensurable" with the old. This term, whose definition remains disputed, originally seemed to imply that a new paradigm constituted a full-blown, all-encompassing worldview that could not be understood or possibly even perceived by those whose allegiance remained with the old paradigm.[57]

Some of these ideas, too, I wish to preserve in my usage of "discourse." A paradigm has coherence; it is based in concrete practices and frequently in technologies of experimentation, and it may centrally include one or more metaphors. A paradigm, once established, falls into the background of knowledge and appears to be little more than common sense, governing the production of truth (in Michel Foucault's phrase) by constituting the obvious. The concept emphasizes the tremendous inertia acquired by established systems of thought, the embeddedness of theory in language, and the large social and cognitive costs of wholesale transitions. The idea that scientific observations are "theory-laden," in the sense that what scientists see is structured by the paradigm they inhabit, also descends from Kuhn. I would like to preserve most of these notions as connotations of "discourse."

But like "ideology," "paradigm" does not fit my purpose here. It has become a term of art in professional history of science, but it has also been popularized to the point of vulgarity, usually in reference to incommensurable gestalts. "Discourse," in my usage, will be neither so hermetic nor so coherent as "paradigm" has often been interpreted to be. Individuals may participate in and be shaped by numerous discourses without being fully determined by any of them. People may have fluent repertoires in alternate, even conflicting, discourses: socially, these discourses may be produced for different purposes in different contexts.[58] Finally, the boundaries of discourses are more ragged and more permeable than those of paradigms. The notion of discourse is much more of an analytical construct than the idea of a paradigm. It allows us to discern a certain order, but this order is not *the* order of things, only one suited to a particular purpose, in a particular context.[59] "Paradigm" is a more totalizing term than "discourse," in my usage, will be.

Another alternative might be the old sociology-of-knowledge concept of *Weltanschauung*, or "worldview." This idea captures the contingent nature of discourse and its relative coherence, and it has the advantage of focusing attention on the subjective reality of the experience produced within a socially constructed system of thought.[60] But the term is too phenomenological, emphasizing the subjective dimension of discourse at the expense of its relationships with technology and other material conditions.

Finally, in the last decade a growing literature in the history and sociology of technology has introduced an array of concepts focused around the idea of "social construction," which I take to mean that technologies are always developed by groups engaged in building, simultaneously, their meaning and their physical form. Of these concepts, at least the following bear strong resemblances to those I develop in this book.

Trevor Pinch and Wiebe Bijker's research program in the social construction of technology signals the power of an analysis of technology guided first and foremost by its role in social groups. They describe how social interpretations of problems fix the meaning and physical form of particular technologies.[61] John Law's "heterogeneous engineering" points to the multiplicity of materials and forces that groups draw upon to put together working technologies.[62] The "actor-network theory" of Bruno Latour and Michel Callon directs attention to the ways science and technology function as networks of power in which the enrollment of active allies (humans, machines, and other "actants") is a primary mechanism.[63] Bijker's idea of "technological frames"—much like that of "paradigms"—refers to the combinations of concepts, theories, goals, and practices used by groups attempting to solve technological problems.[64] Steven Shapin and Simon Schaffer define various "technologies," including the material, the literary, and the social, that seventeenth-century scientists employed to establish a "form of life" and a social space wherein experiments could count as establishing facts.[65] Thomas Hughes has pointed to the role of entrepreneurial "system builders" in creating large technological systems whose scales help them achieve a "momentum" presenting the appearance of a "seamless web" of autonomous technology.[66] Peter Taylor's ideas of "heterogeneous constructionism" and "distributed causality" have significantly expanded the sophistication of the social constructivist program.[67]

These are deep, and deeply important, conceptions of scientific and technological change. However, with the exception of Shapin and Schaffer's Wittgensteinian concept of a "form of life," all of these terms are built to a purpose that is not my own. They help us to understand technological change as a social process, but to do so they focus on the technology itself: innovation, invention, design. As Pinch and Bijker themselves have noted, few studies have managed fully to engage the relationship between the meanings of scientific facts or technological artifacts and their sociopolitical milieu.[68] This is part of my concern in this book, but it is not the whole of it. Instead, my goal is to balance problems in the social construction of technology with their converse, which is to say the *technological construction of social worlds*. The term "discourse" points strongly to the sociopolitical dimensions of technology but at the same time, in my usage, directs attention to the material elements shaping the social and political universe; it is a broad term, in short, for the heterogeneous *media* in which the process of social construction operate.

Having distinguished discourse from these alternatives, let me now develop a positive definition.

Discourse in its narrowest sense refers to the act of conversation (as distinguished from language itself). The analytic use of this term descends from sociological studies of speech in context, sometimes called "discourse analysis." In the larger sense I will employ here, though, discourse goes beyond speech acts to refer to the entire field of *signifying* or *meaningful practices*: those social interactions—material, institutional, and linguistic—through which reality is interpreted and constructed for us and with which human knowledge is produced and reproduced. *A discourse, then, is a way of knowledge, a background of assumptions and agreements about how reality is to be interpreted and expressed, supported by paradigmatic metaphors, techniques, and technologies and potentially embodied in social institutions.* This usage emerges from, though it is not identical with, that of French critical theorists such as Roland Barthes, Michel Foucault, and Jacques Derrida. While "discourse," too, has suffered abuse at the hands of those who would make it explain everything (and so explain nothing), I think that it is still fresh and active enough to fill the role I have in mind. To establish this role more precisely, let me briefly sketch its intellectual ancestry.

Wittgenstein: Language-Games and Meaning as Use
My concept of discourse has a great deal in common with the later Wittgenstein's idea of a "language-game." *A* Wittgensteinian language-game is the set of linguistic and nonlinguistic means that constitute some domain of human social practice. *The* language-game, as a

whole, consists of "language and the actions into which it is woven," according to the *Philosophical Investigations*.[69]

Wittgenstein sees language in the ordinary sense as part of a wider background of practices, materials, and institutions. His semantic theory emphasizes the primacy of training over explanation in the acquisition of language. Especially as children, people come to understand or acquire the meanings of words as part of patterns of action in their lives. They are taught which words to say in innumerable situations, and the first of these uses have to do with the practical satisfaction of needs and desires. Thus people initially experience language not as representation but as action. It is one more thing they can *do*, like reaching for something, crying, or jumping up and down, to get what they want. Once a basic vocabulary is established by training, new language can be learned by explanation. Only at this point can language begin to seem primarily representational. This is the force of Wittgenstein's slogan "meaning is use."[70]

Wittgenstein's most developed example of this phenomenon involves ostensive definition, that is, defining a word by pointing to the object it names. He notes that ostensive definitions are possible only once a "place" in a language-game has been established for the words they define. Thus, pointing to the brake pedal of a car and saying "That's the brake" would only make sense as a definition if the recipient of the explanation already understands automobiles, driving, stopping, starting, and so on. The rest of the context within which "That's the brake" makes sense must be acquired through action—by driving with other people, watching movies involving cars and drivers, and so on. Furthermore, the act of pointing to something itself has a conventional meaning. Infants must be trained to recognize pointing as part of the process of definition. Not only the word defined by ostension, but the pointing itself and the object indicated by pointing, are components of the language-game in Wittgenstein's view: "it is most natural, and causes least confusion, to reckon the samples among the instruments of the language."[71]

Language-games are profoundly public and conventional in nature. People learn to speak in contexts of action that are themselves to some degree habitual, traditional, and institutionalized. Indeed, a sound can only function as a word by virtue of its use in a community. If I label something with a sound I invent, that sound does not quite count as a word until I employ it in a communicative context. This necessarily involves making it public by teaching someone else how it is to be used, that is, in what pattern of action it has a place. Actions, too, can be apprehended by language only once they become patterned and public, for similar reasons.

It is not possible that there should have been only one occasion on which someone obeyed a rule. It is not possible that there should have been only one occasion on which a report was made, an order given or understood, and so on.—To obey a rule, to make a report, to give an order, to play a game of chess, are customs (uses, institutions).

To understand a sentence means to understand a language. To understand a language means to be master of a technique.[72]

Thus language itself operates as a tool. "Language is an instrument. Its concepts are instruments," Wittgenstein says.[73] He means that words are part of concrete actions, just as actions are part of language.

Wittgenstein's ultimate conclusion is that the process of grounding knowledge comes to an end within language-games—not in a reality external to the social world.

But isn't it experience that teaches us to judge like this, that is to say, that it is correct to judge like this? But how does experience teach us, then? We may derive it from experience, but experience does not direct us to derive *anything* from experience. If it is the ground of our judging like this, and not just the cause, still we do not have a ground for seeing this in turn as a ground. No, experience is not the ground for our game of judging. Nor is its outstanding success.

"An empirical proposition can be tested" (we say). But how? and through what? What counts as its test? . . . —As if giving grounds did not come to an end sometime. But the end is not an ungrounded presupposition: it is an ungrounded way of acting.[74]

Language-games make use of all kinds of things, including experience, evidence, and real objects, but there is no ultimate justification for these uses, since justification is itself an "ungrounded way of acting." What "make" propositions true or false are the public practices of justification, verification, etc., of a particular community, not the properties of objects they "describe." (Description, too, is a language-game, part of a cultural discourse.)

Ultimately, for Wittgenstein, language-games are elements of "forms of life," larger, more general, mutually reinforcing patterns of action, language, and logic. In *Leviathan and the Air-Pump*, Shapin and Schaffer offer an extended example. They use Wittgenstein's concept to describe the "experimental life" constructed by Robert

Boyle and his colleagues at the Royal Society in the seventeenth century. Boyle and his followers established what (if anything) an experiment proved, what protocols had to be followed for an event to count as an experiment, what sorts of witnesses were necessary to validate the matters of fact demonstrated by experiment, and other fundamental practices and logics of scientific experimentation. To do so they built what Shapin and Schaffer describe as three kinds of *technologies*: material (Boyle's air-pump as a paradigmatic device), social (the laboratory as a limited public space and its membership as valid witnesses), and literary (forms of description of experiments that allowed readers to function as "virtual witnesses" who could themselves validate an experiment). In short, they constructed the whole form of life, the linked set of language-games and practices, that still underlies science.[75]

Wittgenstein's lessons that language is often if not always a form of action, that meaning is grounded in practice rather than representation, and that the great bulk of human activity occurs within habitual, instinctual, traditional, and institutionalized patterns of action underlie my usage of the term "discourse." These ideas establish a basis for thinking of tools, and the languages and metaphors they generate, together as a single unit of analysis. The tool-like uses of computers and their roles as models, metaphors, and experiences are connected as part of an interrelated set of language-games. We cannot understand their operation as tools in isolation from the way they are taken up in discourse about them, just as we cannot understand discourses *about* computers apart from the devices and the practices that employ them.

Foucault and the Idea of Discourse

The notion of a language-game composed of heterogeneous elements is remarkably similar to Foucault's concept of discourse. But Foucault focuses on a factor Wittgenstein generally ignores: competition among discourses, motivated by power relationships among human groups.

In a society such as ours, but basically in any society, there are manifold relations of power which permeate, characterize, and constitute the social body, and these relations of power cannot themselves be established, consolidated nor implemented without the production, accumulation, circulation and functioning of a discourse. There can be no possible exercise of power without a certain economy of discourses of truth which operates through and on the basis of this association.[76]

Foucault conceives of discourses as the sites where the objects of knowledge are constructed. In a sense, for Foucault the idea of a discourse replaces the more traditional notions of "institution," "convention," and "tradition." Discourses are the Wittgensteinian *forms of life* which institutions and traditions structure for their inhabitants. A form of life is not—or is not only—a form of experience. Discourses create and structure experience, but they are themselves primarily conventional, material, and linguistic, rather than experiential.

In analyzing discourses, Foucault focuses on particularities. He resists reducing discourses to ideologies, or reflections of a "base" in the economy of wealth, seeing instead a multiplicity of "economies" that overlap and vie with each other for dominance. When Foucault describes a discourse as an economy, he means that like the economy of wealth, social institutions constitute self-elaborating and above all *productive* systems with their own elements and logic. This metaphor of an economy is meant in the almost literal sense of a structure of production and exchange of useful things. Like Wittgenstein, Foucault explicitly differentiates the economy of discourse from "a system of representations."[77] He rejects semiotic or linguistic models because they seem to reduce knowledge to the possession of meaningful symbols, whereas knowledge is for him the result of continuous micropolitical struggles.[78]

The economics of discourse is also not a semiotics because the unity of its objects of knowledge is not given by a language or a system of rationality, but created *ad hoc*. Foucault calls sexuality, for example, a "fiction... [an] artificial unity [of] anatomical elements, biological functions, conducts, sensations, and pleasures."[79] The sense of a constantly regenerated and changing discourse differentiates Foucault's concept from the more monolithic stability of Wittgenstein's "forms of life." In a sense, Foucault gives a diachronic view of objects Wittgenstein would have characterized synchronically. He describes discourse as "a series of discontinuous segments whose tactical function is neither uniform nor stable."[80] It is a collection of fragments grouped and interconnected around a "support." The support is the object at once studied and invented by the discourse that surrounds it. I will use this concept to describe the role of computers in closed-world and cyborg discourses.

As an example, Foucault asks us to consider the nineteenth-century campaign against children's masturbation. "This campaign entailed... using these tenuous pleasures as a prop, constituting them as secrets (that is, forcing them into hiding so as to make possible their discovery), tracing them from their origins to their effects.... What was demanded of it was to persevere, to proliferate... rather than to disappear for good."[81] The onanistic child thus became a support or artificial center (in a sense not unlike the "exemplars" of Kuhn's paradigms) not only of a theory of sexuality, but of a whole set of nonlinguistic practices as well, such as the architecture of bathrooms, the enforcement of laws, and the production of books and pamphlets. Such figures as the electronic control center (the War Room, for example) and the cyborg soldier are *supports*, in this sense, for closed-world discourse. The figures of the intelligent machine and the Turing test serve this function for cyborg discourse. Cyborg imagery and problems of control overlap and connect both discourses.

The metaphor of a discursive economy also ties the self-elaborating logic of discourse to the reality of social power. Here Foucault's best example is a "mutation" that occurred in Europe, between the seventeenth and the nineteenth centuries, in the way social control was paradigmatically exercised: from *punishment* by force to *discipline* through training and surveillance—a more subtle but far more pervasive method. For Foucault, Jeremy Bentham's Panopticon, a circular prison constructed so that every inmate is always physically visible to guards in a central tower, was paradigmatic. People who think they are being watched tend to do what they think they are supposed to do, even when they are not. People whose physical actions and emotional responses have been shaped by discipline (soldiers, workers, prisoners) tend to adopt the mindset of the disciplinary institution. Sophisticated, ubiquitous technologies and techniques such as the Panopticon—and computerized recordkeeping has at least the potential to create immensely wide-ranging and insidious panoptic techniques[82]—have increased the ability of institutions to control people without touching them, using the subtle pressures of internalized discipline. In this way, argues Foucault, modern power is more productive than repressive in nature.

If power is productive, what does it produce? First, it generates active compliance rather than passive obedience. But also, for Foucault, power produces truth: true knowledge, warranted by a set of techniques and rules for the creation and evaluation of statements as true and false. More simply, power determines what can *count* as true and false. This is the force of Foucault's concept of "power/knowledge": true knowledge is an effect of power relationships, since power sets the norms for acceptable statements and also sets in motion the process of generating and evaluating those statements—but also itself produces power, since true knowledge enables its possessors to achieve their practical goals. (Thus, in Operation Igloo White, closed-world discourse generated both repressive power—the surveillance and the bombing itself—and productive power—the development of remote sensing techniques, support for the U.S. involvement in Vietnam through the appearance of success, new North Vietnamese tactics, and so on. It also generated specific forms of new knowledge: sensor data and analysis techniques, statistical analyses of Ho Chi Minh Trail traffic, North Vietnamese knowledge of American tactics, and so on.[83]) In this Foucault goes beyond Wittgenstein, who contents himself with pointing out the conventional character of signification and justification, to try to answer the question of *how* these conventions are themselves produced and enforced.

Finally, the constant exchanges of language and knowledge in which a discourse is enacted actually help to constitute individual subjects and describe and mold the social body. Foucault plays upon the different meanings of "subject," as in the "subjects" of the king, "subjection" to torture or surveillance, and "subjectivity" itself, noting their more than trivial interconnections. Experiences, feelings, habits, and customs may be among the products of discourse. In a sense I will develop in chapter 5, discourses create *subject positions* inhabitable by individuals.[84]

Discourse: Technology as Social Process

A discourse, then, is a self-elaborating "heterogeneous ensemble" that combines techniques and technologies, metaphors, language, practices, and fragments of other discourses around a support or supports. It produces both power and knowledge: individual and institutional behavior, facts, logic, and the authority that reinforces it. It does this in part by continually maintaining and elaborating "supports," developing what amounts to a discursive infrastructure. It also continually expands its own scope, occupying and integrating conceptual space in a kind of discursive imperialism. Like a paradigm, much of the knowledge generated by a discourse comes to form "common sense."

As applied to computers in the postwar world, my concept of discourse accepts neither the billiard-ball imagery of technological "impacts" on society nor the too-frequent conspiracy imagery of technological "choices" as governed by dominant social groups. Instead it views technology as one focus of a *social process* in which impacts, choices, experiences, metaphors, and environments all play a part.[85] This vantage point will allow us to explore the politics of material change and the politics of representation as linked elements of the politics of culture.

Objects of knowledge, like other products of human activity, are produced under historically specific conditions from raw materials that are themselves historical products, including practices, objects, symbols, and metaphors. Science and engineering *normally* proceed not so much by the application of well-codified methods to well-defined problems as by what Claude Lévi-Strauss called *bricolage*, or "tinkering."[86] The models, metaphors, research programs, and standards of explanation that make up a scientific paradigm are assembled piece by piece from all kinds of heterogeneous materials. To see science and engineering as tinkering—as discourse—is to blur and twist the sharp, neat lines often drawn between them and the knowledges and practices that constitute other human endeavors such as politics, commerce—or war.

With these conceptual tools ready to hand, we can now explore how computers became a crucial infrastructural technology—a crucial Foucaultian support—for Cold War closed-world discourse.

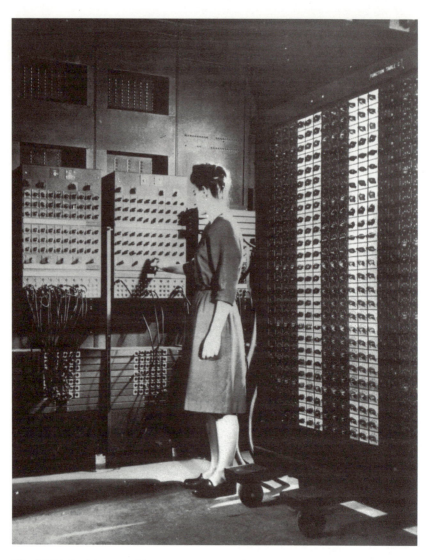

The ENIAC with one of its programmers, circa 1945. Courtesy Charles Babbage Institute.

2
Why Build Computers?: The Military Role in Computer Research

On the battlefield of the future, enemy forces will be located, tracked, and targeted almost instantaneously through the use of data links, computer assisted intelligence evaluation, and automated fire control. With first round kill probabilities approaching certainty, and with surveillance devices that can continually track the enemy, the need for large forces to fix the opposition physically will be less important.... [A]n improved communicative system... would permit commanders to be continually aware of the entire battlefield panorama down to squad and platoon level.... Today, machines and technology are permitting economy of manpower on the battlefield.... But the future offers even more possibilities for economy. I am confident the American people expect this country to take full advantage of its technology—to welcome and applaud the developments that will replace wherever possible the man with the machine.... With cooperative effort, no more than 10 years should separate us from the automated battlefield.[1]

—*General William Westmoreland, former Commander-in-Chief of U.S. forces in Vietnam, 1969*

For two decades, from the early 1940s until the early 1960s, the armed forces of the United States were the single most important driver of digital computer development. Though most of the research work took place at universities and in commercial firms, military research organizations such as the Office of Naval Research, the Communications Security Group (known by its code name OP-20-G), and the Air Comptroller's Office paid for it. Military users became the proving ground for initial concepts and prototype machines. As the commercial computer industry began to take shape, the armed forces and the defense industry served as the major marketplace. Most historical accounts recognize the *financial* importance of this backing in early work on computers. But few, to date, have grasped the deeper significance of this military involvement.

At the end of World War II, the electronic digital computer

technology we take for granted today was still in its earliest infancy. It was expensive, failure-prone, and ill-understood. Digital computers were seen as calculators, useful primarily for accounting and advanced scientific research. An alternative technology, analog computing, was relatively cheap, reliable (if not terribly accurate), better developed, and far better supported by both industrial and academic institutions. For reasons we will explore below, analog computing was more easily adapted to the control applications that constituted the major uses of computers in battle. Only in retrospect does it appear obvious that command, control, and communications should be united within a single technological frame (to use Wiebe Bijker's term) centered around electronic digital computers.[2]

Why, then, did military agencies provide such lavish funding for digital computer research and development? What were their near-term goals and long-term visions, and how were these coupled to the grand strategy and political culture of the Cold War? How were those goals and visions shaped over time, as computers moved out of laboratories and into rapidly changing military systems?

I will argue that military support for computer research was rarely benign or disinterested—as many historians, taking at face value the public postures of funding agencies and the reports of project leaders, have assumed. Instead, practical military objectives guided technological development down particular channels, increased its speed, and helped shape the structure of the emerging computer industry. I will also argue, however, that the social relations between military agencies and civilian researchers were by no means one-sided. More often than not it was civilians, not military planners, who pushed the application of computers to military problems. Together, in the context of the Cold War, they enrolled computers as supports for a far-reaching discourse of centralized command and control—as an enabling, infrastructural technology for the closed-world political vision.

The Background: Computers in World War II

During World War II, virtually all computer research (like most scientific research and development) was funded directly by the War Department as part of the war effort. But there are particularly intimate links between early digital computer research, key military needs, and the political fortunes of science and engineering after the war. These connections had their beginnings in problems of ballistics.

One of the Allies' most pressing problems in World War II was the feeble accuracy of antiaircraft guns. Airplanes had evolved enormously since World War I, gaining speed and maneuverability. Defense from devastating Nazi bombing raids depended largely on ground-based antiaircraft weapons. But judging how far ahead of the fast-moving, rapidly turning planes to aim their guns was a task beyond the skills of most gunners. Vast amounts of ammunition were expended to bring down a distressingly small number of enemy bombers. The German V-I "buzz bombs" that attacked London in 1944 made a solution even more urgent. The problem was solved by fitting the guns with "gun directors," a kind of electromechanical analog computer able to calculate the plane's probable future position, and "servomechanisms," devices that controlled the guns automatically based on the gun director's output signals.[3]

Building the gun directors required trajectory tables in which relations between variables such as the caliber of the gun, the size of the shell, and the character of its fuse were calculated out. Ballistics calculations of this sort have a long history in warfare, dating almost to the invention of artillery. Galileo, for example, invented and marketed a simple calculating aid called a "gunner's compass" that allowed artillerymen to measure the angle of a gun and compute, on an ad hoc basis, the amount of powder necessary to fire a cannonball a given distance.[4] As artillery pieces became increasingly powerful and complex, precalculated ballistics tables became the norm. The computation of these tables grew into a minor military industry. During World War I, young mathematicians such as Norbert Wiener and Oswald Veblen worked on these problems at the Army's Aberdeen Proving Ground. Such mathematicians were called "computers."[5]

In World War II, with its constant and rapid advances in gunnery, Aberdeen's work became a major bottleneck in fielding new artillery and antiaircraft systems. Both Wiener and Veblen—by then distinguished professors at MIT and Princeton, respectively—once again made contributions. Wiener worked on the antiaircraft gunnery problem at its most general level. His wartime studies culminated in the theory of cybernetics (a major precursor of cognitive psychology). Veblen returned to Aberdeen's ballistics work as head of the scientific staff of the Ballistics Research Laboratory (BRL). Just as in World War I, Veblen's group employed hundreds of people, this time mostly women, to compute tables by hand using desk calculators.

These women, too, were called "computers." Only later, and gradually, was the name transferred to the machines.[6]

But alongside them Aberdeen also employed the largest analog calculator of the 1930s: the differential analyzer, invented by MIT electrical engineer Vannevar Bush.

Vannevar Bush: Creating an Infrastructure for Scientific Research

Bush invented the differential analyzer at MIT in 1930 to assist in the solution of equations associated with large electric power networks. The machine used a system of rotating disks, rods, and gears powered by electric motors to solve complex differential equations (hence its name). The BRL immediately sought to copy the device, with improvements, completing its own machine in 1935 at Aberdeen. At the same time, another copy was constructed at the University of Pennsylvania's Moore School of Engineering in Philadelphia, this one to be used for general-purpose engineering calculation. The Moore School's 1930s collaboration with the BRL, each building a differential analyzer under Bush's supervision, was to prove extremely important. During World War II, the two institutions would collaborate again to build the ENIAC, America's first full-scale electronic digital computer.

Bush was perhaps the single most important figure in American science during World War II, not because of his considerable scientific contributions but because of his administrative leadership. As war approached, Bush and some of his distinguished colleagues had used their influence to start organizing the scientific community for the coming effort. After convincing President Roosevelt that close ties between the government and scientists would be critical to this war, they established the National Defense Research Committee (NDRC) in 1940, with Bush serving as chair. When the agency's mandate to conduct research but not development on weapons systems proved too restrictive, Bush created and took direction of an even larger organization, the development-oriented Office of Scientific Research and Development (OSRD), which subsumed the NDRC.[7] The OSRD coordinated and supervised many of the huge science and engineering efforts mobilized for World War II. By 1945 its annual spending exceeded $100 million; the prewar *total* for military R&D had been about $23 million.[8]

Academic and industrial collaboration with the military under the

OSRD was critically important in World War II. Research on radio, radar, the atomic bomb, submarines, aircraft, and computers all moved swiftly under its leadership. Bush's original plans called for a decentralized research system in which academic and industrial scientists would remain in their home laboratories and collaborate at a distance. As the research effort expanded, however, this approach became increasingly unwieldy, and the OSRD moved toward a system of large central laboratories.

Contracts with universities varied, but under most of them the university provided laboratory space, management, and some of the scientific personnel for large, multidisciplinary efforts. The Radio Research Laboratory at Harvard employed six hundred people, more of them from California institutions than from Harvard itself. MIT's Radiation Laboratory, the largest of the university research programs, ultimately employed about four thousand people from sixty-nine different academic institutions.[9] Academic scientists went to work for industrial and military research groups, industrial scientists assisted universities, and the military's weapons and logistics experts and liaison officers were frequent visitors to every laboratory. The war effort thus brought about the most radical disciplinary mixing, administrative centralization, and social reorganization of science and engineering ever attempted in the United States.

It would be almost impossible to overstate the long-term effects of this enormous undertaking on American science and engineering. The vast interdisciplinary effort profoundly restructured scientific research communities. It solidified the trend to science-based industry—already entrenched in the interwar years—but it added the new ingredient of massive government funding and military direction. MIT, for example, "emerged from the war with a staff twice as large as it had had before the war, a budget (in current dollars) four times as large, and a *research* budget ten times as large—85 percent from the military services and their nuclear weaponeer, the AEC."[10] Eisenhower famously named this new form the "military-industrial complex," but the nexus of institutions is better captured by the concept of the "iron triangle" of self-perpetuating academic, industrial, and military collaboration.[11]

Almost as important as the institutional restructuring was the creation of an unprecedented *experience* of community among scientists and engineers. Boundaries between scientific and engineering disciplines were routinely transgressed in the wartime labs, and scientists found the chance to apply their abilities to create useful devices profoundly exciting. For example, their work on the Manhattan Project bound the atomic physicists together in an intellectual and social brotherhood whose influence continued to be felt into the 1980s. Radiation Laboratory veterans protested vigorously when the lab was to be abruptly shut down in December 1945 as part of postwar demobilization; they could not believe the government would discontinue support for such a patently valuable source of scientific ideas and technical innovations. Their outcry soon provoked MIT, supported by the Office of Naval Research (ONR), to locate a successor to the Rad Lab in its existing Research Laboratory of Electronics.[12] Connections formed during the war became the basis, as we will see over and over again, for enduring relationships between individuals, institutions, and intellectual areas.

Despite his vast administrative responsibilities, Bush continued to work on computers early in the war. He had, in fact, begun thinking in 1937–38 about a possible electronic calculator based on vacuum tubes, a device he called the Rapid Arithmetical Machine. Memoranda were written and a research assistant was engaged. But Bush dropped the project as war brought more urgent needs. His assistant, Wilcox Overbeck, continued design work on the machine, but he too was finally forced to give up the project when he was drafted in 1942. Most of Overbeck's work focused on tube design, since Bush was concerned that the high failure rates of existing vacuum tubes would render the Rapid Arithmetical Machine too unreliable for practical use. Possibly because of this experience, Bush opposed fully electronic computer designs until well after the end of World War II.[13]

Bush did, however, perfect a more powerful version of the differential analyzer, known as the Rockefeller Differential Analyzer (after its funding source) at MIT in 1942. This device could be programmed with punched paper tape and had some electronic components. Though committed to analog equipment and skeptical of electronics, he kept abreast of the Moore School's ENIAC project, and the universe of new possibilities opened up by computers intrigued him.[14]

Thus it so happened that the figure most central to World War II science was also the inventor of the prewar period's most important computer technology. Bush's laboratory at MIT had established a tradition of analog computation and control engineering—

not, at the time, separate disciplines—at the nation's most prestigious engineering school. This tradition, as we will see, weighed against the postwar push to build digital machines. Simultaneously, though, the national science policies Bush helped create had the opposite effect. The virtually unlimited funding and interdisciplinary opportunities they provided encouraged new ideas and new collaborations, even large and expensive ones whose success was far from certain. Such a project was the Moore School's Electronic Numerical Integrator and Calculator (ENIAC), the first American electronic digital computer.

The ENIAC Project

Even with the help of Bush's differential analyzer, compiling ballistics tables for antiaircraft weapons and artillery involved tedious calculation. Tables had to be produced for every possible combination of gun, shell, and fuse; similar tables were needed for the (analog) computing bombsight and for artillery pieces. Even with mechanical aids, human "computers" made frequent mistakes, necessitating time-consuming error-checking routines. The BRL eventually commandeered the Moore School's differential analyzer as well. Still, with two of these machines, the laboratory fell further and further behind in its work.

"The automation of this process was . . . the *raison d'être* for the first electronic digital computer," wrote Herman Goldstine, co-director of the ENIAC project. The best analog computers, even those built during the war, were only "about 50 times faster than a human with a desk machine. None of these [analog devices were] sufficient for Aberdeen's needs since a typical firing table required perhaps 2,000–4,000 trajectories. . . . The differential analyzer required perhaps 750 hours—30 days—to do the trajectory calculations for a table."[15] (To be precise, however, these speed limitations were due not to the differential analyzer's *analog* characteristics, but to its electromechanical nature. Electronic equipment, performing many functions at the speed of light, could be expected to provide vast improvements. As Bush's RDA had demonstrated, electronic components could be used for analog as well as digital calculation. Thus nothing in Aberdeen's situation dictated a *digital* solution to the computation bottleneck.)

The Moore School started research on new ways of automating the ballistics calculations, under direct supervision of the BRL and the Office of the Chief of Ordnance. In 1943 Moore School engineers John Mauchly and J. Presper Eckert proposed the ENIAC project. They based its digital design in part on circuitry developed in the late 1930s by John Atanasoff and Clifford Berry of the Iowa State College. (Atanasoff and Berry, however, never pursued their designs beyond a small-scale prototype calculator, conceiving it, as did most engineers of their day, more as a curiosity of long-term potential than as an immediate alternative to existing calculator technology.)[16] The BRL, by this point desperate for new assistance, approved the project over the objections of Bush, who thought the electronic digital design infeasible.

The ENIAC represented an electrical engineering project of a completely unprecedented scale. The machine was about 100 times larger than any other existing electronic device, yet to be useful it would need to be at least as reliable as far smaller machines. Calculations revealed that because of its complexity, the ENIAC would have to operate with only one chance in 10^{14} of a circuit failure in order to function continuously for just twelve hours. Based on these estimates, some of ENIAC's designers predicted that it would operate only about 50 percent of the time, presenting a colossal maintenance problem, not to mention a challenge to operational effectiveness.[17]

When completed in 1945, the ENIAC filled a large room at the Moore School with equipment containing 18,000 vacuum tubes, 1500 relays, 70,000 resistors, and 10,000 capacitors. The machine consumed 140 kilowatts of power and required internal forced-air cooling systems to keep from catching fire. The gloomy forecasts of tube failure turned out to be correct, in one sense: when the machine was turned on and off on a daily basis, a number of tubes would burn out almost every day, leaving it nonfunctional about 50 percent of the time, as predicted. Most failures, however, occurred during the warm-up and cool-down periods. By the simple (if expensive) expedient of never turning the machine off, the engineers dropped the ENIAC's tube failures to the more acceptable rate of one tube every two days.[18]

The great mathematician John von Neumann became involved with the ENIAC project in 1944, after a chance encounter with Herman Goldstine on a train platform. By the end of the war, with Eckert, Mauchly, and others, von Neumann had planned an improved computer, the EDVAC. The EDVAC was the first machine to incorporate an internal stored program, making it the first true computer in the modern sense.[19] (The ENIAC was programmed externally, using switches and plugboards.) The plan for the EDVAC's logical design served as a model for nearly all future computer control structures—often called "von Neumann architectures"—until the 1980s.[20]

Initially budgeted at $150 thousand, the ENIAC finally cost nearly half a million dollars. Without the vast research funding and the atmosphere of desperation associated with the war, it probably would have been years, perhaps decades, before private industry attempted such a project. The ENIAC became, like radar and the bomb, an icon of the miracle of government-supported "big science."

The ENIAC was not completed until the fall of 1945, after the war had ended. The ballistics tables ENIAC was built to compute no longer required urgent attention. But the ENIAC was a military machine, and so it was immediately turned to the military ends of the rapidly emerging Cold War. The first problem programmed on the machine was a mathematical model of a hydrogen bomb from the Los Alamos atomic weapons laboratories. The ENIAC, unable to store programs or retain more than twenty ten-digit numbers in its tiny memory, required several weeks in November 1945 to run the program in a series of stages. The program involved thousands of steps, each individually entered into the machine via its plugboards and switches, while the data for the problem occupied one million punch cards. The program's results exposed several problems in the proposed H-bomb design. The director of Los Alamos expressed his thanks to the Moore School in March 1946, writing that "the complexity of these problems is so great that it would have been impossible to arrive at any solution without the aid of ENIAC."[21]

This event was symbolic of a major and portentous change. The wartime alliance of academic and industrial science with the military had begun as a temporary association for a limited purpose: winning a war against aggressors. Now it was crystallizing into a permanent union.

At the formal dedication ceremony on February 15, 1946, just before pressing a button that set the ENIAC to work on a new set of hydrogen bomb equations, Major General Gladeon Barnes spoke of "man's endless search for scientific truth." In turning on the ENIAC, he said he was "formally dedicating the machine to a career of scientific usefulness."[22] Barnes, like many others in the aftermath of World War II, failed to find irony in the situation: that the "scientific truth" the ENIAC began to calculate was the basis for ultimate weapons of destruction.

Directing Research in the Postwar Era

As the postwar Truman administration began to tighten the strings of the virtually unlimited wartime purse, expectations in many quarters were that, like the armed forces, the enormous scientific and engineering network assembled for the war effort would be demobilized and thrown on its own resources.

For a number of reasons, the looming fiscal constraints never materialized. Postwar federal expenditures for R&D remained far higher than before the war, with most of the money channeled through the armed forces.

In [fiscal year] 1938 the total U.S. budget for military research and development was $23 million and represented only 30 percent of all Federal R&D; in fiscal 1945 the OSRD alone spent more than $100 million, the Army and Navy together more than $700 million, and the Manhattan Project more than $800 million.... In the immediate postwar years total military expenditure slumped to a mere seven times its prewar constant-dollar level, while constant-dollar military R&D expenditure held at a full 30 times its prewar level, and comprised about 90 percent of all federal R&D. In the early 1950s total military expenditure soared again, reaching 20 times its prewar constant-dollar level, while military R&D reattained, and before the end of the decade much surpassed, its World War II high.[23]

Industrial R&D expenditures soared as well. By the late 1940s the total amount of industrial R&D roughly equaled that sponsored by the federal government, but as figure 2.1 shows, a decade later government R&D spending was nearly double that of industry.

This trend, and the politics it reflected, resulted from three concurrent developments in postwar American politics. First, in the rapid transition from World War II to the Cold War, the war's key events served as anchoring icons for postwar policies. Wartime institutions became blueprints for their postwar counterparts. Second, the emerging politico-military paradox of a peacetime Cold War generated a perceived need for new technology, justifying vast military investments in research. Finally, fierce public debates about postwar federal support for science and technology had ended in stalemate. Plans for

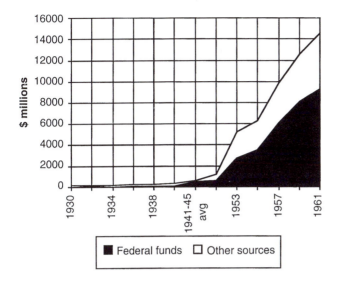

Figure 2.1
U.S. R&D by source, 1930–1961. Data: Kenneth Flamm, *Targeting the Computer: Government Support and International Competition* (Washington, D.C.: Brookings Institution, 1987), 7.

a National Science Foundation suffered long delays, and military agencies were left to fill the resulting vacuum. Let us explore each of these developments in turn.

Transference and Apocalypse
World War II was "the good war," a war not only against greedy, power-hungry aggressors but against an inhumane, antidemocratic ideology. This nearly universal sentiment was vindicated and vastly amplified by postwar revelations of the horrors of the Nazi concentration camps.

Soviet maneuverings in Eastern Europe, as well as the openly expansionist Soviet ideology, provided grounds for the transition into a Cold War. Stalin was rapidly equated with Hitler. The closed nature of Soviet society added a sinister force to mounting rumors of purges and gulag atrocities. In the eyes of many Americans, communism replaced fascism as an absolute enemy. It was seen (and saw itself) not just as one human order among others, but as an ultimate alternative system, implacably opposed to Western societies in virtually every arena: military, political, ideological, religious, cultural, economic. The absoluteness of the opposition allowed the sense of an epic, quasi-Biblical struggle that surrounded the fight against Nazism and fascism not only to survive but to thrive.

This transference of attitudes from World War II to the Cold War included a sense of a global, all-encompassing, *apocalyptic* conflict. In the 1940s and 1950s the partitioning of Europe, revolutionary upheavals across the postcolonial world, and the contest for political and ideological alliances throughout Europe, Asia, and the Middle East encouraged American perceptions that the world's future balanced on a knife edge between the United States and the USSR. Truman's declaration of worldwide American military support for "free peoples who are resisting attempted subjugation by armed minorities or by outside pressures" codified the continuation of global conflict on a permanent basis and elevated it to the level of a universal struggle between good and evil, light and darkness, freedom and slavery.[24]

The only combatant nation to emerge from the war intact, the United States had simultaneously left behind the economic depression and the political isolationism of the 1930s. The magnitude of this change cannot be overemphasized, since we have become used to a very different world.[25] The United States was a world *industrial* power before the Great Depression, but with a few brief exceptions had played only a minor role in world political and military affairs during a period when the European colonial empires still ruled the globe. As late as 1939, the U.S. army numbered only 185,000 men, with an annual budget under $500 million. America maintained no military alliances with any foreign country.[26] Six years later, at the end of World War II, the United States had over 12 million men under arms and a military budget swollen to 100 times its prewar size. In addition,

> the U.S. was producing 45 percent of the world's arms and nearly 50 percent of the world's goods. Two-thirds of all the ships afloat were American built.... The conclusion of the war ... found the U.S. either occupying, controlling, or exerting strong influence in four of the five major industrial areas of the world—Western Europe, Great Britain, Japan, and the U.S. itself. Only the Soviet Union operated outside the American orbit.... The U.S. was the only nation in the world with capital resources available to solve the problems of postwar reconstruction.[27]

The old colonial empires were bankrupt and on the verge of disintegration, the imperial pretensions of Japan had been smashed, and the Soviet Union, though still powerful, had suffered staggering losses. Postwar public sentiment for a return to the isolationism of the

1930s was strong, as was fear of renewed economic depression. The Truman administration initially tended to honor these worries with its heavy focus on a balanced budget and a rapid military demobilization. But the transference of World War II's apocalyptic struggles into the postwar world, the sense of America's awesome power, the fear of future nuclear war, and the need to reestablish war-torn nations as markets for American goods—to stave off the feared depression—combined to render isolationism untenable. The postwar geopolitical situation thus catapulted the United States into a sudden and unaccustomed role as world leader.

America's leaders in the postwar world had been weaned on the isolationist worldview. Except for a brief period after World War I, the United States had never before played a controlling role in world affairs. Thus the war itself provided the only immediately available models for action. Key events of World War II became basic icons in the organization of American foreign policy and military strategy:

- The 1938 Munich accords, in which Great Britain and France handed over parts of Czechoslovakia to Hitler in a futile attempt to stave off war, symbolized the danger of appeasement.
- The Maginot Line—a chain of massive fortresses along the French-German border that the Germans had avoided by the simple maneuver of invading through Belgium—represented the foolhardiness of a defense-oriented grand strategy.
- Pearl Harbor, where unopposed Japanese aircraft destroyed or disabled a significant portion of the U.S. Pacific fleet, signified the perpetual danger of surprise attack.
- Radar (and MIT's Radiation Laboratory, which led wartime radar research) and the Manhattan Project came to represent the power of organized science to overcome military odds with ingenuity.
- The atomic bomb itself, credited with the rapid end to the war in the Pacific, became the enigmatic symbol of both invincible power and global holocaust.

The unfolding political crises of the Cold War *were invariably interpreted in these terms.*[28] For example, the Berlin blockade was perceived as another potential Munich, calling for a hard-line response rather than a negotiation. Truman interpreted Korea through the lens of World War II: "Communism was acting in Korea just as Hitler, Mussolini and the Japanese had acted.... I felt certain that if South Korea was allowed to fall, Communist leaders would be emboldened to override nations closer to our own shores."[29] Critics of the 1950s characterized continental air defense as a Maginot Line strategy for starry-eyed technological optimists. U.S. forward basing of nuclear weapons, in positions vulnerable to surprise air attack, was likened to the risk of another Pearl Harbor.[30] The Manhattan Project was invoked endlessly to rally support for major R&D projects such as the space program. Finally, the growing nuclear arsenal was a reminder of Hiroshima, both horror and symbol of ultimate power, and it was simply assumed (for a while) that no nation would be willing to stand up to a weapon of such destructive force.[31]

Thus in many respects the Cold War was not a new conflict with communism but the continuation of World War II, the transference of that mythic, apocalyptic struggle onto a different enemy.[32]

American Antimilitarism and a High-Technology Strategy
The authors of the U.S. Constitution feared professional armies as dangerous concentrations of unaccountable state power. They saw the career officer corps, which in European armies maintained the hereditary linkage between royalty, gentry, and control of the armed forces, as a linchpin of aristocracy. In addition, military social structure, with its strict hierarchy and its authoritarian ethic, seemed the antithesis of a participatory democracy.[33]

But having won their independence in a revolutionary war, the founders naturally also understood the importance of military power in international politics. The constitutional provision for a citizen army linked the right to participate in government with the responsibility to protect it by force of arms. Every citizen was a potential soldier, but every soldier was also a citizen; thus, in principle, loyalty to military institutions was subordinated to loyalty to state and civil society. In practice, until World War II, this also meant that armed forces were mustered only for war and were greatly reduced once war ended. American political discourse still reflects this ambivalence toward military forces and the conflictual relationship between democratic ideals and military principles of authority.[34]

American antimilitarism, then, is not at all the same thing as pacifism, or principled objection to armed force itself. Instead, antimilitarism is an instance of what political scientist Samuel Huntington has called the "anti-power ethic" in American society, the enormous value this society has always placed on political limits to power, hierarchy, and authority.[35]

In the postwar years a number of factors contributed to a changed perception of the need for a powerful armed force in peacetime. The absolute Allied victory supported a vast new confidence in the ability of military force to solve political problems. The occupation of Germany and Japan meant an ongoing American military presence on other continents. The relative insignificance of American suffering in the war produced an inflated sense of the ease of military victory—the idea that the United States, at least, could buy a lot for a little with military power and new technology. Also, with America's full-blown emergence as a world economic power came new interests across the globe, interests that could conceivably require military defense. Finally, the rapid transition from World War II into the Cold War left little time for a retrenchment into prewar values: the apocalyptic conflict simply continued.

Furthermore, technological factors such as the bomb and the maturation of air warfare now made it possible to conceive of a major military role for the United States outside its traditional North American sphere of influence. Historically, ocean barriers had separated the United States from the other nations possessing the technological wherewithal to mount a serious military challenge. These were now breached. Airplanes and, later, guided missiles could pose threats at intercontinental range. In effect, the very concept of national borders was altered by these military technologies: the northern boundary of the United States, in terms of its defense perimeter, now lay at the limits of radar vision, which in the 1950s rapidly moved northward to the Arctic Circle.

Antimilitarism, because it required that the number of men under arms be minimized, also helped to focus strategic planning on technological alternatives. The Strategic Air Command came to dominate U.S. strategic planning because it controlled the technological means for intercontinental nuclear war. It was the primary threat America could wield against the Soviet Union, yet it required mainly money and equipment, not large numbers of troops. The Army's massive manpower seemed less impressive, less necessary, and more of a political liability in the face of the minimally manned or even automated weapons of the Air Force. As the Soviet Union acquired long-range bombers, nuclear weapons, and then ICBMs, the role of the Air Force and its technology in both defense and offense continued to expand.

The Cold War marked the first time in its history that America maintained a large standing army in peacetime. But its geographical situation of enormous distance from its enemies, combined with its antimilitarist ethic, ensured that the institutional form taken by a more vigorous American military presence would differ from the more traditional European and Soviet approaches of large numbers of men under arms. Instead of universal conscription, the United States chose the technological path of massive, ongoing automation and integration of humans with machines. First Truman and then Eisenhower, each balancing the contradictory goals of an expanding, activist global role and a contracting military budget, relied ever more heavily on nuclear weapons. By the end of the 1950s high technology—smaller bombs with higher yields, tactical atomic warheads for battlefield use, bombers of increasingly long range, high-altitude spy planes, nuclear early warning systems, rockets to launch spy satellites, and ICBMs—had became the very core of American global power.

Support for Research and Development
In his famous 1945 tract *Science: The Endless Frontier,* composed at President Roosevelt's request as a blueprint for postwar science and technology policy, Vannevar Bush called for a civilian-controlled National Research Foundation to preserve the government-industry-university relationship created during the war. In his plea for continuing government support, Bush cited the Secretaries of War and Navy to the effect that scientific progress had become not merely helpful but utterly essential to military security for the United States in the modern world:

This war emphasizes three facts of supreme importance to national security: (1) Powerful new tactics of defense and offense are developed around new weapons created by scientific and engineering research; (2) the competitive time element in developing those weapons and tactics may be decisive; (3) war is increasingly total war, in which the armed services must be supplemented by active participation of every element of the civilian population.
To insure continued preparedness along farsighted technical lines, the research scientists of the country must be called upon to continue in peacetime some substantial portion of those types of contribution to national security which they have made so effectively during the stress of the present war.[36]

Bush's MIT colleague Edward L. Bowles, Radiation Laboratory "ambassador" to government and the military, advocated an even tighter connection. Bowles wrote of the need to "systematically and deliberately couple" scientific and engineering schools and industrial organizations with the military forces "so as to form a continuing, working partnership."[37]

Bush also argued that modern medicine and industry were also increasingly dependent on vigorous research efforts in basic science. The massive funding requirements of such research could not be met by the cash-poor academic community, while the industrial sector's narrow and short-term goals would discourage it from making the necessary investment. Consequently the new foundation Bush proposed would have three divisions, one for natural sciences, one for medical research, and one for national defense.

Bush's efforts were rebuffed, at first, by Truman's veto of the bill establishing the National Science Foundation (NSF). The populist president blasted the bill, which in his view "would . . . vest the determination of vital national policies, the expenditure of large public funds, and the administration of important government functions in a group of individuals who would be essentially private citizens. The proposed National Science Foundation would be divorced from . . . control by the people."[38]

With major research programs created during the war in jeopardy, the War Department moved into the breach, creating the Office of Naval Research in 1946. In a pattern repeated again and again during the Cold War, national security provided the consensual justification for federally funded research. The ONR, conceived as a temporary stopgap until the government created the NSF, became the major federal force in science in the immediate postwar years and remained important throughout the 1950s. Its mandate was extremely broad: to fund basic research ("*free* rather than directed research"), primarily of an unclassified nature.[39]

Yet the ONR's funding was rarely, if ever, a purely altruistic activity.[40] The bill creating the office mentioned the "paramount importance [of scientific research] as related to the maintenance of future naval power, and the preservation of national security"; the ONR's Planning Division sought to maintain "listening posts" and contacts with cutting-edge scientific laboratories for the Navy's possible use.[41] Lawmakers were well aware that the ONR represented a giant step down the road to a permanent federal presence in science and engineering research and a precedent for military influence. House Committee on Naval Affairs Chairman Carl Vinson opposed the continuing executive "use of war powers in peacetime," forcing the Navy to go directly to Congress for authorization.[42]

By 1948 the ONR was funding 40 percent of *all* basic research in the United States; by 1950 the agency had let more than 1,200 separate research contracts involving some 200 universities. About half of all doctoral students in the physical sciences received ONR support.[43] ONR money proved especially significant for the burgeoning field of computer design. It funded a number of major digital computer projects, such as MIT's Whirlwind, Raytheon's Hurricane, and Harvard's Mark III.[44] The NSF, finally chartered in 1950 after protracted negotiations, did not become a significant funding source for computer science until the 1960s (in part because computer science did not become an organized academic discipline until then). Even after 1967, the only period for which reliable statistics are available, the NSF's share of total federal funding for computer science hovered consistently around the 20 percent mark, while Department of Defense obligations ranged between 50 and 70 percent, or 60 to 80 percent if military-related agencies such as the Department of Energy (responsible for atomic weapons research) and NASA (whose rockets lifted military surveillance satellites and whose research contributed to ballistic missile development) are included.[45]

The Military Role in Postwar Computer Research

With the war's end, some corporate funding became available for computer research. A few of the wartime computer pioneers, such as ENIAC engineers Mauchly and Eckert, raised commercial banners. The company they formed developed the BINAC, the first American stored-program electronic computer, and then the UNIVAC, the first American commercial computer.[46]

But military agencies continued, in one way or another, to provide the majority of support. The Army (via the Census Bureau) and Air Force (via the Northrop Corporation's Snark missile project) were Eckert and Mauchly's major supporters. Bell Laboratories, the largest independent electronics research laboratory in the country, saw the percentage of its peacetime budget allocated to military projects swell from zero (prewar) to upwards of 10 percent as it continued work on the Nike missile and other systems, many of them involving analog computers.[47] Many university-based computer researchers continued under ONR sponsorship. Others became involved in a private company, Engineering Research Associates (ERA), which developed cryptological computers for its major customer, the Navy, as well as later commercial machines based on its classified work. (When ERA's

ATLAS became operational, in 1950, it was the second electronic stored-program computer in the United States.)[48]

With military and Atomic Energy Commission support, John von Neumann began his own computer project at the Institute for Advanced Study (IAS). The so-called IAS machine, completed in 1952, became one of the most influential computers of the immediate postwar period. Several copies were built at defense research installations, including the Rand Corporation and the Los Alamos, Oak Ridge, and Argonne National Laboratories.[49]

How much military money went to postwar computer development? Because budgets did not yet contain categories for computing, an exact accounting is nearly impossible. Kenneth Flamm has nevertheless managed to calculate rough comparative figures for the scale of corporate and military support.[50] Flamm estimates that in 1950 the federal government provided between $15 and $20 million (current) per year, while industry contributed less than $5 million—20 to 25 percent of the total. The vast bulk of federal research funds at that time came from military agencies.

In the early 1950s the company-funded share of R&D began to rise (to about $15 million by 1954), but between 1949 and 1959 the major corporations developing computer equipment—IBM, General Electric, Bell Telephone, Sperry Rand, Raytheon, and RCA—still received an average of 59 percent of their funding from the government (again, primarily from military sources). At Sperry Rand and Raytheon, the government share during this period approached 90 percent.[51] The first commercial production computer, Remington Rand's UNIVAC I, embodied the knowledge Eckert and Mauchly had gained from working on the military-funded ENIAC and later on their BINAC, which had been built as a guidance computer for Northrop Aircraft's Snark missile. Though much of the funding for Eckert and Mauchly's project was channeled through the Census Department (which purchased the first UNIVAC I), the funds were transferred to Census from the Army.[52]

Flamm also concludes that even when R&D support came primarily from company sources, it was often the expectation of military procurements that provided the incentive to invest. For instance, IBM's first production computer (the 701, also known as the "Defense Calculator"), first sold in 1953, was developed at IBM's expense, but only with letters of intent in hand from eighteen Department of Defense customers.[53]

Consequences of Military Support

What sort of influence did this military support have on the development of computers? In chapters 3 and 4 we will explore this question in great detail with respect to the Whirlwind computer, the SAGE air defense system, the Rand Corporation, and the Vietnam War. Here, however, I will sketch some more general answers through a series of examples.

First, military funding and purchases in the 1940s and 1950s enabled American computer research to proceed at a pace so ferocious as to sweep away competition from Great Britain, the only nation then in a position to become a serious rival. At the end of World War II the British possessed the world's only functioning, fully electronic digital computer (Turing's Colossus), and until the early 1950s its sophistication in computing at least equaled that of the United States. The Manchester University Mark I became, in June 1948, the world's first operating *stored-program* electronic digital computer (i.e., the first operating computer in the full modern sense of the term). The Cambridge University EDSAC, explicitly modeled on the EDVAC, preceded the latter into operation in June 1949, "the first stored-program electronic computer with any serious computational ability."[54] The firm of Ferranti Ltd. built the first successful commercial computer, also called the Mark I, and eventually sold eight of these machines, primarily to government agencies active in the British atomic weapons research program. The first Ferranti Mark I became operational in February 1951, preceding the Eckert/Mauchly UNIVAC by a few months.

With its financial resources limited by the severe demands of postwar reconstruction, the British government failed to pursue the field with the intensity of the United States. British researchers and producers were in general left to more ordinary commercial and technical resources. By the time large-scale commercial markets for computers developed in the early 1960s, British designs lagged behind American models. Unable to keep up, the fledgling British computer industry declined dramatically: though British firms totally dominated the British market in the 1950s, by 1965 more than half of computers operating in Britain were U.S.-made.[55]

Second, the military secrecy surrounding some of both British and American research impeded the spread of the new technology. Most academic researchers felt advances would come faster in an atmosphere of free exchange of ideas and results. They pressed to reestablish such a climate, and in many cases—such as that of the IAS computer, whose technical reports and plans were widely disseminated—they succeeded. But the wartime habits of secrecy died hard, and in the course of the Cold War tensions between military and commercial interests rose. In August 1947 Henry Knutson of the ONR's Special Devices Center informed Jay Forrester, director of the MIT Whirlwind computer project, that "the tendency is to upgrade the classification [of military-funded research projects] and that all computer contracts are now being reconsidered with the possible view of making them confidential."[56] Much of the Whirlwind work was, in fact, classified. (Indeed, in the 1950s MIT spun off the Lincoln Laboratories from its university operations because of the huge volume of classified research on air defense, including computers.) In the late 1940s, Forrester sometimes had trouble recruiting researchers because so many people refused to work on military projects.[57] John Mauchly, to cite another kind of postwar security issue, was accused of being a communist sympathizer (he was not) and was denied a clearance.[58]

Though many of the military-sponsored computer projects were not classified in the direct sense, informal self-censorship remained a part of postwar academic research culture. As Paul Forman has argued, "strictly speaking there was in this [post-World War II] period no such thing as unclassified research under military sponsorship. 'Unclassified' was simply that research in which some considerable part of the responsibility for deciding whether the results should be held secret fell upon the researcher himself and his laboratory." Forman cites the ONR's Alan Waterman and Capt. R. D. Conrad, writing in 1947, to the effect that "the contractor is entirely free to publish the results of his work, but . . . we expect that scientists who are engaged on projects under Naval sponsorship are as alert and as conscientious as we are to recognize the implications of their achievement, and that they are fully competent to guard the national interest."[59]

Third, even after mature commercial computer markets emerged in the early 1960s, U.S. military agencies continued to invest heavily in advanced computer research, equipment, and software. In the 1960s the private sector gradually assumed the bulk of R&D funding. IBM, in particular, adopted a strategy of heavy investment in research, reinvesting over 50 percent of its profits in internal R&D after 1959. The mammoth research organization IBM built gave it the technical edge partly responsible for the company's dominance of the world computer market for the next two decades. To compete, other companies eventually duplicated IBM's pattern of internal research investment.

Despite the extraordinary vitality of commercial R&D after the early 1960s, the Pentagon continued to dominate research funding in certain areas. For example, almost half of the cost of semiconductor R&D between the late 1950s and the early 1970s was paid by military sources. Defense users were first to put into service integrated circuits (ICs, the next major hardware advance after transistors); in 1961, only two years after their invention, Texas Instruments completed the first IC-based computer under Air Force contract. The Air Force also wanted the small, lightweight ICs for Minuteman missile guidance control. In 1965, about one-fifth of all American IC sales went to the Air Force for this purpose. Only in that year did the first commercial computer to incorporate ICs appear.[60] ICs and other miniaturized electronic components allowed the construction of sophisticated digital guidance computers that were small, light, and durable enough to fit into missile warheads. This, in turn, made possible missiles with multiple, independently targetable reentry vehicles (MIRVs), which were responsible for the rapid growth of nuclear destructive potential in the late 1960s and early 1970s.[61] ICs were the ancestors of today's microprocessors and very-large-scale integrated circuitry, crucial components of modern cruise missiles and other "smart" weaponry.

Another instance was the nurturance of artificial intelligence (AI) by the Advanced Research Projects Agency (ARPA, later called DARPA, the Defense Advanced Research Projects Agency), which extended from the early 1960s until the final end of the Cold War. AI, for over two decades almost exclusively a pure research area of no immediate commercial interest, received as much as 80 percent of its total annual funding from ARPA. ARPA also supported such other important innovations as timesharing and computer networking. In 1983, with its Strategic Computing Initiative (SCI), DARPA led a concerted Pentagon effort to guide certain critical fields of leading-edge computer research, such as artificial intelligence, semiconductor manufacture, and parallel processing architectures, in particular directions favorable to military goals. (We will return to ARPA and its relationship with AI in chapters 8 and 9.)

Thus the pattern of military support has been widespread, long-lasting, and deep. In part because of connections dating to the

ENIAC and before, this pattern became deeply ingrained in postwar institutions. But military agencies led cutting-edge research in a number of key areas even after a commercial industry became well established in the 1960s. As Frank Rose has written, "the computerization of society... has essentially been a side effect of the computerization of war."[62]

Why Build Computers?

We have explored the origin of military support, its extent, and some of its particular purposes. Now we must return once again to the question posed by this chapter's title, this time at the level of more general institutional and technical problems. Why did the American armed forces establish and maintain such an intimate involvement with computer research?

The most obvious answer comes from the utilitarian side of the vision captured in General Westmoreland's "electronic battlefield" speech: computers can automate and accelerate important military tasks. The speed and complexity of high-technology warfare have generated control, communications, and information analysis demands that seem to defy the capacities of unassisted human beings. Jay Forrester, an MIT engineer who played a major role in developing the military uses of computing, wrote that between the mid-1940s and the mid-1950s

> the speed of military operations increased until it became clear that, regardless of the assumed advantages of human judgment decisions, the internal communication speed of the human organization simply was not able to cope with the pace of modern air warfare.... In the early 1950s experimental demonstrations showed that enough of [the] decision making [process] was understood so that machines could process raw data into final weapon-guidance instruction and achieve results superior to those then being accomplished by the manual systems.[63]

Computers thus improved military systems by "getting man out of the loop" of critical tasks. Built directly into weapons systems, computers assisted or replaced human skill in aiming and operating advanced weapons, such as antiaircraft guns and missiles. They automated the calculation of tables. They solved difficult mathematical problems in weapons engineering and in the scientific research behind military technologies, augmenting or replacing human calculation. Computers began to form the keystone of what the armed forces now call "C³I"—command, control, communications, and intelligence (or information) networks, replacing and assisting humans in the encoding and decoding of messages, the interpretation of radar data, and tracking and targeting functions, among many others.

I will argue that this automation theory is largely a retrospective reconstruction. In the 1940s it was not at all obvious that *electronic digital* computers were going to be good for much besides exotic scientific calculations. Herman Goldstine recalled that well into the 1950s "most industrialists viewed [digital] computers mainly as tools for the small numbers of university or government scientists, and the chief applications were thought to be highly scientific in nature. It was only later that the commercial implications of the computer began to be appreciated."[64] Furthermore, the field of analog computation was well developed, with a strong industrial base and a well-established theoretical grounding. Finally, analog control mechanisms (servomechanisms) had seen major improvements during the war. They were readily available, well-understood, and reliable.

Howard Aiken, the Harvard designer of several early digital computers, told Edward Cannon that "there will never be enough problems, enough work, for more than one or two of these [digital] computers," and many others agreed.[65]

Analog vs. Digital: Computers and Control

Most modern computers perform three basic types of functions: calculation, communication, and control.[66] The computers of the 1940s could not yet do this; they were calculators, pure and simple. Their inputs and outputs consisted exclusively of numbers or, eventually, of other symbols punched on cards or printed on paper. In most of the first machines, both decimal numbers and instructions had to be translated into binary form. Each computer's internal structure being virtually unique, none could communicate with others. Neither (with the exception of printers and card punches) could they control other machines.

Deep theoretical linkages among the three functions were already being articulated in the communication and information theories of Norbert Wiener and Claude Shannon. But these theoretical insights did not dictate any particular path for computer development. Nor did they mandate digital equipment. The idea of combining the three functions in a single machine, and of having that machine be an electronic digital computer, came not just from theory—both Shannon and Wiener, for example, were also interested in other types of machines[67]—*but from the evolution of practical design projects in social and cultural context.*

The idea of using digital calculation for control functions involved no special leap of insight, since the role of any kind of computer in control is essentially to solve mathematical functions. (Indeed, the RCA engineer Jan Rajchman attempted to construct a digital fire-control computer for antiaircraft guns in the early 1940s.)[68] But unlike then-extant digital machines, analog computers integrated very naturally with control functions because their inputs and outputs were often exactly the sort of signals needed to control other machines (e.g., electric voltages or the rotation of gears).[69] Thus the difficult conversion of data into and out of numerical form could often be bypassed. In addition, because many electrical devices, including vacuum tubes, have analog as well as digital properties, the increasing shift from electromechanical to electronic control techniques had little bearing on the question of digital vs. analog equipment. In fact, some of the wartime analog computers, such as Bush's RDA and the Bell gun directors discussed below, used electronic components.[70] Finally, the wartime investment in digital computing represented by ENIAC shrank into insignificance when compared with the wartime program in radar and control systems research, which were primarily analog technologies, with the result that far more engineers understood analog techniques than grasped the new ideas in digital computing.

Many of the key actors in computer development, such as Bell Laboratories and MIT, had major and long-standing investments in analog computer technologies. For example, in 1945, as the ENIAC was being completed, Bell Labs was commissioned to develop the Nike-Ajax antiaircraft guided missile system for the Army. Bell proposed a "command-guidance" technique in which radar signals would be converted into missile guidance instructions by ground-based analog computers.[71] Likewise, one of MIT's major wartime research groups was the Servomechanisms Laboratory, which built analog control devices for antiaircraft gun directors and other uses.[72]

With a vigorous tradition of analog computation and control engineering already in place after the war, work proceeded rapidly on general-purpose electronic analog computers. A number of mammoth machines, such as RCA's Typhoon, were constructed under both corporate and military sponsorship. Mina Rees, then director of the ONR's Mathematical Sciences Division, noted in a 1950 public report on federal support for computer research that the ONR continued to fund a variety of analog machines. She pointed to the robust health of the analog computer and control industry as one reason the ONR's analog program was not even larger. "There is," she pointed out, "vastly more analog than digital equipment that has been built without government support, but... the government and its contractors make extensive use of the equipment." Rees also praised the "broad point of view that recognizes merit in both the analog and the digital aspects of the computer art."[73] Analog engineers thought their computers could compete directly with digital devices in any arena that did not demand enormous precision.

These machines and the social groups centered around them (such as industrial research laboratories, university engineering schools, and equipment manufacturers) constituted a major source of resistance to the emerging digital paradigm, especially when it came to using the new machines for purposes other than mathematical calculation. In the words of one participant,

Analog computer experts felt threatened by digital computers. World War II, with its emphasis on automatic pilots and remotely controlled cannon, fostered the analog computer-servo engineering profession.... Many analog computer engineers were around following the war, but so great was the newly realized demand for control devices that the colleges began training increasing numbers.... [O]nly a relatively few servo engineers were able to make the transition to digital machines.... In 1945... we confidently expected that factories would have become softly humming hives of selsyn motors, amplidyne generators and analog computers by the year 1960.[74]

Even as late as 1950, among the groups then developing digital machines, the heritage of World War II analog equipment proved difficult to overcome. When a Rand team seeking a programmable digital machine toured the country's major computer projects, "what [they] found was discouraging." Many of the groups working on reliability and high-speed computing were exploring "modifications of radar technology, which was largely analog in nature.... They were doing all kinds of tweaky things to circuits to make things work. It was all too whimsical."[75]

In addition to social inertia, easy availability, and an acculturated preference for analog technology, there were many other reasons why sophisticated engineers might reject electronic digital computers for most purposes during the 1940s and early 1950s. First, the electronic components of the day were not very reliable. As we have seen, most scientists scoffed at the idea that a machine containing vast numbers of vacuum tubes could ever function for more than a few minutes at a time without breaking down. Thus to contemplate using electronic digital machines for control functions, in real time and in situations where safety and/or reliability were issues, seemed preposterous to many. Second, early electronic computers were huge assemblies, the size of a small gymnasium, that consumed power voraciously and generated tremendous heat. They often required their own power supplies, enormous air conditioners, and even special buildings. Miniaturization on the scale we take for granted today had not emerged even as a possibility. Third, they were extremely expensive (by the standards of analog equipment), and they demanded constant and costly maintenance. Finally, early electronic computers employed exotic materials and techniques, such as mercury delay line memory and the cantankerous electrostatic storage tube, which added their own problems to the issues of cost and reliability.

Even once it became clear (in the late 1940s) that electronic digital computers would work, could be made reasonably reliable, and could operate at speeds far outstripping their mechanical and electro-mechanical counterparts, another issue prevented them from being seriously considered for control functions. As George Valley, one of the leaders of the SAGE project, pointed out in a 1985 retrospective, "relatively few wanted to connect computers to the real world, and these people seemed to believe that the sensory devices would all yield *data*. In fact, only some sensors—such as weighing machines, odometers, altimeters, the angle-tracking part of automatic tracking radars—had built-in counters. Most sensory devices relied on human operators to interpret noisy and complex signals."[76] The problem lay in designing sensory devices that produced direct numerical inputs for the computer to calculate with. Analog control technologies did not require such conversions, because they represented numerical quantities directly through physical parameters.[77]

In 1949, according to Valley, "almost all the groups that were realistically engaged in guiding missiles... thought exclusively in terms of analog computers."[78] A notable exception was the Northrop Snark missile project, which engaged Eckert and Mauchly to build the BINAC digital computer, completed in 1949, for its guidance system. However, the BINAC did not work well, and Northrop engineers afterward moved toward special-purpose digital differential analyzers—and away from stored-program general-purpose computers—for the project.[79]

As late as 1960 Albert Jackson, manager of data processing for the TRW Corporation, could write with authority—in a textbook on analog computation—that electronic analog computers retained major advantages. They would always be better at control functions and most simulations, as well as faster than digital devices.

> The [general-purpose] electronic analog computer is a very fast machine. Each operational unit can be likened to a digital arithmetical unit, memory unit, and control unit combined. Since as many as 100 to 500 of these units will be employed in parallel for a particular problem setup, it can be seen why an analog computer is faster than a digital machine, which seldom has more than one arithmetical unit and must perform calculations bit by bit or serially. Because of their high speed, electronic analog computers have found wide application as *real-time* simulators and control-system components.

Only in the 1980s did efficient *digital* parallel processing become possible, motivated in part by precisely this issue of real-time control. Jackson continued:

> In conclusion, analog computers have found and will continue to find wide application to problems where the knowledge of the physical situation does not permit formulation of a numerical model of more than four significant digits or where, even if such a model could be designed, the additional time and expense entailed in digital computation would not be warranted because of other factors.[80]

Clearly, in the decade following World War II digital computers were a technology at the early phase of development that Trevor Pinch and Wiebe Bijker describe as, in essence, a solution in search of a problem. The technology of digital computation had not yet achieved what they call "closure," or that state of technical development and social acceptance in which large constituencies generally agree on its purpose, meaning, and physical form.[81] The shape of computers, as tools, was still extremely malleable, and their capacities remained to be envisioned, proven, and established in practice. Thus the use of digital devices to create automated, centralized military command-control systems was anything but foreordained.

Computers Take Command

The utilitarian account of military involvement in computer development also fails to explain one of the major paradoxes of military automation. Computers were used first to automate calculation, then to control weapons and guide aircraft, and later to analyze problems of command through simulation. The final step in this logic would be the eventual automation of command itself; intermediate steps would centralize it and remove responsibilities from lower levels. Military visionaries and defense intellectuals continually held out such centralization as some kind of ultimate goal, as in General Westmoreland's dream of the electronic battlefield. By the mid-1980s, DARPA projects envisioned expert systems programs to analyze battles, plot strategies, and execute responses for carrier battle group commanders. The Strategic Computing Initiative program announcement claimed that in "the projected defense against strategic nuclear missiles ... systems must react so rapidly that it is likely that almost complete reliance will have to be placed on automated systems" and proposed to develop their building blocks.[82] DARPA's then-director Robert Cooper asserted, in an exchange with Senator Joseph Biden, that with sufficiently powerful computers, presidential errors in judgment during a nuclear confrontation might be rendered impossible: "we might have the technology so he couldn't make a mistake."[83]

The automation of command clearly runs counter to ancient military traditions of personal leadership, decentralized battlefield command, and experience-based authority.[84] By the early 1960s, the beginning of the McNamara era and the early period of the "electronic battlefield," many military leaders had become extremely suspicious of the very computers whose development their organizations had led. Those strategists who felt the necessity and promise of automation described by Jay Forrester were opposed by others who saw that the domination of strategy by preprogrammed plans left no room for the extraordinarily contingent nature of battlefield situations. In 1964, Air Force Colonel Francis X. Kane reported in the pages of *Fortune* magazine that "much of the current planning for the present and future security of the U.S. rests on computerized solutions." It was, he wrote, impossible to tell whether the actual results of such simulated solutions would occur as desired, because

we have no experience in comparing the currently accepted theory of predicting wars by computer with the actual practice of executing plans. But I believe that today's planning is inadequate because of its almost complete dependence on scientific methodology, which cannot reckon with those acts of will that have always determined the conduct of wars.... In today's planning the use of a tool—the computer—dictates that we depend on masses of data of repeated events as one of our fundamental techniques. We are ignoring individual experience and depending on mass experience instead.[85]

Also in the early 1960s occasional articles in the armed forces journal *Military Review* began warning of "electronic despotism" and "demilitarized soldiers" whose tasks would be automated to the point that the men would be deskilled and become soft.[86] Based on interviews with obviously disaffected commanders, *U.S. News & World Report* reported in 1962—under the banner headline "Will 'Computers' Run Wars of the Future?"—that "military men no longer call the tunes, make strategy decisions and choose weapons. In the Pentagon, military men say they are being forced to the sidelines by top civilians, their advice either ignored or not given proper hearing.... In actual defense operations, military commanders regard themselves as increasingly dependent on computer systems."[87] While these reports certainly exaggerated the actual role of computers in military planning and especially in military operations at the time, their existence shows that the view of computers as a solution to military problems faced internal opposition from the start. They also demonstrate how deeply an ideology of computerized command and control had penetrated into U.S. military culture.

The automation theory alone, then, explains neither the urgency, the magnitude, nor the specific direction of the U.S. military effort in computing. Rather than explain how contests over the nature and potential of computers were resolved, the utilitarian view writes history backwards, using the results of those contests to account for their origins.

Nor does the utilitarian view explain the pervasive military *fascination* with computers epitomized by General Westmoreland's speech in the aftermath of Vietnam. "I see," he proclaimed, "an Army built into and around an integrated area control system that exploits the advanced technology of communications, sensors, fire direction, and the required automatic data processing—a system that is sensitive to the dynamics of the ever-changing battlefield—a system that materially assists the tactical commander in making sound and timely decisions."[88] This is the language of vision and technological utopia, not practical necessity. It represents a dream of victory that is bloodless for the victor, of battle by remote control, of speed approaching the instantaneous, and of certainty in decision-making and command. It is a vision of a closed world, a chaotic and dangerous space rendered orderly and controllable by the powers of rationality and technology.

Why build computers? In this chapter I have tried to show that not only the answers, but also the very question, are complex. Their importance to the future of U.S. military power was by no means obvious at the outset. To understand how it became so, we must look closely at the intricate chains of technological advances, historical events, government policies, and emergent metaphors comprising closed-world discourse. For though policy choices at the largest levels determined research directions, in some cases quite specifically, defining digital computation as relevant to national priorities was not itself a policy issue. Instead it involved a complicated nexus of technological choices, technological traditions, and cultural values. In fact, digital computer research itself ended up changing national priorities, as we will see in the following chapters.

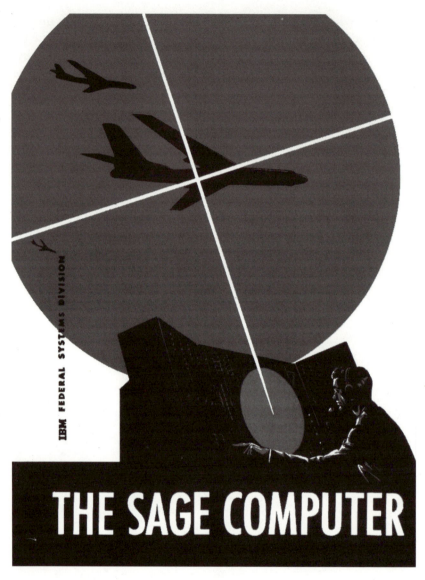

Cover of an IBM instruction manual for the SAGE computer system, circa 1959. Courtesy Charles Babbage Institute.

3
SAGE

By almost any measure—scale, expense, technical complexity, or influence on future developments—the single most important computer project of the postwar decade was MIT's Whirlwind and its offspring, the SAGE computerized air defense system.

In Project Whirlwind many of the questions framed in chapter 2 began to find their answers. Whirlwind started out as an analog computer designed to be part of a control system. It metamorphosed into a digital machine but retained its original purpose, thus linking digital computing to control functions. Originally funded by the Office of Naval Research, the project almost expired during a prolonged crisis over its military justification. It was saved when the Air Force embarked on a search for new air defense technologies after the 1949 Soviet atomic bomb explosion. Whirlwind was chosen, by civilian scientists, as the central controller for the hugely ambitious SAGE continental air defense system. This choice saved the project and led to a vast array of technical developments, such as analog/digital conversion techniques, real-time digital computing, and extremely high reliability, that would be essential to the viability of computers in military control systems.

SAGE was the first large-scale, computerized command, control, and communications system. Although it was obsolete before it was completed, it unleashed a cascading wave of command-control projects from the late 1950s onwards, tied largely to nuclear early warning systems. These systems eventually formed the core of a worldwide satellite, sensor, and communications web that would allow global oversight and instantaneous military response. Enframing the globe, this web formed the technological infrastructure of closed-world politics.

Whirlwind and the Trek from Analog to Digital Control

Whirlwind was conceived late in 1944, in the MIT Servomechanisms Laboratory, as the Airplane Stability and Control Analyzer (ASCA), an analog device intended for use in flight simulators. By 1946, the project had been reoriented toward construction of a general-purpose digital computer. Exploring this transition will highlight the simultaneously technical, social, and institutional character of technological choice.

In the 1940s, flight simulators were servo-operated, electromechanical devices that mimicked an airplane's attitudinal changes in response to movements of its controls. They allowed pilots in training to practice flying in a safe and relatively inexpensive environment. A sufficiently accurate simulation could also allow engineers to study alternative sets of characteristics before building a prototype of a new design. In 1943–44 Captain Luis de Florez, director of the Navy's Special Devices Division, realized that a *general* simulator, one that could be programmed to simulate any desired set of characteristics, could in theory vastly reduce the time and expense of both aircraft development and pilot training.[1] In principle, the flight simulator was what is now known as a "dual-use" technology, equally applicable to training military and civilian pilots. But the urgency of the war made it, in practice, a military technology, and commercial potential was not a factor in justifying the project.

In 1944 Jay Forrester was an advanced graduate student at MIT. As one of Gordon S. Brown's two assistants, he had helped found the Servomechanisms Laboratory in 1940. He was present when Captain de Florez discussed the idea of a general simulator with Brown's group, and when the Special Devices Division issued a contract for the ASCA in December 1944, Forrester took charge of the project.[2]

The Servo Lab was then the most important center of analog control research in the United States, and Forrester spent his first year working on an analog computer for the ASCA. The complexity of the calculations involved—requiring simultaneous solutions of a hundred or more differential equations—frustrated his efforts, but it is important to emphasize that this was *not* because analog techniques were unable, in principle, to solve the equations. Forrester needed to overcome two other problems.

First, the speed of the *electromechanical* analog equipment in terms of which Forrester had been trained to think—the servomechanisms and differential analyzers of the Vannevar Bush era—was too slow. To make a simulator feel realistic, its controller would need to solve the necessary equations virtually instantaneously, that is, without a noticeable delay between the pilot's actions and the machine's response. Computational delays of even significant fractions of a second, as were typical of electromechanical devices, would be intolerable. This was the problem of "real-time" control. In principle, at least, this problem was not unsolvable; electronic analog computation could have achieved the requisite speeds.

A second, more intractable difficulty was the limited accuracy of analog techniques. Because they employ measured physical quantities rather than counts of discrete units, analog devices unavoidably introduce increasingly large errors as their complexity rises. In 1945 Forrester and some associates paid a visit to MIT colleague Frank Verzuh. Verzuh had worked on Bush's Rapid Arithmetic Machine and the various differential analyzers before and during the war, but he was now helping to design the Rockefeller Electronic Calculator, a small digital computer. He "told Jay . . . he would have to use digital techniques," because the best MIT differential analyzer achieved only five significant figures, whereas the ASCA would require as many as ten.[3] Though a research effort on Whirlwind's eventual scale would surely have led to major improvements in analog accuracy, in late 1945 Forrester began to explore digital techniques.

Forrester's interest in digital possibilities was piqued by three further encounters. First, his former fellow graduate student Perry Crawford, who had written a master's thesis on applying digital computation to the automatic control of antiaircraft guns, strongly suggested that Forrester look into digital methods. Then, in late 1945, Forrester attended the Conference on Advanced Computation Techniques, whose major theme was ENIAC research. Finally, he visited the Moore School to learn about the "Pennsylvania technique" and read the ENIAC designers' widely circulated "First Draft of a Report on the EDVAC."[4] Even together, however, these did not amount to some kind of digital conversion experience. The choice, at this point, was anything but clear-cut: Forrester spent the better part of the following year weighing analog and digital methods against each other.

By mid-1946 Forrester had abandoned the analog approach and reoriented the ASCA project toward a general-purpose digital machine, the Whirlwind, that would have the flight simulator as just one of its possible applications. This move from a special-purpose to a general-purpose machine did not correspond precisely with a shift to general or theoretical goals. In keeping with Servo Lab culture, Forrester and Everett remained strongly oriented toward applications.[5] In fact, the Navy continued to view the project in terms of the flight simulator, and work on cockpit design and other features of the eventual ASCA proceeded.

This practical program in simulator design separated Whirlwind from almost all other digital computer projects of this era because it required a device that could be used as a real-time control mechanism.[6] This was a far from obvious goal for a digital computer, given the technology of the day. As we saw in chapter 2, analog computing and control technologies were well developed, with sophisticated theoretical underpinnings and many real-time applications, whereas electronic digital computers had serious problems with component reliability, size, power consumption, and expense, and the logic theory underlying their operation was still quite new. The full implications of the Turing machine's generality remained to be realized, and there was still much controversy over the relative value of general-purpose programmable machines versus special-purpose, task-oriented devices for specific needs.

Furthermore, electronic analog computation presented an alternative that could, in principle, resolve the speed problems of electromechanical machines. Good analog engineers could develop work-arounds to correct for the machines' inherent accuracy limitations. (This approach was being aggressively pursued in RCA's Typhoon, Philbrick Research's Polyphemus, and other projects of the 1940s and early 1950s.[7]) Most other computer projects of the 1940s saw digital machines as giant calculators for scientific computation.[8] Many believed that only a few would ever be needed, and even Forrester at one time apparently thought that the entire country would eventually be served by a single mammoth computer.[9]

By 1948, the ONR's interest in a supersophisticated and by then extremely expensive flight simulator was on the wane. With military budgets declining, the Navy was forced to streamline its research programs. The Special Devices Center's funding for fiscal 1948 was cut from $11 million to $5 million. Meanwhile Forrester, increasingly less interested in the simulator application and more determined to build a high-speed, highly reliable general-purpose digital computer, had openly abandoned work on the simulator cockpit in June 1948.[10] Though he tried to maintain Navy interest by describing Whirlwind as a "fire-control computer," it became clear to the ONR that this was really no more than another general-purpose machine. Since the Defense Department was funding at least twelve other projects for general-purpose digital computers, Whirlwind's justification became increasingly murky. The agency began to demand immediate and useful results in return for continued funding.

This dissatisfaction was due in part to the management of Whirlwind by the ONR's theoretically oriented Mathematics Branch, where the value of a real-time control machine was not well recognized, and in part to Whirlwind's truly enormous expense. Whereas the cost range of computers like the Harvard Mark III and the UNIVAC lay in the hundreds of thousands of current dollars, start-to-finish (most between $300 and $600 thousand), the Whirlwind group was planning to spend $4 million or more. From the Navy's point of view, this money was going to support the useful but hardly defense-critical technology of flight simulation.

MIT requested $1.5 million for Whirlwind in fiscal 1949. This figure would have consumed nearly 80 percent of ONR's mathematics research funds, or almost 10 percent of the *entire* ONR budget for contract research.[11] The actual grant for that year was $1.2 million—still an amazing level of investment, by any standard, in a single project. (Whirlwind's ultimate cost of about $5 million was over five times that of any other computer built during this period—ten to twenty times that of most.[12])

Computers for Command and Control
As the conflict over funding approached a critical phase, Forrester began to cast about for a new, more urgent, and more fundamental military justification. He was in a good position to do this for a number of reasons. First, during the war he had spent time on an aircraft carrier on a combat mission, and so had direct experience with one version of the air defense problem.[13] Second, his laboratory entertained a steady stream of visitors from both industry and military centers, each of whom brought questions and ideas about how a machine like the Whirlwind might be used to automate their operations. Forrester's notebooks indicate that between 1946 and 1948 these visitors raised dozens of possibilities, including military logistics planning, air traffic control, damage control, life insurance, missile testing and guidance, and early warning systems.[14] Perhaps the most significant of these contacts was Perry Crawford,

now at the Special Devices Center. As Forrester recalled in interviews, "it was... Crawford who pushed the whole idea of combat information and control with digital computers."[15] Crawford "circulated through the Navy and through the Washington scene, explaining the ideas to people and developing the necessary backing and funding so that the Navy was in a position to support the early development of the work."[16]

Third, the Whirlwind staff was composed largely of graduate students whose studies had been punctuated by military experience: "people coming back for their master's degrees who had completed an engineering undergraduate degree and anywhere up to perhaps four or five years in military service," Forrester recalled.[17] As a group, such students brought with them more concrete ideas about applications than might students on a more traditional career path. Fourth, Forrester feared the looming prospect of a nuclear-armed USSR and, like many of his peers, hoped his work could make a significant contribution to national defense.[18]

Forrester and his group had in fact been considering the issue of military applications all along. In early 1946, when he had first reported to the Navy on his emerging plan to switch from analog to digital techniques, he had included several pages on military possibilities. There he speculated that ultra-fast, real-time digital computers could replace analog devices in "offensive and defensive fire control," and he foresaw highly automated Combat Information Centers with "automatic defensive" capabilities that would be necessary for "rocket and guided missile warfare."[19] He also mentioned the probable utility of such computers in carrying out other military research and general applications in science and engineering. At a symposium Forrester, commenting on Crawford's paper about computers for missile guidance, predicted the use of computers "as complete control systems in certain defensive and offensive applications" such as "triangulation computations on approaching aircraft" and automatic tracking, targeting, and destruction of incoming ballistic missiles.[20] In October 1947, Forrester, Crawford, and Whirlwind co-leader Robert Everett had published two technical reports (designated L-1 and L-2) on how a digital computer might be used in antisubmarine warfare and in coordinating a naval task force of submarines, ships, and aircraft.[21] That year, in frequent meetings at its Sands Point headquarters, Crawford and other SDC personnel had encouraged Forrester and Everett to continue developing, refining, and planning the blue-sky systems-control ideas of their L-1 and L-2 reports.[22]

The following year, as continuation of ONR support became increasingly uncertain, MIT president Karl Compton requested from Project Whirlwind a report on the future of digital computers in the military. The group produced a comprehensive, compelling vision of computers applied to virtually every arena of military activity, from weapons research and logistics to fire control, air traffic control, antiballistic missile defense, shipboard combat information centers, and broad-based central command-control systems. It presented a plan for a crash 15-year, $2 billion (current) program leading to computerized, real-time command-control systems throughout the armed forces, projecting development timetables and probable costs for each application.[23]

From this point on, Forrester's commitment to the goal of real-time military control systems increasingly differentiated Whirlwind from other digital computer projects. As he recalled later, "from 1948 on, we were seeking machines to go into real-time control systems for military operations. Our circuits had to be extremely reliable compared to anything that previously had been thought necessary or possible. Our applications required very high speed, so we were working at speed ranges that were two and three orders of magnitude above [the Harvard and IAS computer] projects, and at a reliability level that was very much higher than the Institute for Advanced Study. Higher also than Aiken's [Harvard] work."[24] These commitments were realized not only in Whirlwind's technical efforts, but in the language of its self-representation.

Mutual Orientation: Constructing the Future

In one sense, Forrester's (and MIT's) increasingly grand attempts to imagine military applications for Whirlwind represented expert "grantsmanship," or deliberate tailoring of grant proposals to the aims of funding agencies. Grant writing is often dismissed as a kind of game. The usual argument is that grant proposals justifying basic research in terms of eventual applications are simply a vehicle to obtain funds that both recipients and agencies know will really be used for something else.

In the case of Whirlwind, however, a much more significant relationship between funding justifications and practical work also obtained, one we might call *mutual orientation*.

The Whirlwind studies of possible military applications of digital computers and the group's contacts with military agencies expanded the Whirlwind group's sense of possibilities and unsolved technical problems. At the same time, they served to educate the funding agency about as yet undreamt-of possibilities for automated, centralized command and control. While the ONR was not ultimately convinced, the thinking and the documents produced in the exchange kept funding going for several years. Later, these efforts proved crucial in convincing the Air Force to take over support for Project Whirlwind.

The source of funding, the political climate, and their personal experiences oriented Forrester's group toward military applications, while the group's research eventually oriented the military toward new concepts of command and control.[25] Forrester's group, MIT administrators, the SDC, and the ONR all directed each other's attention toward new arenas of concerns and solutions, centered around the articulation of the goals and meanings of a pre-paradigmatic technology. By forcing this articulation, conflicts among the groups' goals—Forrester's high-speed digital research ambitions, MIT's military-based empire-building, the SDC's long-range applications approach, the ONR's budgetary concerns and bureaucratic politics—generated a steady stream of new formulations and an increasingly coherent vision.[26]

Outside the unique circumstances of 1949 and 1950, this vision might have languished. But in the event, Whirlwind's discourse of computerized military control systems lay waiting, ready-made, for a second round of mutual orientation. This time, it would take part in the realignment of Air Force culture and strategy toward its fully modern incarnation as an automated, centralized, computerized command-control system.

To understand how Whirlwind helped reorient the Air Force, we must first understand how the Air Force reoriented Whirlwind. The following section explores how the issue of air defense was understood in the late 1940s. At that time, for a variety of reasons, the Air Force itself had dismissed continental air defenses as impractical. After the USSR's 1949 atomic test and the outbreak of the Korean War in 1950, the Air Force suddenly found itself hard pressed to justify this position. It initiated crash programs designed as much to assuage public anxiety as to provide genuine area defenses. Entering upon this scene, Whirlwind became caught up in a vast web of concerns: political problems of nuclear fear, strategic and tactical problems of air warfare, technical and cultural issues of central control, and, through them, the emerging discourse of the closed world.

Cold War Politics, Strategic Doctrine, and Air Defense

The strategic task of the postwar Air Force, largely self-defined, pivoted on the new weapon—as did its role within the armed forces as a whole.

Before 1947 the Army and Navy were separate services, each with its own cabinet-level representative. During World War II the then Army Air Force (AAF) played such a significant strategic role that it began to seek a place as a third service. The three agencies often saw themselves as competing for military assignments, resources, and prestige.

The AAF seized on the bomb in 1945 as a means to expand its military role. So-called strategic bombing, or area bombing of cities with the aim of killing or disabling the employees of war industries and destroying civilian morale—as opposed to attacking industrial targets directly—was a central strategy of the Allied air forces during World War II (especially in Asia), occasional official pronouncements to the contrary notwithstanding. In addition, World War II-era aerial bombardment had very low accuracy, especially when bombers flew high (as they often did) to avoid antiaircraft fire. This meant that even when industrial or military installations were the intended targets, bombing generally destroyed wide areas around them as well. Thus area bombing was the *de facto* strategy even when not *de jure*.

The postwar *Strategic Bombing Surveys* of Great Britain and the United States showed that this strategy had been relatively ineffective, significant mainly in disrupting enemy fuel and supply lines near the war's end. But the atomic bomb's apparent success in securing Japan's abrupt and complete surrender swept aside their highly skeptical conclusions about air power.[27] Postwar plans, despite the surveys, relied on general attacks against cities and assumed that Hiroshima-like devastation would lead automatically to the enemy's surrender. Nuclear weapons, which unavoidably destroyed everything within miles of ground zero, fit perfectly into this strategic doctrine. Almost without debate, city bombing became the nuclear strategic policy of the new Air Force.

"Prompt Use"

By 1946 the Air Force had drafted a nuclear war plan that called for fifty bombs to be dropped on Russian cities—despite the fact that even a year later the United States had only thirteen bombs, only one of which could have been prepared for use in less than two weeks. In 1947 the National Defense Act elevated the Air Force to the status of an independent service and began, but did not complete, the process of uniting all three services under the new cabinet office of the Secretary of Defense (OSD).[28] In 1948 NSC-30, one of the early directives of the National Security Council (also created by the 1947 National Defense Act), authorized Air Force planners to assume the availability of increasing numbers of nuclear weapons and to establish a policy of "prompt use."

In essence, this was a doctrine of preemptive strike. The Air Force planned an all-out nuclear attack against the USSR in any situation where it appeared the USSR might be about to launch a strike of its own.

The reasoning behind this policy grew in part from the cowboy ethic of Air Force culture. Between the wars, World War I AAF commander Billy Mitchell had mounted an enormous media campaign to promote air power as a kind of ultimate weapon that could make ground warfare obsolete. World War I newsreels more or less commissioned by Mitchell showed airplanes under his command sinking enemy ships. Mitchell's airmen called this activity "air defense" since it involved destroying the sources of enemy fire, a usage which continued until World War II, causing some understandable confusion about the difference between defense and offense in air warfare.[29] It was not until about 1941 that the official Army Air Corps definition of air defense excluded "counter air force and similar offensive operations which contribute to security rather than air defense."[30] As late as 1952, during an interview on national television, General Hoyt Vandenberg reiterated the idea that destroying the sources of enemy fire—in this case, enemy air bases—was the most fundamental tactic of air defense.[31]

Mitchell continued to proselytize after World War I with mock air raids on American cities and articles in the major general-interest magazines *Collier's*, the *Saturday Evening Post*, and *Liberty*. In these forums, during 1924–25, Mitchell challenged the sanctity of civilian lives in modern warfare. He argued that since enemy cities produced munitions and other military matériel, and since their inhabitants directly and indirectly supported this military role, cities (and their populations) were legitimate military targets. Once its industrial centers were bombed with high explosives and tear gas, Mitchell believed, any enemy would be forced to capitulate. Wars could be won from the air. The tireless Mitchell also disseminated his views through Walt Disney films and eventually published a book, *Winged Defense*.[32] His aggressive public attacks on Secretary of War John Weeks and other officers who disagreed with his views eventually led to his court-martial and conviction for insubordination.

Mitchell's flamboyant, swashbuckling image became a basic icon of Air Force culture. (A popular 1955 film, *The Court-Martial of Billy Mitchell*, lionized him as a military prophet.)[33] As we have seen, the very doctrine of strategic bombing that led to his court-martial became official Air Force strategy during World War II. Mitchell viewed air forces as an ultimate war-winning power that required nothing from conventional armies but to be given free rein. This vision became the dream the Air Force pursued to its apotheosis in the Strategic Air Command (SAC), under the flamboyant, cigar-chewing General Curtis LeMay.

The policy of prompt use originated in this culture of the offensive. LeMay reportedly once told an assembled group of SAC pilots that he "could not imagine a circumstance under which the United States would go second" in a nuclear war. Yet because it so deeply contradicted the ideology of America as a nation armed only for its own defense, the policy remained a high-level secret, kept so effectively that according to Gregg Herken "it is likely that few in the government or at Rand [an Air Force think tank] actually knew enough details of Air Force war planning to appreciate the extent to which American nuclear strategy by the mid-1950s was based upon the premise that the United States would land the first blow with the bomb."[34]

There were also, however, significant strategic reasons for the policy of prompt use of nuclear weapons. First, World War II experience with aerial combat and air defense had shown that it was extremely difficult to defend (in the ordinary sense) even relatively small areas against a determined air attack. Second, radar technology of the 1940s could neither see beyond the horizon nor detect low-flying airplanes. Even an ideal radar system using then-current technology could have provided at best one to two hours' advance warning. Worse, attackers flying below 1,000 feet could have evaded it altogether. Third, the enormous length of the U.S. perimeter made complete radar coverage a gigantic and extremely expensive undertaking. Finally, and perhaps most important, commanders generally estimated that even excellent air defenses could prevent only about 10 percent of attacking airplanes from reaching their targets—30 percent, according to the most hopeful, at an absolute maximum.[35] But if the invaders carried nuclear weapons aimed at cities, even a kill ratio of 90 percent would be unacceptably low.

Thus the principle that "the best defense is a good offense" applied in spades to the issue of defense against nuclear-armed bombers. Atomic bombs seemed to produce an even more overwhelming advantage for that offense—which, to Air Force thinking, *was* the defense. By 1950 fifty bombs had been built and many more were on the way. Air defense programs developed at a desultory pace, receiving only minimal commitments of funds and attention. In fact, the Air Force pushed *against* air defense, fearing it would pull resources and commitments away from the Strategic Air Command.

"A Dangerous Complacency": Resisting Air Defense

In August 1947 a panel of officers of the Air Staff reflected the prevailing view within the forces that the AAF neither could nor should plan to provide air defense of the entire United States. Because of its size, they believed, such a commitment might endanger the national economy. Worse, however, it would "leave little room for the air offensive"; this "would be disastrous since real security lay in offensive capability."[36] The panel recommended only point defense of strategic targets.

A Rand Corporation report, commissioned by Air Force science advisor Theodore von Karman, agreed. Rand invoked a favorite Air Force icon: "such an investment [in expensive, near-obsolete, ineffective air defense systems] might ... foster a dangerous 'Maginot Line' complacency among the American people."[37] Thomas K. Finletter's 1947 Air Policy Commission, appointed by Truman to create an integrated air strategy, insisted that the Air Force by 1953 should equip itself with the best, most modern defensive electronics, jet fighters, and ground-based weapons. But in language strikingly similar to Rand's, the commission opposed a total radar coverage system because it might "divert us—as the Maginot Line diverted France—from the best defense against an atomic attack, the counter-offensive striking force in being."[38]

The opposing position was represented by Maj. Gen. Otto Weyland, Assistant Chief of Air Staff for Plans, who pointed out in an exchange with General Earle Partridge (his counterpart in Operations, Commitments, and Requirements) that the AAF now faced a policy contradiction. The agency sought the chief responsibility for air defense but had assigned virtually no equipment or personnel to that task. Weyland argued that at least some minimal system must be built and maintained to demonstrate the AAF's commitment.[39]

Some measures were in fact already under way. Even before the war's end, Army Ordnance and the Air Force had commissioned Bell Laboratories to study continental air defense against high-altitude, high-speed bombers. The result was the Nike antiaircraft missile project. Ground-based analog computers and radar, along the lines of Bell's analog gun directors, would guide the Nike missiles to their targets, where they would be detonated by remote control. Nike R&D was not finished until 1952, and installation was not completed until about 1954.[40] Even then, the Nike-Ajax was a point (as opposed to an area) defense system, and it was controlled by the Army. In a period of intense interservice competition, the Air Force saw the Nike-Ajax project as worse than no defense at all, because it might lead not only to a "dangerous complacency" but to Army control of continental air defense.

Increasingly bellicose Cold War politics, both global and domestic, ultimately mooted the debate. The Air Force role in this process was to amplify fears of Soviet aggression while constructing a military container—in the form of forward SAC bases and nuclear weapons—to hold back the Red tide.

By 1948 Air Force intelligence—contrary to the estimates of both Army and Navy intelligence services and to those of the Central Intelligence Agency—had come to believe strongly in the possibility of imminent Soviet attack. This view bordered on the bizarre. Such an attack would have required (a) the Tu-4 Bull long-range bombers demonstrated for the first time in a 1948 Soviet air show and not produced in any quantity until the following year, (b) a suicide-mission strategy, since the Tu-4 could hold enough fuel to reach the United States from the USSR, but not to return, and, most absurdly, (c) the USSR's willingness to risk American atomic retaliation at a time when it possessed only conventional weapons. The Air Force leadership, grounding its faith in the demonizing discourse of the Cold War, thought the kamikaze strategy a real possibility and apparently suspected, on the thinnest of evidence, that the necessary elements of this strategic scenario might be much more advanced than they seemed.[41]

The strange 1948 emergency alert provides good evidence of the strength of these implausible assumptions. In March of that year, USAF Headquarters ordered the existing skeleton emergency air defense system onto 24-hour alert. The alert lasted nearly a month, until it was suddenly canceled in mid-April. It apparently resulted from reports by Lt. Gen. Ennis C. Whitehead, AF commander in the Far East, of a series of "strange incidents and [Soviet] excursions" over Japan, combined with a change in Soviet European military alignments after the communist coup in Czechoslovakia.[42]

Whatever its causes, this event had the effect of drawing attention to the severe limitations of continental air defense at a time when the so-called Radar Fence Plan was stalled in Congress. This plan, one of several (unimplemented) interim air defense plans proposed between 1945 and 1950, would have used the obsolete World War II-era radars to build a national network including 411 radar stations and 18 control centers, staffed by 25,000 Air Force personnel and 14,000 Air National Guardsmen. The Radar Fence would have cost $600 million and was to become operational in 1953—the earliest date the USSR's first atomic weapons were expected. Despite the emergency alert, Congress balked at the plan's cost. Only in March 1949 did it finally approve an air defense bill, the much smaller Lashup radar system comprising only 85 radar stations and costing a mere $116 million.[43]

The budgetary tide—and the political fortunes of the Air Force—turned hard in September 1949, when the Soviets exploded an atomic bomb years ahead of the schedule forecast by U.S. intelligence. The Truman administration immediately began planning for a two-sided nuclear war.

In the spring of 1950, the National Security Council warned that the Soviets were actually ahead of the United States in the arms race. NSC-68 looked to 1954 as the "year of maximum danger," when Soviet forces would have enough bombs to disarm the United States in a surprise attack. The report recommended spending 20 percent of the nation's gross national product on a massive defense buildup. The outbreak of war in Korea the following year provided the crisis necessary to implement NSC-68's recommendations.[44] By January 1951 Truman had set in place a vast range of new policies for the successful prosecution of a much escalated Cold War. Emergency war powers and renewal of selective service were rushed through Congress. Truman's $50 billion defense budget roughly conformed to NSC-68's guidelines. He increased Army troop strength by 50 percent to 3.5 million men, built new bases in Morocco, Libya, and Saudi Arabia, raised aid levels to the French in Vietnam, initiated proceedings for bringing Greece and Turkey into NATO, and opened discussions with Gen. Francisco Franco in which American aid was eventually traded for military bases in Spain. He also doubled the size of the Air Force, to ninety-five air groups.[45] In less than four years, annual spending for nuclear strategic forces would more than quadruple from 1950 levels.[46]

While its other assumptions about Soviet capabilities and intentions remained implausible, the fact that only Air Force intelligence had predicted a 1949 Soviet nuclear weapon had the effect of vindicating its other views and magnifying its influence on strategic decision-making. The case for air defense suddenly acquired much greater force. Civilians, especially those in the state of Washington (near Boeing Aircraft and the Hanford nuclear facilities), began to clamor for protection. To preserve its credibility, the Air Force would have to come up with something more politically effective than reassurances about the overwhelming power of its offensive forces—especially, perhaps even paradoxically, since the prompt-use policy must remain secret. General Hoyt Vandenberg, now Air Force Commander in Chief, told the Joint Chiefs of Staff in a November meeting that "the situation demanded an urgency and priority similar to the Manhattan District Project."[47]

For the short term, the Air Force initiated a crash program in early warning, stepping up the schedule for the Lashup system. In addition, plans were approved to proceed hastily with the radar "fence" along the polar approaches to the United States, despite the technical problems discussed above.

To get around the problem of low-altitude radar blindness, a colossal network of visual observation posts was established, staffed by civilian volunteers. During the Korean War the Air Force recruited these volunteers with inflammatory—and disingenuous—radio advertisements such as the following:

Who will strike the first blow in the next war, if and when it comes? America? Not very likely. No, the enemy will strike first. And they can do it too—right now the Kremlin has about a thousand planes within striking distance of your home.[48]

At its peak in 1953 the Ground Observer Corps operated more than 8,000 observation posts, twenty-four hours a day, using over 305,000 volunteers. Commanders generally recognized the GOC as unreliable and too slow to provide significant warning. Nevertheless, it continued to function until 1959. The GOC was another buttress for the wall of the container America was building, another support for closed-world discourse. Its function, like so much of the macabre apparatus of nuclear war, was primarily ideological: a genuine defense being impossible, a symbolic one was provided instead.

For the long term, the Air Force turned to scientists for new ideas. Happening almost by chance upon Forrester's crisis-torn computer project, the architects of the long-term solution found a technology neatly packaged together with a ready-made, highly articulated vision of central command and control using digital techniques. They resurrected it from near oblivion and transformed it into the core of the SAGE continental air defense system. Whirlwind, injected with almost unlimited funding and imbued with the intense urgency of nuclear fear, suddenly became a central pillar in the architecture of the closed world's defensive dome.

From Whirlwind to SAGE

In December 1949 the Air Force established the Air Defense System Engineering Committee, headed by MIT professor George E. Valley. The Valley Committee, as it was known, worked for two years, beginning its program with a study of the radar-based air defense of Great Britain during World War II. It emerged with a comprehensive plan for the air defense of North America, a plan that became reality as the Semi-Automatic Ground Environment—SAGE.

Valley's was not the only plan, however, and digital computers were not the only technology. Fierce debates raged inside the Air Force over the issues of centralization and automation. Parallel to the centralized digital system, research proceeded on a less automated, decentralized analog control technique. In the end the MIT scientists and engineers won out, converting the Air Force simultaneously to air defense, central control, and digital computers. They did so not, or not only, by creating a superior technology, but by generating a discourse that linked central automatic control to metaphors of "leakproof containers" and "integrated systems"—Maginot Lines (as Air Force traditionalists called them) for a new technological age.

According to Valley's memoir, he immediately comprehended the enormity of the mathematical problem for a wide-area defense. To triangulate the positions and velocities of aircraft, sightings from two or more radar units would have to be integrated through calculations.

The earth's curvature meant that hundreds, if not thousands, of radars would be required to detect low-flying aircraft.... There was no conceivable way in which human radar operators could be employed to make [the necessary] calculations for hundreds of aircraft as detected from such a large number of radars, nor could the data be coordinated into a single map if the operators used voice communications. The ... computations were straightforward enough.... It was doing all that work in real time that was impossible.[49]

Apparently independently, Valley came up with the idea of using digital computers for this purpose. A month later, in January 1949, Jerome Wiesner told him about the Whirlwind project.

The timing of Valley's encounter with Whirlwind was serendipitous. The ONR was just coming under the influence of a report by the Ad Hoc Panel on Electronic Digital Computers convened by the Committee on Basic Physical Sciences of the Defense Department's Research and Development Board. The panel's major recommendation—that the entire Defense Department effort in digital computation be centralized under a new committee—was never implemented, but its criticisms of Whirlwind were strong and influential. According to its analysis, completing Whirlwind would require about 27 percent of the $10 million DoD computer research budget, which was then supporting thirteen machines from eight suppliers.[50] Without a more urgent end use, the panel held, Whirlwind's expense could not be justified.[51]

By March 1950 the ONR had cut the Whirlwind budget for the following fiscal year to $250 thousand. Compared with the $5.8 million *annual* budget Forrester had at one point suggested—as a comfortable figure for an MIT computer research program leading to military and other control applications as well as state-of-the-art scientific and engineering problem-solving capability—this was a minuscule sum.[52] According to James R. Killian, Jr., then president of MIT, "Project Whirlwind would probably have been canceled out had not George Valley ... come up with the pressure to use Whirlwind as part of the SAGE system."[53]

The Air Force had handed Valley a blank check. Though he

initially heard mostly negative things about Whirlwind from others at MIT, Valley made contact with Forrester, who immediately offered him the 1947 L-1 and L-2 reports on digital computers as central controllers for naval warfare. Forrester and Everett also showed him some plans for air traffic control, but the naval-warfare documents impressed Valley more. Air traffic control was qualitatively different from air defense problems: in the first case pilots are cooperating with controllers and providing information; in the second they are doing the opposite.

Valley approached Whirlwind with some hesitation. He thought of the digital computer application as little more than a way to prove a point about the potential of digital techniques in the control field, then entirely dominated by analog methods, and he took seriously Whirlwind's reputation as an overblown behemoth. He therefore started by proposing that ADSEC rent the Whirlwind for only one year, and he continued to explore other possibilities. (These included another digital-computer-based, but decentralized and lower-speed, system proposed by Northrop Aircraft, former sponsors of the ill-fated BINAC.)[54] Fortunately for Forrester, Whirlwind was finally reaching full operational status—by the end of the year it was regularly running scheduled programs—and with his charisma, energy, and enormous intellectual capability as an additional influence, Valley soon became a believer.

Valley's timing was also lucky in that the Air Force Cambridge Research Center (AFCRC), for reasons having nothing to do with computation, had recently developed methods for digital transmission of data over telephone lines. Their goal was to compress radar information, which was then transmitted over expensive microwave channels, into a bandwidth small enough for telephone transmission. To do this they had created a system known as Digital Radar Relay (DRR). Making such transmissions reliable—in a system designed from the start for analog voice signals—was the key issue.[55] The DRR research, begun just after World War II, had taken four years to complete. Its availability solved one of the many analog-to-digital conversion problems faced by the eventual SAGE.

The budding air defense program thus intersected neatly with newly available digital technologies, including Whirlwind. Simultaneously, Forrester's vision of centralized control systems intersected with the Air Force's recent *tactical* innovation of ground control command.

The original tactics of air-to-air combat sprang from World War I dogfight-style pursuit, with individual pilots identifying their own targets and engaging them on a one-to-one basis. This evolved into a group pursuit strategy in which each group leader chose his own targets from the air, eventually aided by information from ground-based radar. In 1935 Army Air Force Capt. Gordon Saville tested a ground-based control technique in which commanders identified incoming targets using radar and directed interception centrally from their headquarters. Now ground controllers would exercise not "air liaison," but "air command."[56] Saville's approach initially met with resistance from pursuit-group leaders, who were used to commanding their own forces. Tests proved it more effective than the traditional system, however, and by the time of World War II ground control of air defense had been generally adopted—albeit reluctantly, and with a residue of decentralized and loosely organized command structures. When the Royal Air Force successfully employed a ground-control approach in the Battle of Britain in 1940, the Saville method's dominance was assured.

Early in 1950 Saville, by then an Air Force general, was appointed the first Deputy Chief of Staff for Development. In this role, with a swollen budget and virtual carte blanche from his superiors for his technologically oriented tactical imagination, he assumed the role of Air Force liaison to the Valley Committee. When Valley convinced him that digital computers offered the core of a solution to the air defense problem, Saville joined Air Force chief scientist Louis Ridenour as the highest-ranking advocate of centralized computer control within the Air Force command.

Converting the Air Force to Air Defense

The Valley Committee's work was soon extended by other groups. The Weapons Systems Evaluation Group (WSEG), in the Office of the Secretary of Defense, conducted an independent study of air defense beginning in early 1950. In Project Charles, at MIT, a committee of distinguished scientists spent the first six months of 1951 looking into the air defense problem and recommended establishing an air defense laboratory (the eventual Lincoln Laboratory). The East River study of summer 1951, under the Air Force and the National Security Resources Board, found civil defense measures not only dependent on adequate early warning (requiring a much improved radar network) but useless without highly effective air defense. It concluded that computerized systems could improve the prospects for air defense. Finally, in 1952 a Summer Study Group of Lincoln scientists and others, led by MIT physicist Jerrold Zaccharias and including Valley associate Albert Hill and physicists Isidor I. Rabi and J. Robert Oppenheimer, evaluated Lincoln's progress and assessed prospects for a full-scale system.

These committees, led in their thinking by Valley's group, constructed a grand-scale plan for national perimeter air defense controlled by central digital computers that would automatically monitor radars on a sectoral basis. In the event of a Soviet bomber attack, they would assign interceptors to each incoming plane and coordinate the defensive response. The computers would do everything, from detecting an attack to issuing orders (in the form of flight vectors) to interceptor pilots. The plan was first known as the Lincoln Transition System, after MIT spun off its huge Lincoln Laboratory to run the air defense project.[57] It was redesignated SAGE (Semi-Automatic Ground Environment) in 1954.

The final report of Project Charles, while pessimistic about "any spectacular solution of the air defense problem," also expressed "considerable optimism about the contribution to air defense that will be made by new basic technology..., [especially] the electronic high-speed digital computer."[58] The Summer Study Group went much further, crystallizing the Lincoln ideas into an overarching vision around the concept of a highly integrated, computerized air defense control system. Coupled with Arctic distant early-warning radars (a concept rejected by Project Charles), the group expected that such a system could achieve kill ratios of 60–70 percent, an estimate far higher than any Air Force prediction.[59]

Throughout the first half of the 1950s, the Air Force traditionalists continued to oppose the plans developed by the study groups. Robert Oppenheimer, in an article in the July 1953 *Foreign Affairs*, noted a comment made to him by a high-ranking officer to the effect that "it was not really our policy to attempt to protect this country, for that is so big a job that it would interfere with our retaliatory capabilities."[60] Air Force culture, with its emphasis on the nuclear offensive, its pilot-oriented cowboy ethic, and its aversion to defensive strategies, saw the cocky civilian engineers as military naïfs, unable to comprehend battlefield logic and antagonistic to its deeply held traditions and beliefs.

Air defense, Project Charles leader and Manhattan Project veteran Jerrold Zacharias claimed, was finally "sold to Truman over the dead body of the Air Force."[61]

Most of the enthusiasm for air defense, especially in this new, high-technology guise, thus came from civilian scientists and engineers. The latter tended toward the messianic in their promotion of the new technical and strategic concepts. According to one participant, physicist Richard Garwin, Zacharias once told him that "if these people don't come to the right conclusion, then I'll dismiss them and begin another study." Many of them believed deeply that a defensive strategy would prove less provocative than a nuclear sword of Damocles; they saw their work as a kind of end run around their government's belligerent Cold War policies. "We all knew the conclusions we wanted to reach," one summer study group scientist admitted.[62]

Forrester's group and its Project Charles/Summer Study backers were ridiculed within the Air Force as "the Maginot Line boys from MIT" who supported a "Great Wall of China concept."[63] General Hoyt Vandenberg called the project "wishful thinking" and noted that

the hope has appeared in some quarters that the vastness of the atmosphere can in a miraculous way be sealed off with an automatic defense based upon the wizardry of electronics.... I have often wished that all preparations for war could be safely confined to the making of a shield which could somehow ward off all blows and leave an enemy exhausted. But in all the long history of warfare this has never been possible.[64]

The Air Force especially feared that emphasis on air defense would reduce SAC budgets. In an appearance before Congress, Vandenberg reiterated the Air Force dogma that "our greatest *defensive and offensive* weapon is our strategic force plus that part of our tactical force that is based within striking range of the airdromes that would be used by the Soviets."[65] But the Soviet hydrogen bomb explosion of 1953, before the American "super" was ready, renewed public fear of a nuclear holocaust, and this, combined with the "can-do" technological mindset of the 1940s and 1950s, generated the momentum needed to push the SAGE perimeter defense project ahead. President Eisenhower ended up supporting both SAC and the continental air defense program under his high-technology, nuclear-oriented New Look defense strategy.

Centralizing Command, Mechanizing Control

Even when they cooperated, major elements within the Air Force continued to distrust the "nebulous" Lincoln plan. The degree of centralization, especially, concerned commanders. "There was significant concern by the military operators over whether a centralized system . . . was the right way to go or whether one ought to have an improved decentralized system operating at the radar sites much as the old system operated," recalled one of the Air Force backers of SAGE.[66]

The centralization issue arose again later, vastly exacerbated by interservice rivalry, over the question of whether to integrate Army antiaircraft batteries into SAGE. In 1956 Lt. Gen. S. R. Mickelsen, chief of the Army Antiaircraft Command, engaged in intense verbal sparring with Continental Air Defense Command head Gen. Earle Partridge, noting that "early warning and target information from Air Force sources will enhance the effectiveness of AA weapons; detailed control will most certainly degrade it."[67] After a protracted conflict, the issue was taken to Secretary of Defense Wilson, who resolved it in favor of centralized control under SAGE. In interesting contrast, the USSR's eventual air defense program favored a decentralized approach, also advocated by early Rand Corporation and Stanford Research Institute studies. According to one military observer, the Soviet results were excellent: it was "organized like a field-army air defense system—no central control; everybody shoots at anything that looks hostile with everything he has. . . . The Soviet system is what you do when you are *serious* about continental air defense."[68]

Furthermore, to officers steeped in a tradition of human command, the idea of a machine analyzing a battle situation and issuing orders was at best suspicious, at worst anathema. Digital computers were still the province of a tiny élite of scientists and engineers, incomprehensible to the average person. Even for the Air Force, the armed service most open to technical innovation, the Lincoln plan entailed an unprecedented scale of automation using an unknown technology.

There was an alternative. Another system, based on analog computers and automatic assistance for manual techniques, was proposed by the Willow Run Research Center at the University of Michigan at about the same time. The proposal involved adapting and automating the British Comprehensive Display System (CDS), in which radar data sent via telephone and teletype were manually plotted. An individual CDS site could track up to a hundred planes, but there was no way for sites to exchange information automatically. The Willow Run group proposed to automate transfer of data between sites and to provide analog devices to assist in the plotting of tracks and interceptor courses in an Air Defense Integrated System (ADIS). Command would be more decentralized—and much less automated—than under the Lincoln program.

The ADIS project acquired a substantial minority backing within the Air Force, which continued to fund Willow Run research until 1953. Even then, the Air Force only canceled the project when MIT threatened to quit if it did not commit to the digital approach. In fact, it was the Willow Run proposal that catalyzed Lincoln's own proposal for the Lincoln Transition System, a partial implementation of the computerized scheme that the laboratory claimed could be operational by 1955. The Air Force, fearful of losing the goodwill of one of its major technical resources, made a final choice in favor of MIT's centralized digital design. ADIS was canceled, and Lincoln/SAGE development began in earnest.

Working feverishly, the Lincoln group had been able to demonstrate the basic elements of the system—tracking aircraft and controlling interception using radar data relayed over telephone lines—in 1952, using the Whirlwind computer. By then IBM had signed on to design a production version of Whirlwind, the AN/FSQ-7. (Actually the FSQ-7 was modeled on Whirlwind II, the successor machine Lincoln Laboratories had designed specifically for air defense use.) Larger demonstrations followed, such as a reduced-scale experimental SAGE sector in 1954. Meanwhile, Lincoln also designed improved radars and, in collaboration with the Canadian Air Force, ringed the far northern perimeter of North America with radar stations (the Distant Early Warning Line, completed in 1957). Thus three radar networks—the Pinetree Line on the Canadian border, the Mid-Canada Line, and the DEW Line—fed the SAGE system a picture of air traffic as far north as the Arctic Ocean. In a 1957 interview, Gen. Earle Partridge, now commander of the newly created North Atlantic Air Defense Command (NORAD), estimated that 200,000 people worked for the air defense network. He predicted that its total cost between 1951 and 1965 would reach $61 billion.[69]

The Air Force held opening ceremonies for the first SAGE sector at McGuire Air Force Base on June 26, 1957. By 1961 the system's 23 sectors were complete and SAGE was fully operational. Its control centers had cost more than a billion dollars to construct.

SAGE's implementation thus marked the final outcome of an extended political battle within the Air Force. As we have seen, to characterize this as a struggle over strategy and technology would be too narrow, even insofar as the debate took place within the Air Force and its civilian advisory groups. But the debate was not, in fact, so limited. It encompassed wider public arenas as well. There, nuclear fear transcended the technical and strategic merits to transform the debate into a contest of ideology.

Between 1952 and 1955, while development proceeded in the face of fierce Air Force efforts to downscale the project, the air defense study groups sought public support. Many of the scientists involved in the groups saw their work as a way to soften or even eliminate a dangerously aggressive, offensive orientation in national strategy. When they could, they used the press to present their views. In 1953, for example, MIT President James R. Killian, Jr., and Lincoln Laboratory director A. G. Hill published an impassioned plea for better air defenses in the *Atlantic Monthly*. They began by noting that one hundred Soviet atomic bombs, successfully delivered, would kill or injure "not just hundreds or thousands but millions of people" and that existing forces could circumvent "only a small percentage" of such an attack. With the caveat that perfect defenses were not possible, they went on to argue that present capabilities could be improved "manyfold.... Not 100 percent, or even 95 percent" of attacking planes would be downed by the improved system they backed, but it would nevertheless produce "a great gain over our existing powers of attrition."[70] The *Boston Globe*, *Christian Science Monitor*, and *Bulletin of the Atomic Scientists* each republished a condensed version of the essay, and House minority leader John McCormack circulated copies throughout the government.[71]

Whatever the scientists' own beliefs about effectiveness, the nonspecialist opinion leaders they influenced generally exaggerated the prospects for defensive technology. In early 1952's "Night Fighters Over New York," for example, the *Saturday Evening Post* reported (even before SAGE) that a "stupendous, history-making system of defense against an enemy's atom bombs...already covers all approaches to the country." Having committed hundreds of millions of dollars to the radar early warning system and the Air Defense Command, the nation would soon see "thousands of these new supersonic terrors [i.e., jet interceptors]...beating up the airwaves in...the most formidable defense network in history."[72] In a series of articles in 1953, syndicated columnists Joseph and Stewart Alsop, probably aroused by conversations with their contacts among the scientists in the air defense study groups, accused the Air Force of dragging its feet on air defense. They asserted that with an additional yearly investment of $4 billion, the United States could have a virtually leakproof air defense shield (i.e. SAGE) within five years.[73]

Such interpretations, with their underlying concept of an impenetrable barrier surrounding the country—just as the Air Force feared, a Maginot Line—helped constitute the discourse of a closed world protected by high technology. Civilian opinion leaders, the incipient corps of military technocrats, and scientists and engineers with an instinctive belief in technological solutions thus allied *against* powerful Air Force traditionalists. The work of producing SAGE was simultaneously technical, strategic, and political. Its ultimately produced not just a new kind of weapon system but a profound reorientation of strategic doctrine.

Technological and Industrial Influences of SAGE

A central thesis of this book is that computer technology and closed-world discourse were mutually articulated. If this is true, closed-world politics shaped nascent computer technology, while computers supported and structured the emerging ideologies, institutions, language, and experience of closed-world politics. Nothing better illustrates this mutual influence than the history of Whirlwind and SAGE.

Technology

Whirlwind and SAGE were responsible for a vast array of major technical advances. The very long list includes the following inventions:

- magnetic core memory
- video displays
- light guns
- the first effective algebraic computer language
- graphic display techniques
- simulation techniques
- synchronous parallel logic (digits transmitted simultaneously, rather than serially, through the computer)

- analog-to-digital and digital-to-analog conversion techniques
- digital data transmission over telephone lines
- duplexing
- multiprocessing
- networks (automatic data exchange among different computers)

Readers unfamiliar with computer technology may not appreciate the extreme importance of these developments to the history of computing. Suffice it to say that much-evolved versions of all of them remain in use today.[74] Some, such as networking and graphic displays, comprise the very backbone of modern computing.

Almost as importantly, Whirlwind's private publication efforts disseminated knowledge of these advances widely and rapidly. Whirlwind had its own reports editor and printing operation. The project distributed some 4,000 short memoranda, biweekly progress reports, engineering reports, and "R" series reports on major achievements to its own staff. In addition a mailing list of about 250 interested outsiders received Whirlwind quarterly reports and some of the more significant engineering and "R" papers. Internal coordination was one purpose of these publications, but another explicit purpose was "maintaining support and an outside constituency."[75]

Many of Whirlwind's technical achievements bear the direct imprint of the military goals of the SAGE project and the political environment of the postwar era. As a result, despite their priority of invention, not all of these technologies ultimately entered the main stream of computer development via Whirlwind and SAGE. Some, such as core memory, almost immediately made the transition to the commercial world. Others, such as algebraic languages, had to be reinvented for commercial use, for reasons such as military secrecy and their purpose-built character. I will mention only three of many possible examples of this social construction of technology.

First, the Cold War, nuclear-era requirement that military systems remain on alert twenty-four hours a day for years, even decades, represented a completely unprecedented challenge not only to human organizations, but to equipment.[76] The Whirlwind computer was designed for the extreme reliability required under these conditions; in the 1950s, this involved a great deal of focused research. The solution Whirlwind's designers came up with was duplexing—an extremely expensive as well as technically difficult method, since it more than doubled the number of components. Whirlwind research also focused heavily (and successfully) on increasing the reliability of vacuum tubes.[77] The results were impressive. Down time for FSQ-7 computers averaged less than 4 hours per year. Well into the 1970s, other computers frequently counted yearly down time in *weeks*.

Second, SAGE was a control system, and control is a real-time operation. This meant much faster operating speeds than any other machine of the period, not only for the central processing units but for input and output devices as well. For example, the DRR methods of converting radar data into digital form found their first practical uses in SAGE. While high processing speed might seem inherently desirable, in the 1950s the (then) extreme speed of Whirlwind was unnecessary for other computer applications. In non-real-time applications, input/output (I/O) bottlenecks, including human preprocessing and interpretation of results, mattered far more than computer speed in determining overall throughput. Whirlwind thus helped to define both the meaning and the uses of "speed" in early digital computing.[78]

Finally, both the transmission of data from radars and the coordination of the SAGE centers employed long-distance digital communication over telephone lines (some of the first modems[79] were built for this purpose). The computers at different SAGE sectors also exchanged some data automatically. The massive integration of a centralized, continental defense control system required such communications. SAGE was thus the first computer network, structured directly by the needs and locations of the military system it controlled.[80]

Industry

As the history of computer technology bears the imprint of SAGE, so does the history of the emerging computer industry. SAGE contributed devices and ideas to commercial computer technology, increased the competitiveness of American manufacturers vis-à-vis their foreign competitors, and swayed the fortunes of individual companies. For example, IBM built fifty-six SAGE computers at a price of about $30 million apiece. At the peak of the project more than 7,000 IBM employees, almost 20 percent of its total workforce, worked on

SAGE-related projects; the company's total income from SAGE during the 1950s was about $500 million. Between SAGE and its work on the "Bomb-Nav" analog guidance computer for the B-52 strategic bomber, more than half of IBM's income in the 1950s came from military sources.[81]

The benefits to IBM went far beyond profits. They also included access to technical developments at MIT and the know-how to mass-produce magnetic core memory—the major form of random access storage in computers from the mid-1950s until well into the 1970s—and printed circuit boards. IBM's SABRE airline reservation system, completed in 1964, was the first networked commercial real-time transaction processing system. Its acronym stands for Semi-Automatic Business-Research Environment, a direct reference to SAGE (Semi-Automatic Ground Environment), whose essential network structure it copied. Many employees of IBM and Lincoln Labs who learned about computers from working on SAGE went on to start important companies of their own. IBM's decision to participate in the SAGE project was probably, according to Kenneth Flamm, the most important business decision it ever made.[82] Many other computer and electronics firms, including Burroughs, Western Electric, and Bell Laboratories, also benefited from SAGE-related contracts. "Above all," Stan Augarten has written, "SAGE taught the American computer industry how to design and build large, interconnected, real-time data-processing systems."[83]

SAGE also had a critical impact on software. Here IBM proved less foresighted; the company might have achieved even greater dominance in the nascent industry had it elected to do the programming for SAGE as well as building its hardware. IBM declined, according to one participant, because "we couldn't imagine where we could absorb two thousand programmers at IBM when this job would be over someday."[84]

SAGE engineer Norman Taylor's analysis of the effects of SAGE on software technology is worth quoting at length:

> The need for real-time software in the true aircraft-movement sense made the work doubly demanding, since the proper software had to be operated in the proper part of core in synchronism with a real-time clock pacing aircraft as they moved.... To the software world, these activities were as basic as the core memory and the Whirlwind I computer were to the hardware world. When these concepts were later expanded to the full FSQ-7 SAGE computer

in the late 1950s, it became clear that the manual tasks of core register assignments, opening and closing interactive routines, calling programs . . . became a mountainous undertaking . . . and thus began the basic thinking of using the computer itself to assign its own addresses for core register assignments (now known as assemblers) and later for the automatic collection and chaining of program subroutines. . . . [These basic ideas] later developed into concepts of compilers and interpreters. Coincident with these operational software problems, Whirlwind became the testing ground for . . . software diagnostic programs to help an operator to detect and diagnose trouble first in hardware malfunction and later in software ambiguities. This work formed the basis of building real-time systems reliable enough for military use, and capable of self-diagnosis and self-switching to an alternate mode whenever reliability was in question.[85]

After Lincoln Labs had written software for the first three sectors, the Rand Corporation was given the job of programming SAGE. Rand assigned 25 programmers to this work, a number that seems modest but in fact represented about one-eighth of all programmers anywhere in the world then capable of doing such work. Rand spun off its SAGE software division (the System Development Division) in 1957, forming a separate Systems Development Corporation. SDC grew to four times Rand's size and employed over 800 programmers at its peak. The SAGE program code, at a quarter of a million lines, was by far the most complex piece of software in existence in 1958, when it was mostly complete. As one of its programmers recalled, "highly complex programs were being written for a variety of mathematical, military, and intelligence applications, but these did not represent the concerted efforts of hundreds of people attempting to produce an integrated program with hundreds of thousands of instructions and highly related functionality."[86] Here again the spinoff effect was large, as SDC programmers left to start companies and join corporate programming staffs.[87]

Despite these many technical and corporate impacts, I would argue that the most essential legacy of SAGE consisted in its role as a support, in Michel Foucault's sense, for closed-world politics. For SAGE set the key pattern for other high-technology weapon systems, a nested set of increasingly comprehensive military enclosures for global oversight and control. It silently linked defense- and offense-oriented strategic doctrines—often portrayed as incompatible opposites—around centralized computer systems. It provided the technical underpinnings for an emerging dominance of military managers over a traditional experience- and responsibility-based authority system. At the same time, ironically, SAGE barely worked.

SAGE as Political Iconography

A SAGE center was an archetypal closed-world space: enclosed and insulated, containing a world represented abstractly on a screen, rendered manageable, coherent, and rational through digital calculation and control.

Each of the twenty-three control centers received and processed not only digitally coded radar data, which it handled automatically, but weather reports, missile and airbase status, flight plans of friendly aircraft, reports of the Ground Observer Corps, and other information transmitted verbally over telephone and teletype, which operators incorporated into the computer's overall situation picture. It also communicated with other centers, automatically coordinating activities across sectors. Each center tracked all aircraft in its sector and identified them as friendly or unknown. "Air-situation display scopes" superimposed information about aircraft over a schematic map of the sector. Display operators watched the picture of the unfolding air situation and decided on responses. The computer generated interception coordinates and relayed them automatically to the automatic pilots of the interceptors. Unless overridden by their human pilots, the interceptors flew to within closing range of the unknown aircraft under fully automatic control. Eventually SAGE controlled many other weapons systems as well, such as the Air Force BOMARC and the Army Nike-Hercules antiaircraft missile.

Each SAGE center lodged in a windowless four-story blockhouse with six-foot-thick blast-resistant concrete walls, occupying two acres of land (see figure 3.1). The building's entire second story was taken up by the AN/FSQ-7 computer—actually two identical computers operating in tandem, providing instantaneous backup should one machine fail (a technique known as "duplexing"). Weighing three hundred tons and occupying 20,000 square feet, the FSQ-7's seventy cabinets contained 58,000 vacuum tubes. Display consoles and telephone equipment required another 20,000 square feet of floor space.[88] Each center had its own electric power plant to run the computer, air conditioning, and telephone switching systems inside.

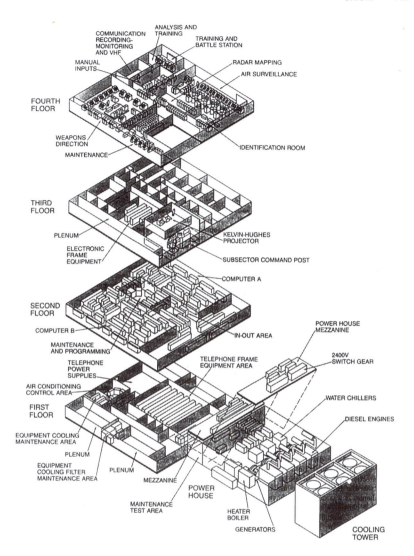

Figure 3.1
Interior of a typical SAGE direction center. Two IBM FSQ-7 computers filled the entire second floor. Drawing by Bernard Shuman, courtesy MITRE Corporation Archives.

Additionally, the dedicated generators insulated the control center from failures of, or attacks on, the commercial power grid. Communications among the centers, however, relied on AT&T commercial telephone lines.[89]

The SAGE centers were the original version of the windowless Infiltration Surveillance Center built a decade later for the Vietnam War.[90] Dim blue light from the consoles illuminated their interiors, known as "blue rooms," where operators used light guns to connect blips on video displays. A 1957 *Life* magazine pictorial on SAGE captured the strange blue glow of the scene within the blockhouse, as well as the eerie calm of battle as an automated process for rational managers. The "huge electronic computer," according to *Life,* could "summarize [data] and present them so clearly that the Air Force men who monitor SAGE can sit quietly in their weirdly lighted rooms watching its consoles and keep their minds free to make only the necessary human judgments of battle—when and where to fight."[91] The abstract electronic architecture of the world represented on their screens, harbinger of the electronic battlefield's virtual reality, was an icon for the political architecture of the closed world.

Strategy and Automated Command

To a casual observer military forces, with their strict hierarchies and authoritarian ethos, epitomize a rigid, rule-bound bureaucracy (and this is, unquestionably, a well-deserved reputation). Scrutinized more closely, however, traditional military hierarchies are anything but mechanical. At every level, individuals *bear responsibilities* rather than *perform functions.* A field officer may be ordered to "take that hill," but the whole point of such an order is that *how* he carries it out is up to him.[92] We may call this system the "command tradition." In the 1950s, within the space of a very few years, the Air Force command traditionalists who had opposed the computerized air defense system either became, or were replaced by, the most vigorous proponents of centralized, computerized warfare anywhere in the American armed services.

One reason this happened was the dawning realization, and then the necessity, that SAGE-style technology could be used for central control of offensive weapons as well as for defense. By the mid-1950s it became obvious that missile warfare would soon augment or even replace airplanes in strategic nuclear war. This rendered the glorious role of the pilot irrelevant. It also decreased response times by an order of magnitude. Only centrally coordinated systems could cope with such speed requirements.

SAGE—Air Force project 416L—became the pattern for at least *twenty-five* other major military command-control systems of the late 1950s and early 1960s (and, subsequently, many more). These were the so-called "Big L" systems, many built in response to the emerging threat of intercontinental ballistic missiles (ICBMs). They included 425L, the NORAD system; 438L, the Air Force Intelligence Data Handling System; and 474L, the Ballistic Missile Early Warning System (BMEWS). SAGE-like systems were also built for NATO (NADGE, the NATO Air Defense Ground Environment) and for Japan (BADGE, Base Air Defense Ground Environment).

Project 465L, the SAC Control System (SACCS), was among the largest of these successors. Its software, at over a million lines, reached four times the size of the SAGE code and consumed 1,400 man-years of programming; SDC invented a major computer language, JOVIAL, specifically for this project. The SACCS was the first major system ever programmed in a higher-level language.[93] In 1962 the SACCS was expanded to become the World-Wide Military Command and Control System (WWMCCS). The WWMCCS, with a global network of communications channels including military satellites, theoretically enabled centralized, real-time command of American forces worldwide. During the Vietnam War this system was actually used by the Johnson administration to direct the air war from Washington (though not in real time). Ultimately the Air Force connected the distant early warning systems originally utilized by SAGE, the BMEWS, and others with computer facilities at the NORAD base under Colorado's Cheyenne Mountain for completely centralized ICBM detection and response.[94]

In chapter 1, I defined three versions of closed-world politics: the West as a world enclosed inside its defenses; the USSR as a closed world to be penetrated or "opened"; and the globe as a world enclosed within the capitalist-communist struggle. In the Big L systems, each of these versions of the closed world found its own embodiment in computerized command and control. SAGE began the process with enclosure of the United States inside a radar "fence" and an air-defense bubble. SACCS continued it with a control system for penetrating the closed Soviet empire. WWMCCS completed it with a worldwide oversight system for total global "conflict management." But while SAGE was still a laboratory experiment, Air Force leaders already conceptualized "the defense of the air space above the Free World [as] a global task"[95]—by which they meant, following pre-World War II strategic doctrine, that intercontinental offensive striking power constituted the best hope of domestic defense. Closed-world discourse was never a simple case of military vs. civilian, liberal vs. conservative, intellectual vs. popular, or defense vs. offense; most positions within the mainstream spectrum of opinion were caught up in its terms, metaphors, experiences, and technologies. Though they often saw themselves as opposites, the builders of both defense- and offense-oriented military systems thus constructed closed-world discourse together.

Despite its importance, the history of SAGE is filled with ironies. It was an Air Force-run project to accomplish a goal most Air Force commanders opposed: air defense against nuclear weapons. It was obsolete before it was complete, rendered militarily worthless by the ICBM and technologically outdated by the transistor and the integrated circuit.[96] Yet it continued to function well into the 1980s. Six SAGE centers were still operating in 1983, using their original vacuum-tube equipment. (By 1984 all had finally shut down, their functions absorbed by more modern elements of the nuclear early warning system.)

Perhaps the most telling irony, from the perspective of closed-world discourse, was that the *automatic* control SAGE promised was then, and remains today, largely an illusion. Whatever the abilities of the computers and their programs, much of the total task still remained to human operators and their organization.[97] Attempts to "program" *this* part of the work—in the form of formal procedures encoded in manuals—always faltered against the unruly complexities not yet enclosed within the system.[98]

> It was impossible to specify in advance all of the contingencies that would be faced in the course of actual operations. Reliance on formal written procedures proved impractical, and unwritten work-arounds soon developed among the human operators of SAGE. Controllers were even reluctant to specify to engineers the exact operating procedures they would employ in particular situations.... For example, small amounts of radar jamming could paralyze SAGE if rule book procedures were followed. Oral agreements between operators could fix this, but these never showed up in official reports.[99]

The closed world was a leaky container, constantly patched and repatched, continually sprouting new holes.

Two problems with automated command thus became dimly visible in the SAGE project. First was the impossibility of providing for, or even articulating, every possible situation in advance, and the consequent need to rely on human judgment. Since the 1950s critics of computer technology (though far outnumbered by optimists who project future solutions) have elevated this incapacity to the status of a philosophical principle.[100] Second was the difference between the formal level of organization, with its explicit knowledge and encoded procedures, and the much larger and more significant informal level, with its situated knowledge and tacit, shifting agreements.[101]

These problems, whatever their visibility, were ignored for reasons that have been elegantly articulated by Paul Bracken. Bracken shows that during the Cold War, command structures evolved from traditional systems based on infrequent periods of full mobilization, with days or weeks of advance warning, to nuclear-era systems based on continuous mobilization, with hours or minutes of warning. This shift required a "vertical integration" of warning, command, and political liaison—essentially, a flattening of the hierarchy of responsibility into an increasingly automatic, and therefore rigid, system able to act on a few minutes' notice. Bracken adduces technological reasons for this shift: the vast increases in the speed of weapons delivery, the amount and complexity of sensor information to be integrated, and the scale of response to be mounted. "To protect itself a nuclear force does the opposite of what a conventional army does. It tries to 'manage' every small threat in detail by centralized direction, reliance on near real-time warning, and dependence on prearranged reactions."[102] Under such conditions, centralized, automated control seemed imperative. In chapter 4 we will see how the promise of automatic control, proffered by SAGE, its descendants, and other emerging uses of computers for strategic analysis, contributed to a realignment of high-level military leadership toward what we may call a "managerial model."

Conclusion

The computerized nuclear warning and control systems both embodied and supported the complex, heterogeneous discourse of closed world politics. Containment doctrine, scientists' and engineers' public pronouncements on strategy, Air Force culture and traditions, public anxiety about nuclear war, and the anticommunist hysteria of the 1950s all participated at least as much as technological changes in the construction of military, rhetorical, and metaphorical containers for the capitalist-communist conflict. Beginning with SAGE, the hope of enclosing the awesome chaos of modern warfare (not only nuclear but "conventional") within the bubble worlds of automatic, rationalized systems spread rapidly throughout the military, as the shift to high-technology armed forces took hold in earnest.

Yet the *military* potential of SAGE was minimal. Many, perhaps most, of those who worked on the project knew this. Such understanding was reflected in another irony of SAGE: the failure to place the control centers in hardened underground bunkers, the only place from which they might have been able actually to control an active defense in a real war. The Air Force located most SAGE direction centers at SAC bases.[103] This decision had only one possible strategic rationale: SAC intended never to need SAGE warning and interception; it would strike the Russians first. After SAC's hammer blow, continental air defenses would be faced only with cleaning up a weak and probably disorganized counterstrike. In any case, SAGE would not have worked. It was easily jammed, and tests of the system under actual combat conditions were fudged to avoid revealing its many flaws.[104] By the time SAGE became fully operational in 1961, SAC bases were unprotectable anyway (because of ICBMs), and SAGE control centers would have been among the first targets destroyed in a nuclear war.

Still, in another important sense, SAGE *did* "work." It worked for the research community, which used it to pursue major intellectual and technical goals. It worked as industrial policy, providing government funding for a major new industry. Perhaps most important, SAGE worked as ideology, creating an impression of active defense that assuaged some of the helplessness of nuclear fear.[105] SAGE represented both a contribution and a visionary response to the emergence of a closed world.

What makes SAGE such an interesting case is its origins within the academic science and engineering community—*not* with military imperatives, though its military funding sources and key geopolitical events spurred it on. Instead the initiative lay with the scientists and engineers, who developed not only machines but a vision and a language of automated command and control. But the construction of SAGE also boosted the redesign and reorientation of an extremely traditional institution—the armed forces—around an essentially technological concept of centralized command. Seen in this light, SAGE was far more than a weapons system. It was a dream, a myth, a metaphor for total defense, a technology of closed-world discourse.

The War Room at the Pentagon, 1959. The wall map displays a "typical" American nuclear first strike. Courtesy Library of Congress, *US News & World Report* collection.

4
From Operations Research to the Electronic Battlefield

We have seen how the rapidly evolving geopolitical concerns of the United States as a nuclear power shaped a grand strategy and a political discourse involving a closed system accessible to technological control.[1] Computers, as tools, supported the technological aspects of that discourse, for example in weapons design, ballistics calculation, and cryptanalysis. In the centralized digital command-control systems of the 1950s, computers also *embodied* the discourse of "containment" and technological closure—its paradoxes and failures as well as its ideals.

This chapter traces the coevolution of computer technology, grand strategy, and closed world politics into the 1960s. We begin with the Rand Corporation, the Air Force think tank where natural scientists, social scientists, and mathematicians worked side by side to anticipate and prepare for the future of war. Rand used systems analysis techniques, born from wartime operations research, to investigate both the mundane problems of weapons procurements and the unknown realm of nuclear strategy. These techniques both benefited from and helped promote Rand's extensive work on computers; Rand's computation center, among the largest in the world, significantly affected the nature and direction of software development in the 1950s. In addition, by legitimating systems analysis, computers helped advance Rand's theory- and simulation-based approach to strategy.

In 1961, Secretary of Defense Robert McNamara established an Office of Systems Analysis staffed with key Rand personnel. In Vietnam, the almost obsessive focus of McNamara and his ex-Rand advisors on quantitative analysis became a serious liability. The formalistic language of statistics, systems, and Rand-style game-theoretic nuclear strategy—all of it reliant on the technological support of digital computers—reduced the problem of war to an issue of algorithms, electronics, and kill ratios. Where SAGE and its successors were built to control global thermonuclear war, Igloo White and similar Vietnam-era projects sought to extend computerized control to the battlefield level, in a kind of information panopticon where nothing and no one could move unobserved. The search for centralized, computerized, quasi-automatic control in war—General Westmoreland's "electronic battlefield"—continued, undisturbed by the failure of the high-tech closed-world infrastructure to contain the peasant guerilla army of North Vietnam.

Operations Research, Systems Analysis, and Game Theory at Rand

Long before the birth of SAGE, theoretically oriented scientists and engineers were framing their own discourse of grand-scale systems, communication, and control.

Alan Turing's theory of digital computers as universal machines, able to solve any precisely formulated problem—any problem that could be systematized, mathematized, modeled, and reduced to an algorithm—achieved widespread currency. New disciplines, such as operations research, cybernetics, information theory, communication theory, game theory, systems analysis, and linear programming, succeeded in devising such algorithms for difficult problems in communication and the control of highly complex systems. Exploited by business, these methods promised to resolve previously intractable problems such as inventory control and work flow in large factories. Management theorists like Herbert Simon began to apply mathematical techniques to decision-making, producing theories of "administrative rationality."[2] Norbert Wiener's cybernetics seemed to offer a comprehensive theory capable of encompassing issues in government and society as well as in science, engineering, and factory production.[3] The Shannon-Wiener theories of information and communication unified a wide range of concepts in language, data analysis, computation, and control.

This extension of mathematical formalization into the realm of business and social problems brought with it a newfound sense of power, the hope of a technical control of social processes to equal that achieved in mechanical and electronic systems. In the systems discourses of the 1950s and 1960s, the formal techniques and tools of the "systems sciences" went hand in hand with a language and ideology of technical control.[4]

Operations research (OR, a.k.a. "operational research" or "operational analysis") was the first and most successful of the emerging systems sciences. Essentially an application of mathematical analysis to the observational data of war, OR found its most famous engagement in antisubmarine warfare during World War II. German U-boat (submarine) attacks on Allied shipping convoys destroyed large numbers of ships, despite both naval and aerial protection. To counter this continuing threat, British and American scientists analyzed U-boat diving patterns and existing Allied reconnaissance methods and their results. Using mathematical techniques, they plotted optimum search strategies for aerial reconnaissance, optimum sizes and patterns for naval convoys, and new fuse settings for depth charges.[5] The latter adjustment instantly raised the number of U-boat kills by a factor of three; German commanders came to believe that the British had invented a powerful new explosive.

Scientists applied similar methods to many other logistical and tactical problems, such as bomber formations, mining of harbors, armoring and arming of airplanes, and emplacement of antiaircraft weapons. Mathematical modeling allowed isolation of significant variables, with sometimes surprising results (e.g., that large convoys, thought more dangerous because easier to locate, were in fact safer than small convoys). By the end of the war, the Army Air Force had established operational analysis divisions in all of its units.[6]

Given its impressive wartime record, military planners of the postwar period naturally sought to deploy and refine OR methods as they prepared for future wars. This trend found an apotheosis of sorts at the Rand Corporation, where mathematical techniques like these became central paradigms for strategic thought.

The Air Force founded Rand (or RAND, an acronym for "Research and Development") in 1946 as a joint venture with Douglas Aircraft. Rand soon separated from Douglas, and its charter as an independent, nonprofit research corporation was signed in 1948. A private research group based in Santa Monica, California, Rand's broad mandate was to study "techniques of air warfare." The organization's annual funding from the Air Force ran in the neighborhood of $10 million. Starting in 1956, Rand diversified its sponsorship with other contracts, mostly from defense and defense-related agencies such as the Atomic Energy Commission and the Advanced Research Projects Agency. Rand straddled the borders between academia, industrial research laboratories, and the military. Many of its staff, with high-level security clearances, had access to some (but not all) top-secret information, and for most of its first decade the majority of Rand reports were classified.[7]

As the ONR did for the Navy, Rand allowed the Air Force to continue the fruitful interdisciplinary team efforts initiated during the war, reflecting the common belief in the military potency of science and the cult of scientists' expertise. Rand's staff included social scientists, economists, and mathematicians as well as physicists and engineers. (Technical research in physics, aerodynamics, and other fields constituted a substantial part of its efforts.) At its peak in 1957, it employed a staff of 2,605.[8] Rand was the center of civilian intellectual involvement in defense problems of the 1950s, especially the overarching issue of nuclear politics and strategy.

Rand's most important contribution was not any specific policy or idea but a whole way of thinking: a systems philosophy of military strategy. Rand supported interdisciplinary studies in operations research, systems analysis, game theory, and other applications of mathematics and logic to problems of politics, strategy, and logistics. The work of such "defense intellectuals" as Bernard Brodie, Herman Kahn, and Albert Wohlstetter—responsible for key nuclear-age concepts of deterrence, limited war, and nuclear war-fighting doctrine—flourished in Rand's hothouse atmosphere. With salaries 50 percent higher than their equivalents at universities and a hands-off, let-them-run-with-it policy toward research, Rand attracted the top talent of the postwar generation. Because of its emphasis on quantitative models, many of Rand's brightest minds were mathematicians. John von Neumann, for example, was a Rand consultant.

Operations research could help optimize the use of *existing* equipment for a given mission. Systems analysis, a technique developed at Rand by Edward Paxson in the late 1940s, went beyond OR to consider a more fundamental, but also more speculative, question: given a mission, how could equipment, logistics, and tactics best be designed to fulfill it?

One of the major variants of this question concerned the intricate interplay of equipment, logistics, strategy, tactics, and *costs*. In the age of nuclear weapons and intercontinental bombers, the problem of how much was enough[9]—how many men, how many bombs and planes, how much air defense, how much research and development—obsessed not only military planners but politicians wrestling with the constraints of still-balanced budgets. Systems analysis could generate and juxtapose a range of options for policymakers to consider. It neatly laid out the costs and benefits of each in money and military effectiveness, measured in terms of targets destroyed—"bang for the buck." By the time the Kennedy administration took office, such methodologies were also known, in a less mystifying phrase, as "cost-benefit analysis."

One of Rand's more sympathetic observers wrote that systems analysis differed from operations research in aiming "at a range of problems to which there can be no 'solution' in a strict sense because there are no clearly defined objectives that can be optimized or maximized. The systems-analysis approach retains, however, elements of rigor and preciseness that offer more reliable assessments of numerous difficult choices than simple intuition or inference from unexamined precept."[10] In practice, especially in the late 1940s and early 1950s, systems analysis depended heavily on quantification, even of objectives not "clearly defined."

The other conceptual scheme behind the Rand strategic paradigm was game theory, invented by von Neumann and economist Oskar Morgenstern. Game theory asked the question of how rational players of "zero-sum" games (games in which one side's loss is the other side's gain, such as poker, chess, or the infamous "prisoner's dilemma") should respond to each other's moves—given that each can also know something about how the other player, if choosing rationally, will respond to his or her response, and so on. The theory expresses how players of such games can balance between maximizing possible gains and minimizing possible losses. It allows precise solutions for all combinations of the probability of some move's success with its potential value to the player.

Rand's *Fourth Annual Report* of March 1950 illustrates how widely game-theoretic analyses already ranged: in the study of "systems for strategic bombardment, air defense, air supply, or psychological warfare, pertinent information developed or adapted through survey, study, or research by Rand [in its mathematics division] is integrated into models.... In this general area of research ... the guiding philosophy is supplied by the von Neumann-Morgenstern mathematical theory of games."[11] By 1960 Rand had also applied game theory to such problems as "radar search and prediction, allocation of defense to targets of unequal value, missile penetration aids, the scheduling of missile fire under enemy pin-down, anti-submarine warfare, and inspection for arms control."[12]

Ultimately the most significant application area was nuclear strategy itself. Since no past experience applied, and since the threat of nuclear attack outstripped all others, predicting the actions and reactions of a nuclear-armed opponent carried a special urgency. Once the commitment to nuclear forces had been made, there was nothing else to do but simulate, predict, and theorize—testing and practice, in any real sense, being out of the question. As a metaphor, the zero-sum game represented yet another symbolic enclosure within which the (il)logic of nuclear politics played itself out.

Systems Analysis, Strategy, and Technology
Rand studies led to a series of shifts in nuclear strategy. One example was the 1952 finding of mathematical logician Albert Wohlstetter that the Strategic Air Command's forward bomber bases could easily be annihilated by a surprise first strike. (The iconography of Pearl Harbor amplified the power of this prospect to disturb; Wohlstetter's inspiration came in part from his wife Roberta, a historian then undertaking a massive study of Pearl Harbor for Rand.)[13] Wohlstetter advocated basing bombers in the United States instead. His own plan, OR-inspired, called for a medium-range bomber force—the kind of force the United States already possessed—relying on forward bases for refueling. SAC commanders were unimpressed, no doubt because they expected to empty their bases long before the Russians arrived.

Wohlstetter's underlying worry was that forward bases would not, in fact, receive attack warnings early enough to launch their planes. The results of his simulation, which eventually disquieted even the steely-eyed SAC, were starkly negative. In what amounted to a Rand crusade, Wohlstetter and his colleagues proceeded to brief ninety-two different sets of military and civilian officials. As the briefing blitz continued, pressure for a genuine response mounted. SAC was forced to act.

Rather than accept Wohlstetter's base-refueling plan, SAC chose the far more expensive, high-technology strategy of U.S.-based *intercontinental* bombers refueled in midair.[14] This required new technologies such as air-to-air refueling and ultra-long-range jet bombers. It also amplified the importance of the distant early warning systems then just programmed for SAGE. The Air Force plan combined strategic considerations with cultural values. As Rand historian Fred Kaplan put it, "Air Force officers, almost all of whom were pilots... didn't care about systems analysis. They liked to fly airplanes. They wanted a bomber that could go highest, farthest, fastest." Even in 1950, Gen. LeMay hoped eventually to command intercontinental bombers, which would need no foreign bases. (LeMay, according to Kaplan, "disliked and distrusted most foreigners.")[15] In a classic case of technological system-building, strategic decisions, technological choices, and Air Force culture became inseparably intertwined.[16]

In his history of Rand, Kaplan argues that the Rand forward-basing studies were a major goad for the U.S. nuclear buildup of the 1950s and 1960s. The Pearl Harbor logic of the Wohlstetter study, with its emphasis on vulnerability and permanent readiness, constituted one of the two major forces behind most of Rand's subsequent strategic analysis. The other was the question of how a nuclear-armed force could establish and maintain credibility as a military threat, since both Truman's "containment" and the Eisenhower/Dulles doctrine of "massive retaliation" seemingly committed the United States either to an all-out orgy of mutual annihilation in response to any Russian advance or to a policy of vacuous rhetorical reactions.[17] Rand produced analyses of how nuclear war might actually be fought as a grand-scale form of conventional war, with winners, losers, and postwar political structures, and how smaller "tactical" nuclear weapons might be used to prosecute "limited" wars. These Rand ideas lay behind flexible-response theory (pegging levels of nuclear response to levels of Soviet provocation), the no-cities doctrine of counterforce (attacking missile silos and other military targets instead of population centers), and other key elements of evolving U.S. nuclear strategy.[18] The Rand analysts' formal models, increasingly assisted by computerized data processing, rapidly became the standard for military thinking about strategy and planning for the future "weapon systems" necessary to actualize it.[19]

Because of their focus on planning, nuclear strategy, and cost comparison, systems analysis and game theory differed in one crucial way from the operations-research methods that gave rise to them. Whereas World War II OR techniques were developed using real combat data, Rand nuclear strategists had virtually no quantitative information upon which to base their calculations of the course of World War III.[20] Systems analysts were by no means unaware of the unreliability inherent in theorizing in the near-complete absence of data. But since they were planning a type of war that had never been fought, they seemingly had no choice but to do their best with theories, models, simulations, and games.

Enormous national military investments based primarily on mathematical models—and the awesome scale of the wars planners contemplated so calmly—created an atmosphere of intense unreality around nuclear strategy, much remarked by participants as well as observers.[21] The Rand thinkers inhabited a closed world of their own making, one in which calculations and abstractions mattered more than experiences and observations, since so few of the latter even existed to be applied. Jonathan Schell notes that the phrase "think tank," of which Rand is indisputably the paradigmatic case, evokes "a hermetic world of thought" that "exactly reflects the circumstances of those thinkers whose job it is to deduce from pure theory, without the lessons of experience, what might happen if nuclear hostilities broke out."[22] Nuclear war existed only as a simulation, a game, a computer model.

To a remarkable extent the Cold War was actually prosecuted through such simulations. Each side based its weapons purchases, force deployments, technological R&D, and negotiating postures on its models of strategic conflict and its projections about the future choices of the other. This is why the Cold War can be best understood in terms of *discourses* that connect technology, strategy, and culture: it was quite literally fought inside a quintessentially semiotic space, existing in models, language, iconography, and metaphor, embodied in technologies that lent to these semiotic dimensions their heavy inertial mass. In turn, this technological embodiment allowed closed-world discourse to ramify, proliferate, and entwine new strands, in the self-elaborating process Michel Foucault has described. In a double sense, systems analysis formalized this discursive connection between technology, strategy, and culture.

First, systems analysis linked choices about strategy directly to choices about technology by asking the question of how a given mission might be accomplished, at what price, with which actual or potential technologies. Its stress on proliferating options for policymakers tended to highlight systems under development or still in the research phase by giving their projected but still unproven capabilities equal weight to those of forces in being, which since they had been tested in practice were less subject to inflated performance claims. Thus it inherently promoted technological change. At the same time it generated what Foucault called a "regime of truth," a set of implicit conventions about what could count as facts and reasons and who was authorized to elucidate them. Civilian strategists, in this "regime" of theories and simulations, gradually gained significant influence over the discourse of their military counterparts, who found themselves unable to bolster their traditional discourse of field experience with any new evidence. In the process, models of rational action displaced Clausewitzian struggle.[23]

Systems analysis also linked technology to strategy in a second way: through its methods, especially computer modeling. Systems analysis did not, strictly speaking, require computers. But as computer scientist Joseph Weizenbaum has noted, the new machines "put muscles on its techniques." In the 1950s, when computers were still very new and rather awe-inspiring, any application automatically inherited their aura of almost erotic scientificity.[24] In the 1950s and 1960s, computers enabled systems analysts to tackle increasingly complex problems; and this complexity, in turn, created an ever greater need for computers. Weizenbaum argues that this quantitative, computer-facilitated increase in the problem-solving power of systems analysis gradually rendered it qualitatively different from its war-era manual counterparts.[25]

Weizenbaum's argument is borne out by Rand's history. The effect of computing at Rand was to increase vastly the abilities, and with them the ambitions, of systems analysts and others concerned with mathematical modeling and simulation. The appearance of "hard" answers achieved by extensive quantitative analysis and simulation lent an air of certainty to results even when based on uncertain assumptions, especially at a moment in American history when the prestige of science and technology had reached an all-time peak. By 1960 Rand estimated the amount of its total effort devoted to "analytical, computer, and simulation techniques" at 18 percent, not including an additional percentage devoted to computers in weapon systems.[26]

Rand and Computer Science

If the computer's effect on Rand was crucial, Rand's effect on computing was even more so. Throughout the 1950s Rand played a direct and major role in computer development, one whose significance has gone largely unnoticed. Much of the corporation's early work in the computer field was and remains classified, but a general picture of its activities can be pieced together from its published documents, oral histories, and studies of other aspects of Rand research.

By 1950 Rand, operating six IBM 604 punch-card calculators and two IBM Card-Programmed Calculators "around the clock," had become "in some sense the world's largest installation for scientific computing" (i.e., machine calculation), according to Rand staffer Fred Gruenberger. But it required still greater capabilities.[27] Von Neumann advised the group to obtain an electronic stored-program machine (i.e., unlike its existing equipment, a true computer). After a tour of computer projects across the continent, Rand's numerical analysis department chose to build a copy of von Neumann's Institute for Advanced Study computer, modified for increased reliability. The Johnniac, named after von Neumann, was operational by 1954.[28] But by then Rand had long been involved in other computer-related projects using equipment available at the major aerospace corporations near its base in southern California.

George W. Brown, a Rand consultant beginning in 1947, believed that Rand's 1950 decision to build the Johnniac was, in fact, a key spur to IBM's decision to commit to digital computer development at about the same time. Rand and the aircraft companies, according to Brown, kept asking for more and better equipment. "IBM regional managers started to see that 'My God, the science and engineering community has needs that we're not going to be able to fill.' When Rand decided to build its own machine, IBM announced the Defense Calculator [the 701] to avoid loss of prestige."[29] Later, claimed Paul Armer, director of Rand's numerical analysis department from 1952 to 1962, Rand "contributed to IBM's dominance of the market . . . by coming up with applications which we developed on IBM gear and then turned over to the Air Force."[30] Though Rand employed many excellent scientists and engineers interested in computers, the organization's "privileged relation to private enterprise because of the DoD connection" led them to avoid replicating private work, recalled Keith Uncapher, an early affiliate of the Rand computer research efforts. The atmosphere of mutual trust thus created, "plus Rand's strong affiliation with the Air Force and later with the DoD, gave the unique capability of bending computer science towards the needs of the Air Force and military."[31]

One of Rand's major projects in this arena was a simulation of the McChord Field Air Defense Direction Center in Tacoma, Washington, part of the pre-SAGE manual air defense system. The project began in 1950 in Rand's System Research Laboratory (SRL), staffed in part by psychologists. The SRL sought to augment the systems-analysis approach:

Operations analysis and system analysis often need to consider the effect of the human factor on system performance. Usually a "degradation factor" is used to qualify the predicted effectiveness. In an effort to better understand the human element in systems, Rand set up the System Research Laboratory to study man's performance in complex man-machine systems.[32]

The Tacoma simulation project sought to explore the psychological and technological dimensions of the "man-machine interface" with an eye to improving training techniques. By 1951 the group had constructed a replica of the McChord Field facility at the SRL. Using Rand's IBM card-programmed calculators, the group designed simulated attacks to appear realistically (but in static form) on the radar team's displays. The lab conducted its first experiments in 1952 using college students. Four six-week simulations, each involving a team of about forty men, showed that practice in a simulator could teach the teams to "distinguish between important and unimportant information input and to shortcut their data-processing procedures" far more efficiently. As one of its designers reported,

The underlying notion behind this research was that it might be possible to obtain the predictable features of a "closed" system by exploiting man's capacity to seek and find problem solutions. That is, if man could be motivated to seek the system's goal, and if he were provided knowledge of operational results, a disparity between actual and desired performance might serve as an error feedback to trigger adaptation of operating practices to improve effectiveness.[33]

System closure was the goal, air defense control the chosen site, proto-computer simulation the chosen method, cybernetic (feedback-controlled) human-machine integration the result.

One of the project's originators was Allen Newell, who had come to Rand from the Princeton mathematics department after an abortive year of graduate study. Newell's intellectual interest—piqued by a previous experience working on logistics for the Pentagon's Munitions Board—was organizational behavior. His initial experiments (with small groups and artificial communications tasks) did not produce enough interesting interaction, so Newell sought out a larger organization to study with more realistic tasks. He found it in the air defense direction center, a project with obvious appeal for the Air Force as well.[34]

The Tacoma experiments were Newell's first experience with a computer of any kind, and he later recalled that it was significant for him that the project was a *simulation* rather than a numerical analysis.

Herbert Simon (then at Carnegie Institute of Technology) was also influenced by the McChord Field simulation; he first began to conceive ideas about the computer as a symbolic processor when he saw Newell's computerized simulations of radar displays. Newell's collaborations with Simon, which became cornerstones of artificial intelligence (AI), began when Simon arrived at Rand in 1952. Simon, in fact, learned to program the IBM 701 at Rand in the summer of 1954. His first conversation with Newell about simulating human problem-solving occurred on the drive to an air defense simulation exercise during the same period.[35]

Newell and Simon, influenced by the work of von Neumann and also of Oliver Selfridge (then building pattern-recognition devices at Lincoln Laboratories, where SAGE was being designed), began to attempt a computer simulation of human thought. They enlisted the help of Rand programmer J. C. Shaw, who had written the assembly language for the Johnniac. Shaw had also helped Newell devise the radar simulator project, but his major contribution lay in co-developing the information processing languages IPL-I and IPL-II, implemented on the Johnniac. Newell and Shaw wrote the former for what was essentially the first AI program, the Newell-Simon-Shaw Logic Theorist, completed in 1956. The Logic Theorist produced proofs of basic logic theorems similar to those of Russell and Whitehead's *Principia Mathematica*, including a new and more elegant proof of Theorem 2.85. In January 1956 Simon walked into his class at Carnegie Institute of Technology (now Carnegie-Mellon University) and announced, "Over Christmas Allen Newell and I invented a thinking machine."[36] That summer, the two attended a conference at Dartmouth, organized by MIT's John McCarthy, that endowed the new field of artificial intelligence with its name. We will revisit AI in chapter 8.

The McChord Field simulation led to the Systems Training Program (STP), an Air Force contract to train air defense specialists. By the mid-1950s Rand employed 500 people, including about 200 psychologists, to produce training materials for air defense personnel.[37] The STP evolved into Rand's Systems Development Division (discussed in chapter 3) in 1955. By this point true electronic computers were available commercially. Rand obtained one of the first models of the IBM 701 (the "Defense Calculator") in August 1953, and the STP began using it immediately. As Rand acquired new machines, the simulations grew in sophistication; an IBM 704, which produced dynamic radar images, required programs some 80,000 lines in length.

As SAGE approached operational status, the STP designed its training materials. Rand spun off the Systems Development Division, including the STP, into the independent Systems Development Corporation in 1956. By 1958, the STP's training programs had reached 125,000 lines of code, about half the size of the SAGE program itself; they eventually grew to nearly 750,000 machine instructions. By 1959, SDC and the Systems Training Program employed more social scientists than almost any other private firm.[38]

Rand's report to the 1957 System Simulation Symposium captured the larger significance of the air-defense systems simulations: its successful use of concepts and technologies of information and control to integrate humans into mechanical systems.

A real-time simulation approach to man-machine systems fairly heavily depends upon the system being largely an information processing or control system—or perhaps more basically, it depends upon the system's being one that deals with symbols—symbolic representations of things instead of things. Certainly business, management in particular, is in large part manipulation of symbols, and this is becoming increasingly the case in this electronic age. There will of course be limitations in simulating such systems, but in conjunction with the proper analog devices, the general flexibility of the modern digital computer can be expected to solve many of the problems which will arise.[39]

Computers linked other technologies to human beings by constituting systems—constituting them conceptually, practically, and metaphorically—as information processors.[40] They created a closed world of semiotic values in which future wars could be imagined, their soldiers trained, and their outcomes deduced. Yet the world of computer simulations was more than a game. For unlike the chess-style war games of previous eras, computer-age commanders could engage in simulations using equipment that not only resembled, but sometimes actually *was*, the equipment used for real war. The closed world within the machine, and the closed world of real strategy it supported, blurred together in an intricately woven, discursively constituted whole.[41]

Robert McNamara, Systems Analysis, and Military Management

Apart from a few successful crusades like Wohlstetter's vulnerability study, Rand influenced actual policy only intermittently. By the mid-1950s Rand's armchair generals had aroused significant antagonism among certain elements of the Air Force, especially SAC. As an independent think tank, it had no direct authority within either the military or the government. In 1958, President Eisenhower's "New New Look" defense policy rejected the Rand-influenced 1957 Gaither Report (which, like NSC-68, painted a terrifying picture of a looming Red menace and recommended a 50 percent hike in the military budget).[42]

All this changed in 1961 when John F. Kennedy assumed the presidency. The electorate chose Kennedy in part for his promise of an aggressive response to Rand-inspired predictions (leaked by Gaither Report commissioners) of a looming "missile gap" between the United States and the USSR.[43] As his Secretary of Defense, Kennedy appointed the cerebral, 44-year-old Robert S. McNamara.

McNamara had begun his career as a professor at the Harvard Business School. When World War II broke out, he was recruited to the Statistical Control Office of the Army Air Corps, where he used OR-style mathematical techniques to plan the logistics of bombing raids, first in Germany and later in the Far East under Gen. Curtis LeMay (later SAC's commander-in-chief). McNamara's techniques raised by 30 percent the flying time logged by LeMay's bomber command. After the war, McNamara and nine other former Statistical Control Office analysts sought work as a group, hoping to apply their new skills to industrial productivity.[44] They ended up at Ford Motor Company. There the McNamara group, known as the "whiz kids," introduced these same techniques into business management, with sometimes astounding success. In 1960 Ford named McNamara its president.

Like many other intellectually oriented managers of the 1950s, McNamara found mathematical modeling techniques far superior to traditional wisdom or intuitive approaches to management based on shop-floor experience. McNamara, according to former strategic analyst Barry Bruce-Briggs, "was purely and simply, from beginning to end, what they call in Detroit a 'bean-counter': an accountant, a man who expects the numbers to add up, and who expresses initiative and creativity by making the numbers add up."[45] Kaplan's assessment is similar:

> His experiences in the war and at Ford provided McNamara with the confidence that he could gain command of any situation and that he could do so more quickly and proficiently than the conventional experts in the field, whether they be auto executives or Air Force generals. McNamara was coldly

clinical, abrupt, almost brutally determined to keep emotional influences out of the inputs and cognitive processes that determined his judgments and decisions. It was only natural, then, that when Robert S. McNamara met the Rand Corporation, the effect was like love at first sight.[46]

The Office of Systems Analysis

McNamara offered Rand economist Charles Hitch the Pentagon comptroller's post after reading Hitch's book *The Economics of Defense in the Nuclear Age* while preparing for his new job. Hitch and McNamara then established an Office of Systems Analysis (OSA) within the Office of the Secretary of Defense (OSD). To direct the OSA, Hitch chose another Rand economist, Alain Enthoven. (Enthoven, who had also worked on SAGE, had just completed a pessimistic report on SAC vulnerability, which he saw as "the most important danger facing the Western world.")[47] Other major Rand figures, such as Harry Rowen, Albert Wohlstetter, and W. W. Kaufmann, were also given jobs or retained as consultants. The Pentagon systems analysis group also came to be called the "whiz kids," after McNamara's Ford group.

McNamara, Hitch, and the OSA approached the Defense Department like the business managers and economists they were. They sought to rationalize the DoD by forcing the services to justify their procurement requests in terms of roles and missions, that is, the exact strategic and tactical purposes each military element would be expected to fulfill under U.S. foreign policy. In their public statements the new OSD and OSA emphasized the importance of managerial techniques, especially cost-benefit analysis based on "data" rather than the judgments of experience. "It is important," Enthoven proclaimed in his 1962 address to an Army symposium on operations research, "that these estimates [of effectiveness and cost of alternatives] be as objective as possible; that is, as free as possible from personal tastes or value judgments. To be so, their derivation needs to be explicit and reproducible rather than based on unanalyzed personal opinion."[48] Enthoven explicitly rejected both experience and history as guides to military policy-making in the post–World War II age of high-technology forces. Instead, he wrote, "modern day strategy and force planning has become largely an analytical process."[49]

The top-heavy reliance McNamara, Hitch, and Enthoven placed on quantitative information became legendary. (Once, when informed by a White House aide that the Vietnam war was doomed to failure, McNamara reportedly shot back "Where is your data? Give me something I can put in the computer. Don't give me your poetry.")[50] The whiz kids frequently defended this attitude by characterizing quantitative assessments as merely a background for value judgments. At the same time, they insisted on quantifying everything that might serve as an input to the analysis. Enthoven, for example, noted that such factors as courage, training quality, and dedication cannot be quantified directly: "although some of these questions call for numerical answers, the answers cannot be produced entirely by calculations."[51] Nevertheless, he held, some "calculative choices" must be made in order to complete the analysis. Similarly, McNamara admitted that "no significant military problem will ever be *wholly* susceptible to purely quantitative analysis." But he immediately went on to argue that "every piece of the total problem that can be quantitatively analyzed removes one more piece of uncertainty from our process of making a choice.... The better the factual basis for reflective judgment, the better the judgment is likely to be. The need to provide that factual basis is the reason for emphasizing the analytical technique."[52]

To the new Secretary of Defense, a major reason for the need to rationalize Pentagon procurements lay in the problem of technological choice.

Our problems of choice among alternatives in strategy and in weapon systems have been complicated enormously by the bewildering array of entirely workable alternative courses which our technology can support. We believe the Nation can afford whatever investment in national security is necessary. The difficult question is "What is required?" It is far more difficult to build a defense program on this kind of foundation than it is to set a budget ceiling and then squeeze into it whatever programs you can.[53]

McNamara initiated five-year plans for weapons research and development, combined with cost-cutting, soon after taking office. Hitch contributed the so-called Planning-Programming-Budgeting System (PPBS), essentially systems analysis applied to the articulation of military strategy with technology and cost. In testimony before Congress, McNamara offered an "oversimplified" example of how such an analysis might proceed:

Whether we should have a 45-boat Polaris program, as the Navy has suggested, or a 29-boat program, as the Air Force thinks, is in part affected by the decision we make on the Air Force Minuteman missile program.... A major mission of these forces is to deter war by their capability to destroy the enemy's warmaking capabilities.... [T]his task presents a problem of reasonably finite dimensions which are measurable.

The first of McNamara's "reasonably finite, measurable" dimensions—the number, type, and location of enemy targets—could be established by the intelligence services through information-gathering. The second dimension, or step, in his analytic process was to determine the number and yield of warheads needed to destroy those targets, a relatively simple calculation from known, testable quantities. At the third step, matters became both more complex and more speculative.

The third step involves a determination of the size and character of the forces best suited to deliver these weapons, taking into account such factors as (1) the number and weight of warheads that each type of vehicle can deliver; (2) the ability of each type of vehicle to penetrate enemy defenses; (3) the degree of accuracy that can be expected of each system ... ; (4) the degree of reliability of each system; (5) the cost-effectiveness of each system, i.e., the combat effectiveness per dollar of outlay.[54]

This example illustrates the methods McNamara's OSD applied to the issue of technological choices. But it also shows how McNamara used systems analysis to deal indirectly with the more significant problem of integrating the Army, Navy, Air Force, and Marine Corps into a single functional whole.

Military Management: Integrating the Armed Services

The National Defense Act of 1947 *nominally* integrated the four military branches, creating the Office of the Secretary of Defense and the Joint Chiefs of Staff as unifying bodies. Since that date, Secretaries of Defense had gradually accumulated increasing power, but none until McNamara had succeeded in fully imposing the layer of civilian control first envisioned by Truman. McNamara completed the unification of the services by obtaining authority over their budgets.

McNamara saw centralization of DoD decision-making as an imperative imposed not only by problems of cost, strategy, and technological choice, but by standard managerial practice. He expanded the office of the Defense Director of Research and Engineering (DDR&E), created by Eisenhower in the wake of Sputnik, in order to centralize research and development decisions. McNamara also established a Defense Intelligence Agency, a Defense Supply Agency for logistical support, a National Military Command at the level of the Joint Chiefs of Staff, and a Defense Communications Agency. All of these were civilian bodies within the OSD, and all were charged with centralizing and coordinating previously separate activities of the individual services.[55] McNamara and his associates saw central organizations like these as necessary to prevent wasteful duplication of effort, not only in operations but also in weapons programs. For example, rather than research, develop, build, and buy a different jet fighter for each service, each with slightly different capabilities, McNamara hoped to save money by producing a single plane to fill the needs of all three.

The McNamara OSD in effect brought Rand and its methods inside the Pentagon through the OSA and its Planning-Programming-Budgeting System (see chapter 1). Together they rationalized and centralized the Department of Defense and imposed upon it a civilian managerial, rather than a military command, form of leadership. In certain respects they merely completed a process begun during World War II, when the military first began to employ corporate managerial structures to handle logistical problems. Indeed, even in the 1950s military sociologist Morris Janowitz noted an emerging split between "heroic" leaders or "gladiators" and a new class of "military managers," whose careers were based not on combat leadership but on organizational abilities.[56] The importation of industrial management methods into military organizations had become such a powerful trend that by 1961 an Army lieutenant colonel could write that "every professional military man in recent years has read or heard the theory expressed that management and command are essentially the same thing. Chances are better than even, in fact, that he has never heard any serious dissent from the proposition that the terms command and management are synonymous, or nearly so."[57]

The shift toward a managerial approach had everything to do with burgeoning technological change, as the ramifications of high-technology military systems unfolded. Martin van Creveld notes the very rapid increase in complexity experienced by the armed services as they incorporated electronic communications and data processing (among other technologies) into their command systems after World War II:

Toward the end of World War II the 9,700,000 men enlisted in the U.S. Services were divided into 1,407 different Military Occupation Specialties (MOS's), the average number trained in each MOS being thus 6,894.... By 1963 the number of enlisted men in all four services was approximately

2,225,000, in 1,559 MOS's, so that the average number of men in each MOS was down to 1,427—and even this figure misrepresents the true situation because many of the 1963 MOS's, especially those connected with the rapidly growing field of electronic gear maintenance and operation, were really agglomerations of several different specialties. . . . [A] calculation that would put the complexity, relative to size, of the 1963 armed forces at . . . four and a half times that of their 1945 predecessors would probably err on the side of caution. . . . [U]nit by unit, the amount of information needed to control the US Forces in 1963 was perhaps twenty times (4.5 × 4.5) larger than in the case of their 1945 predecessors.[58]

In an era in which business schools were theorizing management as a problem of information processing, while Rand strategists increasingly formulated command as information processing and war itself as a problem of communication, the rise of a military managerial style was perhaps inevitable.

Command and Control

Centralized military management remained—and remains—in tension with the command tradition, which presumes nested spheres of responsibility within which detailed planning and control devolve to the *lowest* levels of authority. Nuclear forces, by contrast, flatten their hierarchies as much as possible and retain authority at the upper levels. They do this because they require instantaneous and massive responses, which must be preprogrammed because their execution must be virtually automatic, and because the consequences of the release of nuclear weapons are too great to be devolved upon lower-level commanders.

Computerization supported this flattening by automating many tasks and permitting rapid, accurate, and detailed central oversight. Similarly, corporate managerial styles of the 1950s—evolving out of the Taylorist and Fordist managerial practices of the 1920s and 1930s—sought to raise detailed control to increasingly higher levels of authority.[59] In fact, as Bracken and others have observed, "command" and "control" (in their traditional senses) are complementary or even opposite rather than synonymous aspects of military organizations.[60] (Whereas traditional military forces have *command* systems, nuclear forces have *control* systems.) But by the early 1960s, military parlance treated the two as virtually identical. A decade later, "command, control, communications, and information" (C^3I) had become a single unified process.[61]

McNamara's search for military "options" other than all-out nuclear confrontation led him to a Rand-inspired "flexible response strategy." This required a yet wider array of nuclear weapons of all sizes, to preserve the possibility of "tactical" strikes, as well as increases in conventional forces, to be able convincingly to confront a Russian conventional assault on Europe. Enthoven later credited the OSA's PPBS with this strategic shift, and he drew direct parallels between the PPBS's formal analytic methods and the resulting strategy: "The themes 'options,' 'flexibility,' and 'choice' have become as fundamental to our military strategy as they have been to our approach to analysis and planning of the defense program. The charge that centralized decision-making leads to an inflexible strategy based on a single set of assumptions is refuted by the historical facts."[62]

Multiplying strategic options also meant complicating the automated command-control systems designed in the 1950s. Now, instead of serving as simple coordinators for a tripwired "massive retaliation," command-control systems had to permit many-faceted, multistage actions more carefully integrated with political interventions.

By this point the untempered optimism of the SAGE designers had given way, if only among computer scientists, to a more sober view of the capacities of large automated systems. In late 1962 the Air Force Electronics Systems Command, together with the Mitre Corporation (a Lincoln Laboratory spinoff responsible for ongoing work on SAGE), sponsored the First Congress on Information Sciences at Hot Springs, Virginia. Conferees from Mitre, Rand, Raytheon, the Air Force, and other agencies voiced disappointment with the performance of existing automated command-control systems such as SAGE. Many expressed despair at the prospects for coordinating the rapidly proliferating DoD computer systems (through standard computer languages, for example) and noted that progress in hardware had far outstripped understanding of the command process itself.

At the same time, they generally agreed that "the trend to centralize which has been associated with DoD management to date is also apt to apply to C&C [command and control] systems." Rand's Norman Dalkey noted that flexible-response strategy would require "techniques for rapidly adjusting to unforeseen conflicts." The "increased capability" such techniques demanded "will be met [partly] by increased automation for both data systems and weapons." Despite well-known problems with radically centralized command, such as its tendency to amplify minor mistakes, Dalkey guessed that "for some years to come, the political advantages to centralization will outweigh the operational disadvantages."63

This statement was not only about the value of centralization to politicians collecting power. Flexible-response strategy required that political leaders continue to communicate during an escalating nuclear exchange. Indeed, such communication was the whole point of the strategy, since it offered (in theory) the possibility of ending such a battle somewhere short of all-out war. Therefore, preserving central command and control—political leadership, but also reconnaissance, data, and communications links—achieved the highest military priority.

Computers as Icons
The role of computers as analytical tools in the McNamara OSD was certainly significant. From the perspective of closed-world discourse, however, this was not as important as their *symbolic* role. Whether the systems analysts required room-size IBM mainframes or whether they used desktop adding machines made no difference; the computer became their icon. For example, a 1962 *U.S. News and World Report* article listed a number of major Kennedy administration procurement decisions under the heading "Where Computers Indicated 'Yes' and 'No'," implying that the decisions had been left entirely to machines.64 The military itself reacted to the intrusions of civilian strategists and computerized formal analysis with an ambiguous mix of disgust and accommodation. The Joint Chiefs of Staff and each service individually created their own systems analysis departments, in part to be able to combat the OSA's analyses with their own.65 But they still felt, as SAC General Tommy Power bluntly put it in 1964, that "computer types who are making defense policy don't know their ass from a hole in the ground."66

Computers became icons for the increasing success of managerial style within the command structure as well. As noted in chapter 2, accusations of "electronic despotism" and computers "running the wars of the future" had become commonplace by the late 1950s. By the time the Office of Systems Analysis became fully operational in 1963, it was already necessary for Enthoven to defend its techniques in the pages of the *Military Review*.

"Computers are replacing military judgment." "Weapon systems are being computed out of existence." "Computers are running the wars of the future."

These and similar statements are being published in prominent American newspapers and magazines.... [T]hese statements are... worse than wrong, they are dangerous.[67]

Enthoven's words reflect the popular iconography that equated DoD centralization and civilian management with computer analysis.

As with SAGE, the McNamara era reflects a discourse process of mutual orientation in which computers played a key part. Civilians, in this case data-oriented managers and economists, sought—using computers to implement the PPBS and the OSA to institutionalize systems analysis—to centralize and rationalize DoD procurements. Their discursive categories—systems, options, data, flexibility, limited war—required the development of program choices that linked strategy with technology and cost. The military, in response to these essentially managerial requirements as well as to rapidly evolving technology, constructed strategic options that depended upon increased centralization of command. The discursive categories of command and control, in turn, motivated the increasing sweep of automated, computerized command systems. The Pentagon managers oriented field commanders toward programmable "options" and integrated technological choice, while the commanders oriented the managers toward computerized command and control systems for the electronic battlefield.

Vietnam

The first half of the 1960s saw the high tide of technocratic optimism in the United States. Riding on the steadily rising wave of 1950s economic growth, federal budgets burgeoned with possibility. Kennedy embarked on a new Cold War contest: the space race, a Manhattan Project aimed at technological prestige. With Sputnik and the space race came a new iconography of global closure: an Earth floating in the black void, encircled by orbiting spacecraft.

The Eisenhower administration had used manned space missions as a cover for the secret military agenda of establishing rights of satellite overflight in international law. The Mercury flights served as legal precedents for spy satellites that would "open up" the USSR to American surveillance. Presidents Kennedy and Johnson adopted similar programs, but they raised the ante far higher, declaring an all-out push to land a man on the moon by the decade's end. Kennedy spoke of a "new frontier" in which high technology would unseal the closed globe to exploration of the limitless universe beyond.[68] But as Walter McDougall has argued, this program, in addition to covering for military missions, embodied two other Cold War purposes. First, it transformed space research into an arena of competition for international prestige. In this arena, the magnificence of American space ventures would demonstrate the power of its technology and vision. Second, the program offered one more opportunity to exercise the most long-term of America's Cold War strategies: the effort to spend communism into extinction.

NASA director James Webb—like McNamara, a kind of ultimate technocrat—wrote in 1963 that the United States was "in the midst of a crucial and *total* technological contest with the Soviet Union." Webb thought of the space program as a kind of model for future social systems adapted to rapid technological change. He called these "prototypes for tomorrow." Invoking the Manhattan Project as a precedent, he envisioned these systems as large-scale, heavily managed, technologically controlled, and technocratically organized. Webb's language echoed the systems discourse of McNamara's DoD: Space Age social organization would require "adaptive, problem-solving, temporary systems of diverse specialists, linked together by coordinating executives in organic flux." Webb's future world "required the forging of a 'university-industry-government complex' for the waging of 'war' on the technological frontier."[69] This infrastructure of space technology—aimed at spinoffs, systems, and technological power for managing domestic culture and global Cold War—belied the ideology of space as a limitless frontier.

A heavy irony lay behind the discursive *décalage* between the frontier imagery and the Cold War competition: most of the swarming satellites and spaceships were sent up only to look *down*. With every launch another orbiting object drew its circle around the planet, marking the enclosure of the world within the God's-eye view from the void. Even in the dizzy technological euphoria of the first moon landing, the barren moonscapes, sterile capsules, and sealed spacesuits emphasized not the bounty of a green frontier but the utter aloneness of the living Earth. After all was said and done, the space program's chief products were not outward- but inward-looking: spy cameras to pierce the Soviet Union's veil, pictures of the Earth drifting alone through space, pictures of the closed world.

Kennedy's military rhetoric—like the frontier iconography of the space program—seemed at first to promise a way out of the Cold War. In fact, however, Kennedy went even further than his predecessors, promising to "bear any burden, pay any price" in the defense of freedom throughout the world. McNamara approved the thousand-strong Minuteman ICBM force in order to "regain" nuclear superiority, though by then he was well aware that the United States already had superiority and that the new missiles would only spur the Russians into a renewed arms race. After Kennedy faced down Krushchev during the 1962 Cuban missile crisis, his portfolio as a Cold Warrior was complete. In the attack of nerves that followed their long stare into the inferno of nuclear war, both Kennedy and Khrushchev lightened their tones, as reflected in the limited test-ban treaty of 1963. Nevertheless, Kennedy remained committed to the eventual "evolutionary" demise of communist societies. By the time of his death in November 1963 he had raised the military budget to $56 billion.

Kennedy focused heavily on Third World upheavals as arenas of capitalist-communist conflict, and part of his legacy was the upward-spiraling course of American involvement in Vietnam. Truman and Eisenhower had already committed U.S. assistance, through financial aid first to the hapless French and then to the authoritarian but noncommunist regimes of South Vietnam. It was Eisenhower who first pictured Southeast Asia as a row of dominoes, each set to topple all the others in a "fall" to communism, but it was Kennedy who enthusiastically undertook to uphold the "dominoes." His Secretary of State, Dean Rusk, in classic Cold War language, continually warned of a possible "Munich" in Asia. Kennedy raised the number of military personnel in South Vietnam from a few hundred to over 15,000. Many of these so-called advisers came from Kennedy's new Special Forces, the Green Berets, conceived and trained precisely for Third World counterinsurgency warfare.

Power at a distance was the new watchword: America would micromanage the global struggle against communist insurgencies, furnishing the equipment, the know-how, and the money while its proxies supplied the men-at-arms. Vietnam would serve—as Walter Rostow and others often put it, in their Third-World-as-laboratory language—as the "test case." McNamara, Rostow, Vice-President Johnson, and other high-level Kennedy administration officials visited the country often. All cheerfully reported the war under control and moving rapidly toward an expeditious end.

When Kennedy was assassinated late in 1963, he was succeeded by his less sophisticated, more bellicose, and more driven vice president, Lyndon Johnson. Despite his rangy Texan image, Johnson was every bit the technocrat that Kennedy had been. He carried through Kennedy's promise to put a man on the moon (though the astronauts landed only in mid-1969, a few months after Johnson's bitter departure from the White House). He pursued a vision of social management for a higher good, the "Great Society," that was perhaps even grander than Kennedy's. Johnson also inherited Vietnam.

The new president retained most of the Kennedy national security apparatus, including McNamara and the OSA, and prosecuted similar policies with ever-increasing vigor. McNamara still regarded Vietnam as "a test case of U.S. capacity to help a nation meet a Communist 'war of liberation.'"[70] In 1964 he sent more "advisers." In the middle of that year, a North Vietnamese "provocation" provided the occasion for the congressional Gulf of Tonkin Resolution, authorizing the President to commit American troops for combat. Now the military, too, would get a "test case"—for its new technology and its doctrines of limited war. Johnson, on the advice of McNamara and others, opened a strategic bombing campaign against the North on March 2, 1965.

Computers and the "Production Model of War"

To understand the role of computers and closed-world discourse in the Vietnam War, it is necessary to understand the war's enormous scale. American jets ultimately dropped more bombs on Southeast Asia than had been dropped by all combatants in *all previous wars combined*. This strategic bombing nevertheless failed—as it had failed in Germany, Japan, and Korea—to win the war. So did the ground troops sent in four months after the bombing began. At the war effort's peak in 1969, between one-third and one-half of all U.S. combat-ready forces were actively involved. Well over half a million troops were stationed inside Vietnam, supported by at least 100,000 troops operating staging areas on Manila and Guam or logistical support efforts in the United States and elsewhere.[71] Even these numbers do not include the tens of thousands of South Korean, Australian, and other Southeast Asia Treaty Organization allied troops—or the hundreds of thousands in the Army of the Republic of Vietnam.

In many ways Vietnam was the apotheosis of closed-world politics. Ideology, not national interest, was at stake. The war embodied the containment doctrine's defense of "free" nations (though the regimes being propped up were widely known to be corrupt and could never have sustained themselves alone). It began as a proxy conflict, with the United States arming and advising South Vietnam and the Russians and Chinese doing the same for the North. It remained a "limited" war, in that nuclear weapons were sometimes suggested but never seriously considered. It was a high-technology war of the first order, prosecuted with the most advanced equipment America could build and engendering an enormous wave of new inventions.

But Vietnam was not the set-piece, central war envisioned by the designers of SAGE, NATO, and the nuclear strategists. In 1965 computers still bulked too large to serve at the front of such a mobile war. (Microprocessors, which permitted the integration of computers weighing just a few pounds directly into weapons such as cruise missiles, were not invented until 1970.) Nevertheless, electronics, in the form of communications equipment, made possible vastly increased centralization of command and control. About a third of the matériel brought into Vietnam consisted of electronic communications equipment. The problems of coordinating small-unit jungle combat led to major problems of specialization and complexity.[72] With field radios, every unit could and often did maintain constant contact with commanders, who frequently directed ground movements from helicopters above the battlefield. McNamara, as noted in chapter 1, managed the air war in detail from Washington. Logistical support—stripped from individual units by Gen. Westmoreland—was also centralized and run from the United States, leading to byzantine requisitioning procedures and interminable delays.

Most of the uses of computers in Vietnam occurred at the rear, where they churned the daily reports generated by nearly everyone involved in the war into statistics. The war's gigantic scale, and the logistical problems of running, from Washington, a war on the other side of the world, contributed to an insatiable demand for information, delivered in the form of computer-processed statistics. Army and Air Force computers were set up in trailers resembling mobile homes so they could be moved from base to base, processing battle reports as they came in from the front.[73]

In his meticulous study of Vietnam, *The Perfect War*, James William Gibson argues that the institutions responsible for the war conceived the problem it was supposed to solve in the mechanistic terms of physical science. Metaphors of falling dominoes, popping corks, and chain reactions were used to describe the diplomatic situation. Communist governments and armies were depicted as demoniac machines, conscripting their people as parts and consuming their energy; Gibson calls this imagery "mechanistic anticommunism." The entire transaction was understood as an accounting procedure in which capitalists scored "credits" and communists "debits." Thus its planners were managers who saw the war as a kind of industrial competition. Gibson names this factory model "technowar" or "the production model of war." It represents "the military mode of strategy and organization in which war is conceptualized and organized as a high-technology, capital-intensive production process."[74] Counterinsurgency, the new technology of limited war, would allow the prosecution of "technowar" in the revolutionary jungles of the Third World.

The production model measured its progress primarily by means of statistics, and for this reason statistical information took on an inordinate importance in Vietnam.[75] Yet of all wars America has fought, the actual course of the Vietnam War was probably the least susceptible to measurement. There were few static geographic fronts to be charted. Technological assets such as factories, bridges, roads, and motor vehicles were much less important in the peasant society of Vietnam than in Europe, where military planners had focused most of their attention since World War II. Weapons factories were located in China and the Soviet Union, where they could not be attacked. The North Vietnamese Army (NVA) moved supplies by bicycle or oxcart or on foot, over ever-changing trails obscured by dense jungle canopies. When American planes blew up their bridges, NVA convoys simply forded the rivers. Communications relied on human couriers, not telephone lines. The guerrilla forces used women, children, and old people—traditionally, in the West, noncombatants—as spies and part-time soldiers. Most wore no uniforms and so were indistinguishable from the civilian population. Thus counting North Vietnamese losses of soldiers and matériel became a difficult problem of interpretation.

Many kinds of statistics—body counts, bomb tonnage, percentage of the population loyal to the South, patrols performed, hamlets "pacified"—were collected to measure the war's progress. The importance attached to this information by the high command led to a system of incentives for reporting, and especially for reporting what commanders (and their civilian superiors) wanted to hear. Promotions and perquisites were based on high "outputs." As a result troops systematically overreported their "productivity."

Body counts are perhaps the most notorious example. Frequently American soldiers were killed in ambushes and night engagements, but despite returning fire, after the smoke cleared their compatriots failed to find any Vietnamese bodies. Such one-sided losses could be deeply embarrassing. Furthermore, counting enemy dead was usually dangerous, since the NVA often left behind snipers and mines at the sites of skirmishes. Troops thus found themselves risking their lives to obtain information of dubious quality (how many of their dead had the enemy recovered during the night?) for their superiors to use to promote their careers. They were pressured to produce high counts—so they inflated them or simply made them up.

Statistics for the air war were also falsified, though less frequently. Numbers of sorties flown by jet fighters, tonnages of bombs and shells expended, and canisters of napalm dropped could all be measured. In a war where defining progress at all was difficult, these statistics offered a reassuring aura of certainty. What could not be measured, except by extrapolations of unknown validity, was the actual effect of the bombing on the North Vietnamese war effort. As in Igloo White, damage to targets was often estimated rather than directly observed. As Gibson shows, these extrapolations of losses tended to assume that the enemy's society—its technology, its human resources, and its political aims—mirrored America's. Destruction of bridges and vehicles scored more points, in this system of assumptions, than creating equitable land distribution or ending government corruption.

Gibson argues that the production model served major ideological purposes.

The production system with its precise reports of how many bodies were found on operations created the *appearance* of highly rational, scientific warfare. Body counts, weapons/kill ratios, charts of patrols conducted, helicopter and jet plane missions flown, and artillery rounds fired—all the indices of war production created at various command levels—presented Vietnam as a war managed by rational men basing their decisions on scientific knowledge. Statistics helped make war-managers appear legitimate to the American public.[76]

Computers, by facilitating the collection and analysis of statistical information, participated in the discourse of "technowar" in the same way, assisting its legitimation.

The OSA's role in the war effort was connected with this statistical legitimation process. As usual, it forced the services to justify their procurements and expenditures of equipment in terms of cost per target destroyed. The OSA did not participate in the detailed management of the war. But its systems approach, with its vast demands for data, did help create what van Creveld has called the "information pathologies" of Vietnam.[77] By this term van Creveld refers both to the communications overloads that resulted from attempts at every level to exert detailed control over lower levels, and the efforts (partly in consequence) to understand the war through the abstract lens of statistics.

To a certain degree, argues Kaplan, the Rand Corporation also influenced the strategic concepts used to plan the war and to define the meanings of "progress" and "victory." The doctrines of limited war were applied directly to counterinsurgency. Each military maneuver was conceived as a political message about the costs of continued war to the other side. Rand theorist Thomas Schelling was consulted about how to "communicate" U.S. intentions via carefully limited bombing punctuated by pauses to allow Vietnamese response, and how exactly the United States could know whether the message had been received. When National Security Adviser McGeorge Bundy notified President Johnson of the war plans being drafted in 1964, he wrote that "the theory of this plan is that we should strike to hurt but not to destroy, and strike for the purpose of changing the North Vietnamese decision on intervention in the South."[78] He emphasized the "deterrent" effect he expected from rapid, massive, highly visible U.S. troop deployments. McNamara's strategy in Vietnam borrowed its essential philosophy from the counterforce/no-cities theory of nuclear strategy. Strategic bombers attacked military targets in an attempt to "communicate" with the North Vietnamese, holding in reserve the possibility of attacks on cities in case they did not respond with the correct reply.[79]

Of course, though American bombers broadcast the message repeatedly and in increasingly horrific terms, the enemy never did receive it—or more accurately, received it but simply did not agree with its terms. The official discourse of the war in its first years was mainly optimistic, despite some forebodings. By the end of 1967, official statistics had been deliberately revised to reduce the number of NVA troops to 299,000—ignoring over 120,000 guerrillas, whom planners simply deleted from the enemy order of battle in order to bolster the discourse of success. After almost three years of virtually continuous bombing of its supply lines, harbors, and industrial centers, the NVA mounted a major offensive in 1968 during Tet (the Vietnamese New

Year). The large scale, fierce drive, and symbolic efficacy[80] of the Tet offensive took military leaders by surprise, but its impact on domestic American politics was even stronger, fueling a burgeoning antiwar movement by confirming the growing convictions of many that the bombing had failed. The end of the war was nowhere in sight.

On the Electronic Battlefield
We have now returned full circle to the story that opened this book. Operation Igloo White, like the SAGE system it so closely resembled, was a product of civilian scientists attempting to offer a high-technology *defensive* strategy to replace the unpopular "retaliatory" (but in fact offensive) bombing campaign. The group that designed the so-called McNamara Line (the sensor-strewn stripe across the Ho Chi Minh Trail in southern Laos) was known as the Jason Division of the Institute for Defense Analysis (IDA). IDA was a Rand-style think tank, dating to the early 1950s, that advised the Weapons System Evaluation Group. Like the Valley Committee and the Summer Study Groups that created SAGE, the Jasons were a blue-ribbon commission of civilian scientists who spent their summers working on military problems. Also like their predecessors, they proposed a computerized, centralized command-control system as the basis of a barrier against invasion—as an enclosure of the "free" world. Their idea was to draw a line of fire between North and South Vietnam that would be so intense and so deadly accurate the NVA could not cross it. One version of the plan called for clearing a ten-mile-wide swath through the jungle, using chemical defoliants, all the way across Vietnam at the demilitarized zone.

Components of the McNamara Line began operating in 1967, and the Igloo White system functioned at least until the official end of the American bombing in 1972. As we saw in chapter 1, its effectiveness was limited at best. Once again official statistics vastly inflated the destruction rates of the "target signatures," as the white video lines representing North Vietnamese convoys were called. In 1972 the North Vietnamese launched a major tank and artillery assault within the South, something that should have been impossible if official claims for Igloo White's success were true.

As Gibson notes, an essential feature of technowar in Vietnam was its inability even to imagine failure.

As the possessor of an advanced technological system of war production, the United States began to view political relationships with other countries in terms of concepts that have their origin in physical science, economics, and management. A deeply mechanistic world view emerged among the political and economic elite and their intellectual advisers.... The Vietnam War should be understood in terms of the deep structural logic of how it was conceptualized and fought by American war-managers. Vietnam represents the perfect functioning of [their] closed, self-referential universe.[81]

The "deep structural logic" of the "war-managers" of Vietnam, their "closed, self-referential universe," was closed-world discourse. It carried each of the characteristics described in chapter 1. Rand-style game-theoretic *techniques* shaped the conceptual basis for overall strategy; statistical analysis provided the measuring tool for the war's "productivity." *Experiences* of the closed, apparently manipulable political world of the 1940s, when the United States was still the sole possessor of atomic weapons, and of the successes of operations research in military logistics and factory management in the 1950s, had shaped the thinking of the men then at the peak of their political and military careers. *Practices* of systems management had achieved wide currency through the ongoing evolution of operations research and systems analysis. The goal of the Vietnam effort was couched in terms of systemic manipulation: the United States would "strike for the purpose of changing the North Vietnamese decision on intervention in the south."

Computer tools had now reached technological maturity and could play powerful roles in data analysis. In Igloo White they served a SAGE-like function as the core of a centralized, automated command-control system; elsewhere they also contributed to command as management.[82] Finally, a *formalistic language* of statistics, systems, and strategy tied these elements of the discourse together. The PPBS, for example, unified the DoD approach to war under categories of costs, benefits, weapon systems, credits, and debits. Numerical analysis superseded military experience as the discourse of decisionmakers. The new language interpreted the war using the categories of games, bargaining, production, and management. It reinforced the view of war as a rational problem, rather than a struggle with its roots in ancient feelings of patriotism, desires for justice, and resentments of foreign intervention that might not respond to a "rational" challenge.

Conclusion

Vietnam, Kaplan reflects, "exposed something seamy and disturbing about the very enterprise of the defense intellectuals. It revealed that the concept of force underlying all their formulations and scenarios was an abstraction, practically useless as a guide to action."[83] It was an attempt to apply the logic of the closed world to war in a green-world jungle. Napalm and defoliants, however, proved incapable of laying bare the forests that concealed the guerrilla army. The frustrations of the most technologically sophisticated war in history brought home—at least to some—the hard lesson that it was "a gross oversimplification to regard Communism as the central factor in every conflict throughout the developed world," as the disillusioned McNamara put it in 1966. Even he, an architect and himself an archetype of closed-world discourse, ultimately regretted the "almost ineradicable tendency to think of our security problem as being exclusively a military problem.... We are haunted by this concept of military hardware."[84]

Yet others drew the opposite lesson. In the "concept of military hardware," whatever its shortcomings in Vietnam, lay the ripening seeds of a bright future whose maturing technologies would finally lift the Clausewitzian "fog of war." Westmoreland's 1969 vision of the "electronic battlefield" was *based* on his Vietnam experience. Drawing on his own experiences of the military promise of high technology, he contemplated a future war zone "under 24 hour real or near real time surveillance of all types ... an Army built into and around an integrated area control system that exploits the advanced technology of communications, sensors, fire direction, and the required automatic data processing."[85] Computers lay at the heart of Westmoreland's vision, which echoed Forrester's grand schemes of twenty years before.

The political purpose of the electronic battlefield was to build a deadly version of what Shoshana Zuboff has called an "information panopticon." Zuboff's panopticons are offices and factories whose central information systems allow their managers to record every employee's activity at microscopic levels of detail. By relying on the recorded database and its statistics rather than personal observation to judge employees, these systems "create the fantasy of a world that is not only transparent but also shorn of the conflict associated with subjective opinion." They reflect a desire for "light without heat," knowledge without confrontation, power without friction.[86] Ideally, panoptic power is self-enforcing: people who know their every act is "on the record" tend to do and say what they think they are supposed to do and say. On the pseudo-panoptic battlefields of the Vietnam War, soldiers subjected to panoptic control—managed by computers—did exactly what workers in panoptic factories often do: they faked the data and overrode the sensors. The Americans made up body counts and fabricated statistics. The NVA tape-recorded truck sounds and carried bags of urine to confuse the McNamara Line's sensors. Crippled by its own "regime of truth," the system faltered and was finally defeated.

Westmoreland believed that on the future electronic battlefield commanders would find at last the transparent, laser-perfect vision to resolve the "endless quest for certainty" van Creveld calls the essence of command.[87] In a world surrounded by a swarm of communications and photographic satellites, with submarine sensors ringing the ocean floor and radars scanning the sky, no movement would go undetected. With instantaneous communication and automatic, computerized control, not a minute's delay would intervene between commands and their execution. Conventional war would achieve the "near-real time" scales and the automatic action pioneered by nuclear C^3I.

Pure information, "light without heat," would illuminate future war. In its bright and tightly focused beam, the army of the information age would finally discover certainty in command, combat without (American) casualties, total oversight, global remote control. Political leaders could achieve the ideal of American antimilitarism: an armed force that would function instantly and mechanically, virtually replacing soldiers with machines. The globe itself would become the ultimate panopticon, with American soldiers manning its guard tower, in the final union of information technology with closed-world politics.

Book jacket, 1949. Courtesy Charles Babbage Institute, University of Minnesota.

5
Interlude: Metaphor and the Politics of Subjectivity

Chapters 6–8 explore the origins of cybernetics, cognitive psychology, and artificial intelligence (AI), three research programs that relied on the computer as model, metaphor, and tool.

According to received histories, in their early days all three of these fields were speculative and theoretical, without much practical import. Their significance lay mainly in their perspective on human nature: they pictured minds as nested sets of information processors capable of being duplicated, in principle, in a machine. In this sense, then, they are all "cognitive" theories. In contrast, I will argue that the *cyborg discourse* generated by these theories was from the outset both profoundly practical and deeply linked to closed-world discourse. It described the relation of individuals, as system components and as subjects, to the political structures of the closed world.

Cognitive theories, like computer technology, were first created to assist in mechanizing military tasks previously performed by human beings. Complete *automation* of most of these activities—such as aiming antiaircraft guns or planning air defense tactics—was not a realistic possibility in the 1940s and 1950s. Instead, computers would perform part of a task while humans, often in intimate linkage with the machines, did the rest. Effective *human-machine integration* required that people and machines be comprehended in similar terms, so that human-machine systems could be engineered to maximize the performance of both kinds of components. Work on these problems during and after World War II brought psychologists together with mathematicians, neurophysiologists, and communications engineers as well as computers.

This chapter focuses on the political significance of computer metaphors in psychological theories. To grasp what it might mean to speak of a politics of metaphor, I explore how metaphors work and how they shape theories. First, I argue that theories play a critical role in the politics of modern culture because they assist in constituting the subject positions inhabited by individuals and the cultural representations of political situations. Second, I claim that metaphor, as a major mode of representation, frequently helps to organize theories of all sorts. I use Lakoff and Johnson's conjectures about metaphors in ordinary language to describe their systematic, wide-ranging structural effects. Third, I examine some of the entailments of computer metaphors in psychology and compare them with the metaphors used in other psychological theories.

Psychological theories describe subjects: how they make decisions, how they communicate, and how they understand their relation to objects. In representing possible subject positions, they simultaneously describe one of the two faces of modern political order, namely the individual. Cyborg discourse links the psychology of cognitive actors to the social realm of closed-world politics: the institutional and ideological architecture those subjects inhabit.

Politics, Culture, and Representation

We start with politics, which we can define as the contest among social groups for power, recognition, and the satisfaction of interests. The contest is acted out in many arenas and with varying degrees of visibility. Institutions of government, such as the military and the legislature, are only the most obvious and most discussed domain. Equally important are organizations and institutions in civil society, such as labor unions, factories, and universities.

But to say that politics is about acquiring power and satisfying interests is not enough. What *is* power? Is it an actual force that can be created and exchanged? Or is it merely an analytical concept for naming the shifting contingencies that lead to victories, useful only in retrospect? Why does the political recognition of persons matter? Is it important in itself, or only in conjunction with power and material gains? What is an interest? How do groups come to recognize collective interests in the first place, and how do they decide what counts as their satisfaction? Do similar interests somehow create coherent, politically meaningful groups from otherwise unconnected individuals, or do preexisting group identities determine individual interests? Do group identities such as gender, race, and family originate in some sort of primordial status, while others such as economic or social class stem from specific political conditions? Or are all group identities in some sense *outputs* of political systems? These questions, perennial problems of political theory, begin to open the space for an interrogation of what I will term the *politics of subjectivity*.

To understand how power is created and employed, recognition expressed and interpreted, and interests fashioned and fulfilled, we must first grasp the relationship between individual subjective desires and the objective political interests of groups.[1] Most political theories attempt to explain one of these as a consequence of the other. Such explanations increase in complexity along with the degree of difference between how people actually act, what they actually want, and what the theory assumes they *ought* to do and want if they are acting rationally and in their own best interests. The simplest political theories are those based in classical economics, which circularly assumes that what people get is what they really want and vice versa. Utilitarian doctrines stemming from the work of Jeremy Bentham fit this mold. For them the problem of democratic politics amounts to little more than arranging for free, unhindered voting. Preferences revealed through the vote constitute the electorate's straightforward judgments about the relative value of the choices it has been offered, just as preferences revealed through the market mark preexisting needs and their valuation.

More sophisticated theories note that politics, like advertising, not only announces the availability of various choices but attempts to transform the recipients of its messages into consumers who will be satisfied by choosing among the options they are offered. Democratic elections, for example, promote not only various candidates but also the idea of voting itself. The exercise of the right to vote becomes a simultaneous experience of difference and unity, one's group identity vis-à-vis other social groups and one's membership in a single community of participants. It reconciles electoral losers to their fate by affording them recognition. Thus the electorate comes to know that it has gotten what it wanted *by learning to want what it gets*. In politics, the vehicle of such transformation is political theory itself. Democratic theory, for example, involves the premise that making the right choice is less important than having the right to choose. In the liberal political tradition, being respected as a Kantian chooser matters more than receiving the Benthamite benefits of some more streamlined system. As political theorist Robert Meister puts it, "to be free" within the modern liberal state "is to be recognized as an individual by an institution that one recognizes as a state."² This requires that we explicitly experience our subjectivity in our roles as voters and participants in state politics, and that we comprehend the state as made up of our collective subjective choices.

The point here is that theories of politics, in the modern world, play a direct and major role in the construction of political subjects. Political theory is actually *about* the relations of subjects to objects. That is, it describes how people decide which things and services they value and how the political system shapes and/or provides their access to these things and services. More subtly, any political theory can be said to describe how the relations of subjects to each other are mediated through their relations to objects. This insight was most thoroughly articulated by Hegel, who argued, as Meister puts it,

that others can control us by controlling the external objects we need and value. [Hegel] also suggests that our awareness of material needs has a social dimension, even to begin with: we need things partly because they are recognized as valuable by others as a means of social control; the satisfaction of our needs is in part, therefore, a way of gaining recognition from others for the autonomy of our desires. For this reason Hegel insisted that the relative value of things is inherently intersubjective: the things we need will define the nature of our dependency, autonomy, and power in relation to others.³

At the core of democratic political theory lies a theory of subjectivity as a limited sovereignty, that is, a domain in which individuals choose for their own reasons—or for no reason at all, since the very basis of the theory is that they need not justify their political choices. At the same time, political recognition takes place within an intersubjective field of articulated reasoning: "Freedom of the individual... is the development of an identity through which one gains recognition for one's choices by learning how to make the reasons for them intelligible to others."⁴

Yet this particular subject position—subjectivity as limited sovereignty and shared rationality—is largely a creature of the theory. Under some political systems (and the theories that justify them) different people's choices receive different treatments, while under others some individuals do not make political choices at all. *Theory creates political subject positions* that individuals inhabit and that form the preconditions for the constitution of collective political actors. This analysis points to the crucial importance of *culture* as another political domain.

Culture consists of the shared, informal world of language, art, narrative, play, architecture, visual imagery, imagination, and so on within which social, epistemological, and ethical realities are constructed for human subjects. It also includes those theories (of anything whatever) that become part of "common sense" and the artifacts that embody these constructions and theories. Culture encompasses the public manifestations of subjectivity and the communicative practices that define a shared "life-world." While few concepts boast more ragged edges, the idea of "culture" does allow us to understand *representation* as an arena of political action. The "politics of culture" refers to the embedding of structures of power and interests in shared representations: concepts, media, and conventional structures of thought.

Is culture an arena of true political power? Looked at broadly, the concept of "power" is usually operationalized in terms of coercive physical force (armies, police, weapons), wealth (which can purchase influence and tools for action), and the institutionalized roles of those who make choices for groups (the president, the chairman of the board). But as Michel Foucault, among others, has shown, under a lens of higher magnification such definitions help very little. As Hegel observed of the emerging democracies of the nineteenth century, in the universe of modern political subjects "what is to be authoritative ... derives its authority, not at all from force, only to a small extent from habit and custom, really from insight and argument."[5] Under democracies, at least, argumentation complements pure force and arbitrary choice as a basic source of world-shaping decisions. Rationality itself has become a source of power; consensual political systems require agreement in thought as well as acquiescence in behavior. Twisting the liberalism of Hegel's point in light of decades of discussion of the politics of representation, we must ask how any given claim *comes to count* as an insight and *from what source* arguments derive their social force.[6]

This problem has been addressed most explicitly in the sociology of knowledge. Recent social studies of science have termed the epistemological standpoint that assumes a relation between power and knowledge an "equivalence postulate." Barry Barnes and David Bloor, for example, describe this position as follows:

Our equivalence postulate is that *all beliefs are on a par with one another with respect to the causes of their credibility*. It is not that all beliefs are equally true or equally false, but that regardless of truth and falsity the fact of their credibility is to be seen as equally problematic. . . . Regardless of whether the sociologist evaluates a belief as true or rational, or as false and irrational, he must search for the causes of its credibility. Is [a belief] enjoined by the authorities of the society? Is it transmitted by established institutions of socialization or supported by accepted agencies of social control? Is it bound up with patterns of vested interest? ... All of these questions ... should be answered without regard to the status of the belief as it is judged and evaluated by the sociologist's own standards.[7]

Instead of looking for fixed, universal laws of logic guaranteeing the connection of particular phenomena to general concepts, sociologists of knowledge seek the learned, contingent principles of thought actually used by human groups and refuse the temptation to judge them against rules of their own. To investigate signification and justification as social practices, we have to explain why cognitive approaches differ without appealing to the "facts" of the world. Barnes and Bloor put the point eloquently: "the general conclusion is that reality is, after all, a common factor in all the vastly different cognitive responses that men produce to it. Being a common factor it is not a promising candidate to field as an explanation of that variation."[8] This returns us to the Foucaultian argument of chapter 1: power is continuously constructed in very ordinary interactions via the production and circulation of discourse. The "micropolitics" of power can appear in the construction of "regimes of truth" as well as in the exercise of force. The cultural arena of *knowledge* is an arena of power as well. (Power, as Foucault argued, is productive as well as repressive.)

To insist on the idea of a cultural politics is to claim that political interactions—the maintenance and the shifting of power among groups and individuals—occur in the representation of situations as well as in the situations themselves. In fact, representations are generally inseparable from the situations they describe, largely because representation is itself a form of action.[9] The institution of slavery, to take an obviously political example, depended on the economic demand for cheap labor and the material control of white masters over black slaves, exerted through physical means. But it could not have existed without certain semiotic "means"—that is, representations—as well, such as the roles of slave and master, the ideological justifications offered to explain the suitability of each race to its role, laws governing the legal institution of slavery, and so on. Or consider contemporary debates about abortion, where the "facts" of the situation—the number of fetal brain cells, putative consciousness or lack thereof, even viability outside the womb—do not, and cannot even in principle, present an unambiguous picture with regard to choices between the competing claims of mother and fetus. Even to discuss fetal "claims" is to accept one of several plausible metaphorical representations of the situation. Larger cultural systems such as class, race, and religion (with their clashing ideologies of freedom, family, and motherhood) play crucial roles in how different groups represent that situation.

Thus material culture and representation maintain a reciprocal, not a reflective, relationship. Representations shape material culture as well as the reverse. Representation can be a political act, and its political significance increases as a given representation becomes embedded in ordinary language—or in scientific discourse, which in the modern world serves as the paradigm of rational "insight and argument."

The idea of a politics of culture and representation will be useful when we turn to the historical construction of certain metaphors for the human mind, especially the metaphor of the mind as a kind of computer. Computer metaphors were constructed and elaborated by scientists working in relatively unconstrained laboratory situations. Here we will find accounts neither of scientists coerced into producing particular theories, nor of theories that merely reflect underlying relations of production. Instead we will encounter practices of what Donna Haraway has called "constrained and contested story-telling." Such story-telling "grows from and enables concrete ways of life," in Haraway's words; scientific "theories are accounts *of* and *for* specific kinds of lives."[10] Power and knowledge, in French *pouvoir/savoir*—being able to do something and knowing how to do it—arise together in these practices and ways of life.

Metaphor is, of course, one mode of representation. It is also a key discursive process, one that relates concepts to each other through shared experiences. The mechanics of metaphor in language is our next point of exploration.

The Power of Metaphor

The linguist George Lakoff, working with the philosopher Mark Johnson, has been the foremost recent exponent of the view that language and thought are essentially structured by metaphor.[11] Lakoff and Johnson's work has shown that, far from being a literary or occasional phenomenon, metaphor is ubiquitous in human language. Furthermore, metaphors form coherent systems that reflect the coherence of certain aspects of experience. At the same time, the elaboration of coherent metaphorical schemes and the development of correspondences among schemes can itself structure experience. A unique feature of their theory is that it does not picture conceptual structure as a reflective representation of external reality. Instead, it views concepts as essentially structured by human life and action, and especially by the human body in its interaction with the world.

Lakoff and Johnson emphasize the way in which aspects of embodiment—the physical experience of having a body and moving around in the world—form the basis for innumerable metaphors in ordinary language. For example, the metaphor HAPPY IS UP, SAD IS DOWN appears in expressions like these:

I'm feeling *up*.

That *boosted* my spirits.

My spirits *rose*.

Thinking about her always gives me a *lift*.

I'm feeling *down*.

I'm *depressed*.

I *fell* into a *depression*.

My spirits *sank*.[12]

They suggest a basis for these metaphors in the physical experience that "drooping posture typically goes along with sadness and depression, erect posture with a positive emotional state." However, the conclusion is not—as might be assumed under an objectivist philosophical scheme—that physical experience determines the metaphorical expression of more abstract terms. Rather, individual metaphors are frequently chosen because they cohere with others to form a larger system.

For example, happiness also tends to correlate physically with a smile and a general feeling of expansiveness. This could in principle form the basis for a metaphor HAPPY IS WIDE; SAD IS NARROW. And in fact there are minor metaphorical expressions, like "I'm feeling expansive," that pick out a different aspect of happiness than "I'm feeling up" does. But the major metaphor in our culture is HAPPY IS UP; there is a reason why we speak of the height of ecstasy rather than the breadth of ecstasy. HAPPY IS UP is maximally coherent with [other core metaphors such as] GOOD IS UP, HEALTHY IS UP, etc.[13]

Thus, while many metaphors ultimately have a physical, experiential basis, just which *aspects* of experience form core structures of metaphorical systems is heavily influenced by other factors, especially culture.

Lakoff and Johnson emphasize that the metaphorical structuring of concepts is always only partial. This means that while metaphors reveal hidden aspects of reality (by providing a frame that highlights them), they also always hide other features. Perhaps the best example comes from the CONDUIT metaphor, in which people discuss the nature and functioning of language using the complex metaphor

IDEAS (OR MEANINGS) ARE OBJECTS.
LINGUISTIC EXPRESSIONS ARE CONTAINERS.
COMMUNICATION IS SENDING.

Language is thus conceived as a conduit through which objects (ideas), packaged in containers (words and phrases), are transferred from a sender to a receiver. CONDUIT metaphors are extremely common:

It's hard to *get that idea across* to him.
Your *reasons came through* to us.
It's difficult to *put my ideas into words*.
Try to *pack more thought into fewer words*.
Your *words* seem *hollow*.[14]

As Lakoff and Johnson note, without prompting many of these expressions do not seem metaphorical at all. For them, this indicates the depth to which the CONDUIT metaphor fundamentally structures our thinking about language. However, the metaphor has a number of "entailments," or metaphorical and logical consequences, that when examined more closely reveal what the CONDUIT metaphor hides. Most notably, the characterization of meanings as objects and words as containers for meanings entails that meanings exist outside contexts and beyond the personal intentions of speakers. This picture of language works well much of the time, when meaning is unambiguous and context doesn't matter. Yet there are many circumstances where context does matter and where no clear meaning can be said to exist apart from the intentions of the speaker within a shared context of use. It was the dominance of exactly this CONDUIT metaphor that prompted Ludwig Wittgenstein's repudiation of the logical-atomist view of language as a picture of reality and forced him to the conclusion that meaning must be understood as embodied in the *use* of language—in a context, for a purpose.

On this theory, certain aspects of human experience actually require metaphorical structuration. These are domains that lack the relatively clear, definite, and readily shared structures we encounter in physical experience. While Lakoff and Johnson hold that physical experience grounds metaphorical structuration, they do not conclude that physical experience is therefore more basic in some ontological or epistemological sense. Instead, it is simply more common and convenient to conceptualize the (less clearly delineated) nonphysical in terms of the (more clearly delineated) physical.[15] More abstract domains of experience such as the mental and the emotional, which initially acquire structure from physical-experiential metaphors, may come in turn to serve as metaphors for each other and for physical experience, though this is less typical.

Important evidence for Lakoff and Johnson's theory of metaphor is that many metaphors exist not in isolation, but are elaborated into complex systems whose coherence emerges from the experiential coherence of the source domain of the comparison. For example, the major metaphor AN ARGUMENT IS A JOURNEY, expressed in such phrases as "we have *arrived at* a disturbing conclusion," relies upon the common human experience of taking actual journeys. Other elements of the journey experience (for example, that a journey defines a path, and the path of a journey is a surface) lead to other ways of using the metaphor, such as "we have *covered a lot of ground* in this argument," since the metaphorical logic entails that THE PATH OF AN ARGUMENT IS A SURFACE.[16] Furthermore, metaphorical systems frequently interact with each other. Thus arguments are also metaphorically structured in terms of BUILDINGS (to *buttress* an argument) and CONTAINERS (to have the right ideas *in* an argument). Elaborations of these metaphors lead to areas of compatibility (as in the expression "as we *go along*, we will *go into* these issues *in depth*," which contains elements of both the JOURNEY metaphor and the CONTAINER metaphor), as well as to areas of difference. For example, it would be senseless to speak of *"buttressing* an argument *in depth,"* since the aspect of the BUILDING metaphor picked out by "buttress" and the aspect of the CONTAINER metaphor picked out by "deep" do not cohere.

Lakoff and Johnson conclude that metaphors have two crucial consequences for the understanding and use of concepts. First, they provide concepts with necessary systematic structure, in a fully Wittgensteinian sense. Second, all metaphors characterize only certain aspects of the concept they delineate. Partially coherent systems of different metaphors for the same concept help cover the gaps left by any given metaphor, but even a well-elaborated system typically obscures some aspects of the experience it describes.

The ultimate conclusion of the Lakoff-Johnson theory of metaphor is that *experience itself* has metaphorical components.

Some natural kinds of experience [i.e., domains of experience that form a single gestalt] are partly metaphorical in nature, since metaphor plays an essential role in characterizing the structure of the experience. Argument is an obvious example, since experiencing certain activities of talking and listening as an argument partly requires the structure given to the concept "argument" by the ARGUMENT IS WAR metaphor. The experience of time is a natural kind of experience that is understood almost entirely in metaphorical terms (via the spatialization of time and the TIME IS A MOVING OBJECT and TIME IS MONEY metaphors).[17]

Metaphor, then, is far more than a rhetorical device. It mediates the relationships among language, thought, and experience. The elaboration of metaphorical schemes is both a central function and a central method of cultural exchange, and it is based in action and experience.

For my purposes here, the most significant feature of this theory is what it tells us about the political power of metaphor. All metaphors are political in the weak sense that they focus attention on some aspects of a situation or experience at the expense of others. A metaphor channels thought and creates a coherent scheme of significance not only by making certain features central, but by establishing a set of connections with other metaphors and openings toward further elaboration. This means that metaphor is not merely descriptive, but also prescriptive. Often, if not always, our representations of situations contain within them indications of appropriate responses and attitudes.

This, in itself, is intriguing but trivial, if only because ordinary language is riddled with metaphorical constructions. It is impossible to notice every aspect of a situation at once, and foolish to treat all aspects as equally important and equally interconnected. Metaphor serves the important communicative purpose of structuring understanding and guiding the limited resources of human attention. Little of interest can be said about the political aspects of metaphor in much of our everyday language, where our focus shifts frequently (and appropriately) depending on our purposes and our partners in conversation.

It is not the sheer fact of metaphorical language, but the larger patterns of metaphor in discourse that are political in a stronger sense. Some metaphors become entrenched so deeply that they guide and direct many other systems of description. These "master tropes"[18] provide what amount to basic structures for thought and experience. They may also actually provide constitutive frameworks for institutions. Some examples are TIME IS MONEY and LABOR IS A RESOURCE, root metaphors in capitalist economic systems but irrelevant, even incomprehensible, in subsistence or barter economies. Another is THE BODY IS A MACHINE, a metaphor crucial to modern Western medical science but altogether absent from many traditional medical systems.

Abstract concepts such as freedom or the mind do not have boundaries as obvious and clearly defined as concrete, experiential concepts such as bodily orientation, containers, fighting, or fever. To a much larger degree they are literally constructed by and within our language about them. Since it is so difficult to talk and think about them directly, and so relatively easy to talk and think about concrete concepts, metaphorical constructions do more than provide a convenient way of understanding a preexistent, Platonic world of the abstract: they play a key role in the construction and use of the concepts themselves. Metaphor is part of the flesh of thought and culture, not merely a thin communicative skin. Therefore the politics of culture is, very largely, a politics of metaphor, and an investigation of metaphor must play an integral role in the full understanding of any cultural object. The mind is such an object, and the computer is such a metaphor.

Computers as Metaphors

Computer metaphors are neither as pervasive nor as obviously politically resonant as some of the others I have discussed. Yet the computer lies at the center of a series of unusually significant discourses about the human mind and about the nature of certain essentially political problems. Since World War II, computer metaphors have been central in the reconstruction of certain conceptual boundaries, both within and outside scientific discourse, such as those between humans and machines, intelligence and intuition, rationality and emotion.

The most famous example of the use of computers as metaphors is the so-called Turing test for machine intelligence discussed in chapter 1. In this test, a human interrogator sits at a terminal connected to two entities, one a computer and the other a person, in another room. The computer is programmed to imitate as closely as possible ordinary human capacities. The interrogator attempts to discover which entity is human by comparing their responses to questions, which may be on any subject and take any form.

Turing meant the test to be taken literally, as a criterion for determining whether a machine could be counted as intelligent. But historically its major effect was to crystallize, in a single image, a metaphorical structure that connects minds with computers via tacit assumptions about both communication and information processing. Under the regime of the test, written natural language must be seen as an adequate representative of human communication. Turing did not require that the computer imitate a human voice or mimic facial expressions, gestures, theatrical displays, laughter, or any of the thousands of other ways humans communicate. What might be called the intelligence of the body—dance, reflex, perception, the manipulation of objects in space as people solve problems, and so on—drops from view as irrelevant. In the same way, what might be called social intelligence—the collective construction of human realities—does not appear in the picture. Indeed, it was precisely because the body and the social world signify humanness directly that Turing proposed the connection via remote terminals.

The Turing test makes the linguistic capacities of the computer stand for the entire range of human thought and behavior. The content of a communication process is thus assumed to be independent of its form; in the same way, the content of intelligent thought is assumed independent of its form. The manipulation of written symbols by computer and human being become processes exactly analogous to, if not identical with, thought. These postulates represent the basic principle of the Turing machine, namely that any precisely specified problem can in principle be solved by a computer.

The Turing test thus uses the computer as a metaphor not only to delineate the nature of intelligence abstracted from any embodiment, but also to describe us to ourselves. It provides a graphic image with which to understand the meaning of human communication and thought. In the test computers serve not only as channels for communication and processors of information, but as metaphors for the structure of communication and the process of information processing. They represent, in a sense, pure subjectivity, abstracted from the physical, experiential, and cultural contexts in which human relations with objects and others ordinarily take place.

Turing conceived his test at a time when computers were still enormous machines understood and used by only a few élite scientists and military personnel. Since then, computers have become far more accessible to a wide range of users. By now, in fact, most of the American middle class has probably had some kind of direct experience with computers in school, at work, or at home. This phenomenon has provided the conditions under which a metaphor can evolve into a living element of language and thought. The computer has become, in Sherry Turkle's eminently useful phrase, an "object to think with."[19]

The experience of having a "mind"—knowledge, perception, consciousness, rationality—is exactly the sort of nonphysical experience whose own structure is too weak to support the demands made upon it by ordinary language, much less by the more rigorous investigations of science. Richard Rorty has argued that Descartes essentially invented the modern concept of "mind" by gluing together a list of heterogeneous elements of thought, action, and experience using an analogy to mathematics. For Descartes, the experience of certainty found in mathematical proof provided the core metaphor by which such diverse phenomena as perception, will, belief, knowledge, denial, love, and imagination could be welded together into a single concept.[20] Reading Rorty's analysis of Descartes through the lens of Lakoff and Johnson, we may conjecture that the Cartesian concept of mind became a problem for philosophy because the mind-as-mathematics metaphor was too weakly structured to bear the burdens of further elaboration. Attempts to extend it failed, and the metaphorical edifice collapsed, piece by piece.

The computer metaphor contributes to the understanding of "mind" its far greater concreteness and vastly more detailed structure. For cognitivism, this metaphor provided a powerful new frame through which groups of previously unrelated phenomena could be viewed as connected, as well as a source of experimental designs. For the eventual subcultures centered around computers, such as hackers and computer scientists, it offered a unified way to grasp life, work, and experience. For the wider world, it eventually came to constitute a cultural background whose terms—like those of psychoanalysis, as Turkle has argued—increasingly pervade the self-understanding of ordinary people.[21]

Entailments of Computer Metaphors
As Lakoff and Johnson have shown, one way to evoke the full range of a metaphor's cultural potentialities is to explore its entailments. What are the entailments of the Turing-test metaphor THE MIND (OR BRAIN) IS A COMPUTER? The most obvious ones are these:

The brain is *hardware*.
The brain is a *rapid, complex calculating machine*.
The brain is made up of *digital switches*.
The mind is *software*.
The mind is a *program* or *set of programs*.
The mind *manipulates symbolic representations*.
The mind is an *information machine*.
Thinking is *computation*.
Perception is *computation*.
Memory is *looking up stored data*.
The function of the mind and brain is *information processing*.

All these claims have in fact been made, in more or less these terms, by cognitivists over the last four decades. They have achieved such currency that some of these ideas, such as the notion that the brain processes information, no longer seem metaphorical at all.[22]

The entailments of the COMPUTER metaphor lead off in a range of directions, some obvious and some less so. For example, the metaphor of the mind as a set of programs, or symbolic instructions that process inputs and control outputs, provides a rich set of analogies that allow us to portray the complex, hidden, abstract processes of thinking and the production of behavior in terms of the relatively simpler and more concrete ones involved in computer programming.

Like much human behavior, most computer programs are not built in or "hard-wired." This implies that behavior and thought patterns can be changed, erased, or replaced. Imperfect computer programs have "bugs"—flawed instructions that cause erratic, unwanted results. Human behavior and thought, too, can "go haywire." The COMPUTER metaphor implies that with diligence "bugs" can be located and corrected. Programs, especially simple ones, are quite rigid, prescribing patterns of action that are not always right for the situations that trigger them. Thus to say that someone "acts like a computer" has the negative connotation that s/he responds in rigidly patterned ways, rather than flexible, appropriate ones.

The computer is most familiar as a calculating machine and a symbol processor. It manifestly does not betray any capacity for emotion or sensitivity to the emotions of human beings. The COMPUTER metaphor also implies, then, that emotion is either irrelevant to the understanding of human thought, or that emotion might somehow be represented as a symbolic process. The computer is a logic machine. Thus the COMPUTER metaphor privileges one mode of human thought at the expense of other, paralogical or tropological modalities.[23] It points toward a reductive explanation of the paralogical, the tropological, and the intuitive in terms of a more rigorous, mathematical or quasi-mathematical logic. In effect, it returns to the Cartesian metaphor of the mind as a mathematical engine, but with a massively elaborated concrete structure that vastly enriches the Cartesian concept.

Other Metaphors for the Mind
Let us compare the entailments of the COMPUTER metaphor with some alternative metaphors current in other epochs.

First, consider the classical animal-machine metaphor, ANIMALS ARE REFLEX MACHINES. If HUMANS ARE ANIMALS as well (a claim that deeply entangles literal and metaphorical connotations), then HUMANS ARE REFLEX MACHINES. This metaphor compares humans to the animals of the paradigmatic behaviorist experiments, such as Pavlov's dogs, Tolman's rats, or B. F. Skinner's pigeons. The Pavlovian picture draws a parallel between the transference of a natural reflex (salivation at the smell of food) onto an arbitrary stimulus (the sound of a bell) and the "mental" process of associating words ("Dinnertime!") with their meanings. For Tolman, the world was a maze much like that navigated by his rats; cognition involved mapping the maze. The Skinnerian picture is similar but places emphasis on conditioned "operant" behavior—semi-random, novel exploratory operations on or in the environment rather than built-in reflexes. Skinner's rats and pigeons learned to press bars or peck at different colors and shapes (operant behavior) in order to receive rewards. For Skinner, human mental processes are essentially operant behaviors (such as babies' babbling) shaped into structured responses (such as adult language) by the differential rewards offered by the environment, including other people; the experience of thought is epiphenomenal.

These metaphors entailed consequences such as the following:

Mental processes are *tacit physical behaviors*.
Mental processes are *controlled* by *the environment*.
Learning is a process of differential *reinforcement*.
Thoughts are *tacit conditioned verbal responses*.

The REFLEX MACHINE metaphor has certain parallels with the COMPUTER metaphor, but it leads in wholly different directions. For example, symbolic activity (such as language, problem-solving, and perception), physical behavior, and emotional responses are all on a par under the REFLEX MACHINE conception. The metaphor directs attention toward the external variables controlling a response rather than toward internal transformations. It suggests that deep insights into human behavior can be gained from the study of animals. The REFLEX MACHINE metaphor directs the experimenter's focus toward how behavior is learned (built up from simple components) rather than toward the structure of (complex) established behavior patterns.[24]

Second, consider a metaphor for the mind drawn from the very different perspective of psychotherapy. Freud founded his system of psychoanalysis upon the metaphor THE MIND IS A HYDRAULIC SYSTEM, a sort of complex, leaky network of plumbing governed by pressures and flows. This entails:

Unconscious thoughts *burst through* or *leak* into consciousness.
Instincts and emotions exert *pressure* on the conscious mind.
Sexual energy *builds up* and must be *released*.

The HYDRAULIC SYSTEM metaphor invites us to view emotion and instinct, rather than rationality or action, as the central features of the mind. A hydraulic mind needs "outlets" for inevitable buildups of pressure. Societies that provide insufficient or ill-designed outlets for their constituents may implode in decadence, war, or internal violence. Individuals may require extensive therapy to be able to "contain" or "divert" their irrational impulses. Civilization itself, as Freud saw it, was a set of structures for diverting the forces of sexuality and aggression into creative channels: "sublimation."[25]

The REFLEX MACHINE and HYDRAULIC SYSTEM metaphors do not simply contradict specific entailments of the COMPUTER metaphor. Rather, each leads off in a different direction. If THE MIND IS A COMPUTER, it may be reprogrammed, while if it is a REFLEX MACHINE, its responses may be modified through new conditioning. While reprogramming and behavior modification are different processes, they have in common the precept of a flexibility of the mental apparatus and the possibility of change and learning. In contrast, the HYDRAULIC SYSTEM model offers the diversion of unchanging instinctual pressures into new channels, rather than wholesale changes in mental structure.

The instinctual sources of psychic energy cannot, on the Freudian view, be altered—only redirected. The analyst provides one vessel into which these flows may be channeled, and psychoanalysis emphasizes the therapeutic relationship as a vehicle for understanding and potential change. The REFLEX MACHINE metaphor concentrates instead on environmental variables as triggers for behavior, suggesting a focus on the social system of rewards as the ultimate "technology of behavior," in Skinner's phrase. Since the notion of operant behavior presumes random creativity, and even reflexes are subject to deliberate restructuring, the REFLEX MACHINE metaphor leads to a view of behavior as infinitely flexible. The COMPUTER metaphor instead draws attention to the internal structure of the mind and its representational schemes. It suggests the possibility of "reprogramming" the mind by setting up new thought patterns or restructuring its "hardware" with drugs, surgery, or implanted microchips. But it also promises to reveal inalterable high-level structures, genetically programmed, such as Chomsky's universal grammar.

Thus the direct entailments of different metaphors for the mind point to radically different sets of questions, ethical positions, and views of human nature. The kinds of questions scientists should ask, the kinds of morality appropriate to the corresponding concepts of human nature, and the sorts of expectations human beings might reasonably have of each other under the three schemes are quite different (although not in every respect). As usual, each metaphor draws attention to certain features of the domain of the mental, obscures others, and through its elaboration may actually generate new forms of experience.

Subject Positions and Cyborg Discourse

Theories of mind are a case of "constrained and contested storytelling" because such theories are necessarily, and simultaneously, representations or constructions of possible subject positions. Whether scientific, quasi-scientific, or popular, theories of mind concern the relation between subject and object: perception, memory, decision-making, motor action. They are also about modalities of intersubjective relations: language, communication, emotion. Just as democratic theory plays a crucial role in constructing political subjectivity, theories of mind are central to the construction of subjectivity more generally. *Like political theories, theories of mind are largely about how people recognize and choose among alternatives.* The phrase "cyborg discourse," introduced in chapter 1, captures the COMPUTER metaphor's creative potential for structuring subjectivity.

In the early 1980s, Sherry Turkle studied Boston-area subcultures centered around computers. Her work documents, in effect, the progressive elaboration of the COMPUTER metaphor among children learning to program computers, video game players, hardware hackers, university-based software hackers, and AI researchers. Against the backdrop of a virtual explosion of computers into popular culture (following the introduction of inexpensive personal computers), Turkle explored how computers provided a new medium for self-understanding.

Mark, one of Turkle's interviewees, was a junior computer-science major at MIT. Through his experience with computers and the concepts of computer science, he gradually developed a detailed and highly sophisticated model of his own mind. Mark's model assumed that the brain was a kind of computer.

"This does not mean that the structure of the brain resembles the architecture of any present-day computer system, but the brain can be modeled using components emulated by modern digital parts. At no time does any part of the brain function in a way that cannot be emulated in digital or analog logic," [Mark says.] ... In Mark's model the computational actors in the brain are simple. Each is a little computer with an even smaller program, and each "knows only one thought." ... [A]ll of the processors have the same status: they are "observers" at a long trough. Everything that appears in the trough can be seen simultaneously by all the observers at every point along it. The trough with its observers is a multiprocessing computer system.[26]

Mark's model describes consciousness as a "passive observer" that sees only some of what gets "dumped" into the trough, and it analogizes the processors to neurons. He concludes that consciousness is epiphenomenal and, therefore, that the notion of free will is an illusion. In his words,

> You think you're making a decision, but are you really? For instance, when you have a creative idea, what happens? All of a sudden, you think of something. Right? Wrong. You didn't think of it. It just filtered through—the consciousness processor just sits there and watches this cacophony of other processors yelling onto the bus and skims off the top what he thinks is the most important thing, one thing at a time. A creative idea just means that one of the processors made a link between two unassociated things because he thought they were related.[27]

The upshot, Turkle concludes, is that "creativity, individual responsibility, free will, and emotion [are] all ... dissolved" in Mark's picture of human nature. The COMPUTER metaphor plays an important and direct role in Mark's self-understanding and in his ethical system, as well as in his more speculative thinking about the nature of the mind.

Mark understands himself as a Turing machine, a mechanical mind stripped of precisely those qualities Turing's test was designed to make irrelevant to the understanding of mind and intelligence. As a political subject, he is a totally rational decision-maker. But freedom in the Hegelian sense is not a value for him; he does not require recognition as a Kantian chooser, since he does not understand his choices as being up to him. Seeing himself as a system composed of many-leveled subsystems, Mark is a true cyborg subject. Political unfreedom would be an inefficient use of his "processors," but it would not offend against any sense of inherent worth.

Mark's is only one of many possible versions of the self to draw upon the COMPUTER metaphor. However, a variety of observers suggest that similar constructions of selfhood are common among subcultures centered around computers.[28] By exploring how such subcultures use the COMPUTER metaphor in articulating key cultural formations such as gender and science, we can discern some of the salient patterns in cyborg subjectivity.

In a key insight, Turkle discovered two very different approaches among children learning to program in the LOGO language at a private school. Those she named "hard masters" employed a linear style that depends on planning, advance conceptualization, and precise technical skills, while "soft masters" relied upon a less structured system of gradual evolution, interaction, and intuition. In her words,

Hard mastery is the imposition of will over the machine through the implementation of a plan. A program is the instrument for premeditated control. Getting the program to work is more like getting to say one's piece than allowing ideas to emerge in the give-and-take of conversation.... [T]he goal is always getting the program to realize the plan. Soft mastery is more interactive... the mastery of the artist: try this, wait for a response, try something else, let the overall shape emerge from an interaction with the medium. It is more like a conversation than a monologue.[29]

Turkle thus establishes a dualism at the heart of cyborg subjectivity, a "hard" self and a "soft" one. This is an evocative, problematic, and paradoxical dichotomy.

"Hard" and "soft" are exceptionally rich words that cover a variety of overlapping conceptual fields. "Hard," according to the *Random House Dictionary*, includes among its fifty-four uses the meanings not soft; difficult; troublesome; requiring effort, energy, and persistence; bad; harsh or severe, unfriendly; and sternly realistic, dispassionate. "Soft" may mean not hard, easily penetrated; smooth and agreeable to touch; pleasant; gentle, warm-hearted, compassionate; responsive or sympathetic to the feelings of others; sentimental; not strong, delicate (the example given is "He was too soft for the Marines"); easy; submissive. The words also have obvious sexual connotations.

What is it about being a Marine that requires a sufficient "hardness" in a man? How is that kind of hardness linked to the hardness of control, planning, and the "imposition of will over [a] machine"? How is a gentle, delicate softness linked to the ideological role of women in war, and how does it connect to the conversation-like artistry in the interactive approach of "soft mastery"? For though Turkle provides both male and female exemplars of each style, she admits that boys tend to opt for the "hard" approach while girls prefer the "soft." How does the computer become a foil in the politics of subjectivity?

The "hard" master of computers is a subject whose major cognitive structures are preconceived plans, specific goals, formalisms, and abstractions, who has little use for spontaneity, trial-and-error, unplanned discovery, vaguely defined ends, or informality. This is also American culture's prevalent image of scientists, generically portrayed as disciplined thinkers who deploy long chains of logical and mathematical reasoning to arrive at their subtle, powerful understanding of nature's ways. Men, too, are supposedly tough-minded, rational, unswayed by emotion—good with maps and mathematics—while women supposedly outdo them in more "intuitive" skills such as nursing and child care. Computers, scientists, and men are "hard" subjects; children, nurses, and women are "soft" ones.

In practice, of course, the image is false. Scientists of both sexes, including computer scientists, may experience their work as a visceral, creative, social, and unpredictable enterprise. For many, formal, linear thinking is only a part of a larger process involving many kinds of thought and practice. Many women are fully capable of all the "hard" tasks of science and computer work; equally, many men are not without a certain "softness."

Regardless of its truth, however, the hard/soft split plays a major ideological role. It is reinforced by the popular media and by professional cultural practices such as the highly impersonal style of scientific journals and textbooks.[30] It can also be found reduplicated within science (there are "hard" sciences like physics and "soft" ones like psychology). Even within disciplines, there are "hard" and "soft" approaches. Contests for legitimacy are staged between those who deploy a "hard" cognitive approach, using a technical language, mathematical or logical formalisms, a technical apparatus (including computers), and the other trappings of the hard style, and those who rely more upon the "soft" resources of nontechnical language, broad heuristics, and nontechnical methods such as clinical practice. Often, experimentalism is not the key distinction; indeed, part of the role of the hard/soft metaphor is to distinguish what *counts* as an "experiment" from mere clinical observation or interaction.

Computer scientists have usually enjoyed a mystique of hard mastery, in this sense. Computer work is associated with vast mental powers, a kind of genius with formalisms akin to that of the mathematician, and an otherworldliness connected with the classical picture of the scientist. Computers symbolize unblinking precision, calculative power, and the ability to synthesize massive quantities of data. At the same time they stand for the rigidities of pure logic and the impersonality of centralized corporations and governments. This reputation was one source of their authority in the construction of closed-world discourse.

These semiotic dimensions of computers have much to do with how they function—with what their use requires of their users. For the COMPUTER metaphor is articulated not only at the level of broad comparisons, but in the detailed practice of computer work, where the machines have come to serve as a medium for thought, like English or drafting tools. Turkle's phrase "objects to think with" captures their triple status as tools, metaphors, and domains of experience—supports, in the Foucaultian sense, for cyborg discourse.

"Objects to Think With"

One way to think with a computer—in their first three decades, almost the only way—is to learn its "language." All existing computer languages consist of a relatively small vocabulary of admissible symbols (from several hundred to a few thousand in the most sophisticated) and a set of simple but powerful rules for combining those symbols to form sequential lists of machine instructions.[31] They are much simpler and more restrictive than any human language, especially insofar as human language tolerates—indeed, relies on—ambiguity and imprecision.[32] In their half-century of evolution, computer languages have come far closer to approximating human language in vocabulary and grammar. But mistakes that would be entirely trivial in an exchange between humans, such as a misspelled word or misplaced punctuation mark, still routinely cause catastrophic failures in computer programs. Attempts to get computers to understand unrestricted spoken or written English have been plagued by precisely this problem.[33] Thus, if a language is a medium for thought, the kind of thinking computer languages facilitate is quite different from the reasoning processes of everyday life.

Computer scientist Jonathan Jacky has observed that each computer language tends to encourage a particular programming style, as do subcultures associated with each one. Certain languages seem more likely to lead toward more organic, "soft" methods of programming than "harder," more structured languages. Thus the Pascal language was deliberately designed to promote a highly structured, "self-documenting" approach to programming, while AI languages such as LISP seem to breed a murkier, more intuitive and interactive approach, with unexpected results a part of the goal.[34] This corresponds to computer programmers' stereotypes of each other—LISP programmers as hackers, sloppy but artistic visionaries; Pascal programmers as precise but uncreative formalists, self-described "software engineers." The ongoing invention and spread of new computer languages is a symptom of the search not only for convenience of interaction, but for styles of thinking—subject positions—congenial to different kinds of users and their projects.[35]

Yet despite these vivid differences, all computer programs work in essentially the same way. They manipulate symbols according to well-defined, sequentially executed rules to achieve some desired transformation of input symbols into output symbols. Rule-oriented, abstract games such as checkers or chess also have this structure. As a result, *all computer programming, in any language, is gamelike*.[36]

Many writers have suggested that "hard" modes of thought, such as highly developed procedural planning, mathematical logic, and formal gaming, seem more familiar and friendly to most men than to most women.[37] They fit well with a culturally defined "masculine" conception of knowledge as an objective, achieved state rather than an ongoing, intersubjective process, and with a "masculine" morality built on abstract principles rather than shifting, contextually specific, emotionally complex relationships.[38] The similarity of such modes to mathematics identifies them with the Western tradition of rationality itself, going back to the ancient Greeks. In this "rationalistic tradition," as Winograd and Flores call it,

emphasis is placed upon the formulation of systematic logical rules that can be used to draw conclusions. Situations are characterized in terms of identifiable objects with well-defined properties. General rules that apply to situations in terms of these objects or properties are developed which, when logically applied, generate conclusions about appropriate courses of action. Validity is assessed in terms of internal coherence and consistency, while questions concerning the correspondence of real-world situations with formal representations of objects and properties, and the acquisition of knowledge about general rules, are bracketed.[39]

Cognitive "hardness"—the ideal of the rationalistic tradition—has a very long history of evaluation as "masculine," and has often served as a kind of master trope in the construction of gender, politics, and science.[40] Computers embody the values of this tradition and its intimate association with science. The overwhelming perception (if not the reality) is that successful programming both demands and helps engender (as it were) a "hard" style of thought. Furthermore, the demands of commercial efficiency seemingly preclude the more unstructured style, although they cannot eliminate it entirely.[41]

Another kind of link between gender identity, science, and computing has to do with the emotional structure of computer work. Programming can generate a strong sensation of power and control. To those who master the required skills of precision, planning, and calculation, the computer becomes an extremely malleable device. Tremendously sophisticated kinds of play are possible, as well as vast powers to transform, refine, and produce information. Control of myriad complex systems, such as machines, robots, factories, and traffic flows, becomes possible on a new scale; all this is "hard" and powerful work. Members of many computer subcultures—hackers, networking enthusiasts, video game addicts—become mesmerized by what Turkle calls the computer's "holding power."[42] The phrase aptly describes the computer's ability to fascinate, to command a user's attention for long periods, to involve him or her personally. Many describe a kind of blending of self and machine, an expanded subjectivity that extends deep into the computer.

What gives the computer this "holding power," and what makes it unique among formal systems, are the simulated worlds within the machine: what AI programmers of the 1970s began to call "microworlds,"[43] naming computer simulations of partial, internally consistent but externally incomplete domains. Every microworld has a unique ontological and epistemological structure, simpler than those of the world it represents.[44] Computer programs are thus intellectually useful and emotionally appealing for the same reason: they create worlds without irrelevant or unwanted complexity.

In the microworld, the power of the programmer is absolute. Computerized microworlds have a special attraction in their depth, complexity, and implacable demands for precision. The programmer is omnipotent but not necessarily omniscient, since highly complex programs can generate totally unanticipated results. Comprehending even a simple program, especially if it contains subtle "bugs," may require extraordinary expertise and ingenuity. This makes the microworld exceptionally interesting as an imaginative domain, a make-believe world with powers of its own. For men, to whom power is an icon of identity and an index of success, a microworld can become a challenging arena for an adult quest for power and control.[45]

Human relationships can be vague, shifting, irrational, emotional, and difficult to control. With a "hard" formalized system of known rules, operating within the separate reality of a microworld, one can have complexity and security at once: the score can always be calculated; sudden changes of emotional origin do not occur. Things make sense in a way human intersubjectivity cannot.[46] Turkle notes that in the cultural environment of MIT, computer science majors of the early 1980s were veritable outcasts, the loners and "lusers" (an in-group pun on computer "users") in a culture of loners. Levy and Weizenbaum describe hacker culture, from the 1960s to the present day, in similar terms.[47] Certainly women can be loners and outcasts. But male gender identity is *based* on emotional isolation, from the demands for competitive achievement at others' expense through the systematic repression of means of emotional release (especially of grief and fear) to the organized violence at the center of the masculine gender role.[48] It seems likely that many men choosing engineering careers replace missing human intimacy with what are for them empowering, because fully "rational" and controlled, relationships with complex machines.

What all this means is that the experience of the computer as a second self *is the experience of the closed world of a rule-based game*. The second self computer users find within the machine is, in general, a "hard," quasi-scientific, male self, an experience of reality in the terms of closed-world discourse. The disembodiment of subjects operating inside the computer has sometimes opened the possibility of new articulations of gender, age, race, and other identities classically inscribed in the human body (as others have argued).[49] Nevertheless, cyborg subjectivity during the Cold War tended overwhelmingly to reinscribe the rationalistic male identity on its new electronic surfaces. Thus the cyborg—as both experience and theory, subject position and objective description—is a profoundly and inherently political identity.

In the following chapters I explore the history and political context of cyborg discourse. My ultimate argument will be that just as political theory has played a crucial part in constructing political subjects, cognitive theories and computing machines assisted in constructing the subjects who inhabited the electronic battlefields of global cold war.

Interpreting human minds as information-processing machines, cyborg discourse created subject positions within a political world enclosed by computer simulation and control. Cyborg discourse collaborated with closed-world discourse on a technical level, generating techniques and theories of human-machine integration while developing the long-term possibility of total automation via artificial intelligence. At the same time, it collaborated in the creation of powerful closed-world metaphors, analyzing the mind as a closed control system subject to technical manipulation. Cyborg discourse integrated experience and action at the level of individuals with the technology and politics of global war.

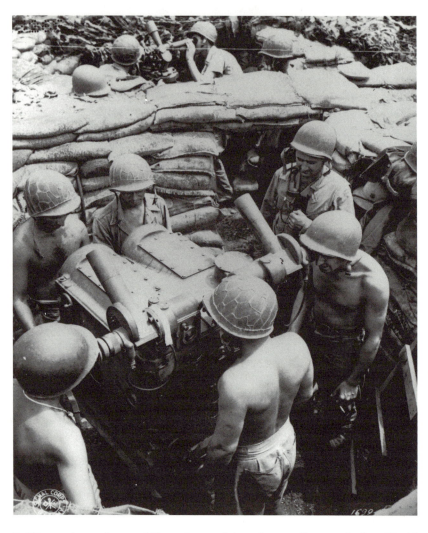

Soldiers using a Sperry M7 analog antiaircraft gun director during World War II. Courtesy Sperry Corporation Collection and the Hagley Museum and Library.

6
The Machine in the Middle: Cybernetic Psychology and World War II

At the Hixon Symposium on Cerebral Mechanisms in Behavior, held at the California Institute of Technology in 1948, the neuropsychiatrist Warren McCulloch presented a paper that began:

> As the industrial revolution concludes in bigger and better bombs, an intellectual revolution opens with bigger and better robots. The former revolution replaced muscles by engines and was limited by the law of the conservation of energy, or of mass-energy. The new revolution threatens us, the thinkers, with technological unemployment, for it will replace brains with machines limited by the law that entropy never decreases. These machines, whose evolution competition will compel us to foster, raise the appropriate practical question: "Why is the mind in the head?"[1]

This problem became "practical" during World War II, when the human being as object of psychological knowledge began to be seen as a servomechanism or analog computer. New forms of technological power based on the amplification and insertion of soldiers' bodies inside electromechanical systems produced new ways of approaching the nature of the mind and led to the construction of new kinds of biopsychological explanatory spaces. These new tools and metaphors rejuvenated human experimental psychology, involving academic psychologists in practical design projects as consultants on the "machine in the middle" of complex human-machine systems.

Psychology as Power/Knowledge

Though it lacks the public visibility of psychotherapy and psychiatry, human experimental psychology is just as tightly interwoven with the fabric of sociopolitical meanings and practices. Psychology is the discipline that constructs and maintains the human individual as an object of scientific knowledge. Contests for models, metaphors, research programs, and standards of explanation in theoretical psychology represent simultaneous struggles for pictures of "human nature" and norms of behavior. The human natural-technical objects[2] constructed by psychology find uses far beyond the restricted discourse of the scientific community. They are incorporated into clinical therapeutic apparatuses, integrated into ergonomic design, used to justify social and political practices, and absorbed by popular mythology.

Psychological theories have often backed interventions that have been profoundly and directly political in nature. In *Civilization and Its Discontents*, Freud wrote that women were unsuited to public life, since "the work of civilization has become increasingly the business of men; it confronts them with ever more difficult tasks and compels them to carry out instinctual sublimations of which women are very little capable."[3] Freud's theories and others like them supported clinical practices that focused on reconciling women to their primal need for a penis and their consequent "natural" dependence on men. Various theories of human intelligence, in the late nineteenth and early twentieth centuries, purported to prove that nonwhite and lower-class humans were inherently less well endowed than Caucasian males. The resulting technologies—"intelligence tests"—in turn reinforced repressive social practices and ideologies. As a case in point, during the successful passage of the anti-ethnic Immigration Restriction Act of 1924, American politicians frequently invoked World War I Army IQ tests, which "proved" that half the tested population had a "mental age" of under thirteen.[4] Foucault has shown how theories and practices of clinical psychiatry and penology, by changing the focus of attention from the bodies to the souls and minds of insane people and prisoners, both represented and incited a spreading "normalization" in nineteenth-century France.[5]

Psychology—both clinical practice and academic theory—constitutes a potent form of Foucaultian power/knowledge: a discourse of truth that maintains "a circular relation with systems of power that produce and sustain it, and to effects of power that it induces and that extend it."[6] Knowledge of the mind as an information-processing device developed in just such a "circular relation" to high-technology military power. The military organization of science during and after World War II played a crucial role in creating conditions under which fruitful encounters among psychology, computer science, mathematics, and linguistics could occur.

Psychology as an academic discipline profited enormously from the war effort, gaining visibility, legitimacy, and funding from its war work on such issues as propaganda analysis, morale-building, and psychological warfare. Its ranks swelled dramatically during and after the war: between the war's beginning and its end, the American Psychological Association's membership grew from 2600 to 4000. By 1960, it counted over 12,000 members, almost five times its prewar peak.[7] As with many other scientific disciplines, during the war the majority of psychologists worked on war-related issues. The federal government directly employed about half of all professional psychologists, either on research contracts, in military service, or in civilian agencies. Many others volunteered time, in teaching, research, or service, toward filling the country's wartime needs.[8]

The work of many psychologists on essentially military problems during the war changed the direction of their interests and the scope of their social and professional relationships. James Capshew, in a detailed survey of psychology in the war years, has concluded that

World War II [was] the most important event in the social history of American psychology.... It aided the expansion of psychology beyond its academic base into professional service roles. As psychologists became involved in war work their research and practice was overwhelmingly reoriented toward applied science and technology. Relationships were forged with new patrons in military and civilian government agencies which provided markets for psychological expertise. Psychologists exploited these new markets by promoting themselves as experts in all aspects of the "human factor" in warfare, ranging from the selection, training, and rehabilitation of soldiers, to propaganda and psychological warfare, to the design of equipment.[9]

Yale psychologist Seymour Sarason concurs with Capshew's analysis and places the change in the context of wider transformations throughout the social sciences:

After World War II the social sciences experienced remarkable growth in terms of numbers, funding, prestige and influence in the halls of public and private power. Social scientists were cocky and confident ... sociologists [were] enamored with grand, abstract theories of the structure and dynamics of society ... psychologists promis[ed] much about their capacity to fathom the basic laws of development and behavior, to prevent individual abnormalities and miseries.... Anthropologists, who became instantly valuable to the government during World War II, became even more so with the war's end, when the nation emerged as the dominant military-social-political force in the world ... and took on administrative supervision of diverse peoples and cultures. Before World War II the social sciences were, except for economics, university-based disciplines having, for all practical purposes, no ties with the political system. World War II changed all that; social scientists became needed, and they wanted to be needed.[10]

The tools of this transformation were political, social, and technological as well as theoretical. World War II-era psychology was *militarized* knowledge production, in the sense that national military goals came to define broad directions for research, particular problems, and the general nature of useful solutions. To say this is simply to state a fact about the postwar condition of American science. I am not arguing that militarized science was somehow illegitimate, nor am I claiming it was ideologically driven—at least no more so than other historical research regimes. Militarization was not a false, but simply a *particular* process of knowledge production.[11] Cyborg discourse emerged from this militarized regime of truth, forming the heterogeneous ensemble of psychological theories, experimental designs, machine interfaces, quasi-intelligent devices, and personal practices that constituted subjectivity in the closed world.

Cognitivism, Behaviorism, and Cybernetics

By the time the cognitive approach to experimental psychology reached maturity in the mid-1960s, it was distinguished by the following characteristics:

1. A fundamental interest in *complex internal "cognitive" processes*, especially perception, memory, mental imagery, and the use of language, and a concurrent emphasis on the use of meaningful data or stimuli in experimentation.

2. Emphasis on *experiments with human subjects*, coupled with the use of subjective reports as experimental data, especially when these could be correlated with observable behavior. Reaction-time experiments were among the most common.

3. A conception of *organisms as active and creative* in the domains just listed. Goals, plans, expectations, and other internal structures were cast in a relation of reciprocal influence with perception, memory, and language.

4. *Rationalist philosophical roots*. Most cognitivists assumed that many mental structures, especially those providing for learning and language, were innate.

5. Some degree of commitment to the *metaphor of computation* in theoretical descriptions. This took such paradigmatic forms as computer modeling, simulation, the use of information and communication theory, and cross-borrowing from artificial intelligence research.
6. Theories attempted to deduce *internal representational systems* from experimental evidence. Cognition became, fundamentally, *symbolic information processing*, or computation on physically represented symbols.[12]

Since cognitivism arose in the midst of a polemical debate with behaviorism, contrasting the two schools briefly will help reveal the significance of cognitive psychology.[13] Most versions of behaviorism were devoted primarily to animal experimentation. They studied only "observable" behavior, usually gross physical movement (actions like bar-pressing and pecking at colored disks). They cast themselves as theories of learning, rather than of thinking and perceiving. They conceived the organism as a relatively passive "black box" subject to a mechanical form of causality from its environment, responding automatically to "stimuli." Behaviorism had empiricist and positivist philosophical roots, often explicit, and it modeled complex processes as aggregates or series of basic, simple "building block" responses. Most of its experimental results came from studies of rats, pigeons, and cats, and its central analogies were to deterministic machines. The major issue for behaviorism, however, was never to specify mechanisms but to predict and control. Applications of behaviorist theory to human beings were strongly oriented toward social control and manipulation, as exemplified in B. F. Skinner's phrase "technology of behavior."[14]

Behaviorism was a mechanistic theory, and psychologists built or designed many machine analogs.[15] Such models were, generally, true machines rather than information processors. (A major exception was the electrical analogy of the telephone switchboard, which was popular from the 1920s on.)[16] Nor were machines of any sort the primary source of models and metaphors for behaviorism; animal behavior filled that role.

Where behaviorism emphasized comparisons between animal and human behavior, and psychoanalysis concentrated on human social and discursive effects, cognitive psychology reconstructed both humans and animals as cybernetic machines and digital computers. The explicit goal of the proto-cognitive theories of the 1940s and 1950s—of what I will call "cybernetic psychology"—was to understand the processes of perception, memory, and language in terms of formalizable transformations of information and feedback circuits or control loops.

Cybernetics: The Behavior of Machines

Norbert Wiener's *Cybernetics* and Claude Shannon's "The Mathematical Theory of Communication," the benchmark documents of information and communication theory, did not appear in print until 1948.[17] By that time, a major symposium at the California Institute of Technology had brought psychologists and cybernetic theorists together.[18] Psychologists had begun working closely with information theorists and electrical engineers as early as 1946, though, as we will see, the groundwork for cybernetic psychology was laid during the war, in the form of social networks, studies of the psychology and psychometrics of communication at Harvard's Psycho-Acoustic Laboratory, and speculation by engineers, mathematicians, neurologists, and a few psychologists in the context of technical military problems. Cybernetic psychology began as an effort to theorize humans as component parts of weapons systems and continued, after the war, to draw crucial models and metaphors from those concerns.

As noted in chapter 2, the antiaircraft gunnery problems of World War II created a need for servomechanisms that could accurately predict an aircraft's future position, allowing automatic aiming of weapons and reducing the effects of human error. In 1941 Wiener, then a mathematics professor at MIT, joined a team of scientists at the Radiation Laboratory who were studying this problem under the sponsorship of the NDRC. Working with engineer Julian Bigelow and Mexican neurobiologist Arturo Rosenblueth, Wiener started thinking about how one might predict an airplane's future course from information about its present location and velocity. Out of this work came a highly general statistical theory of prediction based on incomplete information. This was the theory of feedback control, which became the basis for the design of servomechanisms.[19] This theory had a double face: it described not only how mechanisms could be made to predict future position, but how their human controllers could do the same thing and potentially, given enough information about the *human* mechanism, why they failed.

By May 1942 the theory had gained enough substance for Rosenblueth to give an extended presentation to a small interdisciplinary conference, known as the Cerebral Inhibition Meeting, organized by the Josiah Macy Jr. Foundation's medical director, Frank Fremont-Smith. The meeting's two topics were hypnosis and "the physiological mechanism underlying the phenomena of conditioned reflex." It was led by behaviorist psychologist Howard Liddell, "an experimenter on the conditioning and behavior of mammals," and Milton Erickson, a hypnotist.[20] Others attending this meeting by invitation included neurologist Warren McCulloch, anthropologists Gregory Bateson and Margaret Mead, psychologist Lawrence K. Frank, and psychoanalyst Lawrence Kubie. With its promise of a major new interdisciplinary research paradigm, Rosenblueth's presentation galvanized the audience, especially the social scientists. Mead later recalled that she broke one of her teeth during the conference but was so rapt she did not notice until it was over.[21]

The most important element of the Rosenblueth-Wiener-Bigelow theory was the concept of "negative feedback," or circular self-corrective cycles, in which information about the effects of an adjustment to a dynamic system is continuously returned to that system as input and controls further adjustments.[22] In 1943 the three published the landmark article "Behavior, Purpose, and Teleology," which emphasized comparisons between servo devices and the behavior of living organisms guided by sensory perception. Essentially, they described goal-oriented "teleological" behavior as movement controlled by negative feedback.

Three aspects of their servo/organism analogies were important historically in the construction of "cognition" as a cybernetic natural-technical object:

• the specification of a "behaviorist" analysis applicable to both machines and living things,

• the redefinition of psychological and philosophical concepts in the terminology of communications engineering, and

• analyses of humans as components of weapons systems as a central source of analogies.

Rosenblueth, Wiener, and Bigelow contrasted behaviorism with "functionalism." They defined the former as the study of input-output relationships in abstraction from the (functionalist) internal characteristics of the entity under examination.[23] While the proto-cyberneticians had no quarrel of principle with the "functionalist alternative," they chose the behaviorist approach partly because it would allow them to apply mathematical formalisms from their radar tracking work. Similarly, Kenneth Craik, a British psychologist who also worked on the radar tracking problem, wrote near the end of the war that "the human operator [of tracking devices] behaves basically as an intermittent correction servo" and suggested that operator tracking errors could be described by the same equations used for the periodicity of the tracking servomechanisms.[24] Such comparisons became important metaphors in postwar psychological research as well as in cybernetics' canonical origin story.[25]

Cybernetic psychology was first introduced, perhaps ironically, as an extension and vindication of behaviorism. Wiener, who built the word from a Greek root meaning "steersman," intended cybernetics to encompass "control and communication in the animal and the machine."[26] As a general mathematical theory of self-regulating mechanisms, cybernetics would transcend the boundary between machines and organisms. It would do this not by rejecting concepts of purposes, goals, and will (as in behaviorist psychology), but by *expanding the category of "machines,"* via the concept of feedback, to include these notions. Thus even in 1943, Rosenblueth et al. claimed that "a uniform behavioristic analysis is applicable to both machines and living organisms, regardless of the complexity of the behavior.... The broad classes of behavior are the same in machines and in living organisms."[27]

Typically, psychological behaviorists described the general form of their research problems as the discovery of a function f relating a stimulus S to some response R, of the form $R = f(S)$.[28] Adequate definitions of S and R presented a difficulty that was usually resolved by adopting extremely general classifications. Stimuli were "any portion of the environment to which the organism is exposed under uniform conditions" and were operationally defined in experimental situations: "the number of different S's said to be present... will depend upon the number of independent experimental operations." Similarly, "any movement or sequence of movements may be analyzed out of behavior and treated as a 'response'." In theory, then, stimulus and response existed as organized units only in virtue of choices of the experimenter. In themselves, they were meaningless; and this implied that any response could be conditioned to any given stimulus. (Of course, real behaviorist experiments never respected this impossible principle but divided stimuli and responses along "obvious" lines of meaningful connection.)[29]

In "Behavior, Purpose, and Teleology," Rosenblueth et al. gave a superficially similar description of organism/environment relations in engineering terms. For stimuli, the cyberneticians offered "input" as "any event external to the object that modifies this object in any manner." Responses become "output": "any change produced in the surroundings by the object." Behavior itself was defined as "any modification of an object, detectable externally."[30] "Input" and "output," originally electrical engineering terms for currents entering and exiting circuits, had been adopted analogically by communications engineers to describe the entry and exit versions of *messages* in a channel. For Rosenblueth et al., behavior now became any transformation of energy *or information* from the environment.[31] The possible transformations were hierarchically distributed as shown in figure 6.1. A cue ball moving across a billiard table typifies passive behavior; active behavior requires some provision of output energy by the object itself. The most interesting classes of behavior involved feedback, which utilizes energy returned to the system as information. Prediction and control "behavior" topped this hierarchy.

The proto-cyberneticians' ultimate concern was to explain human voluntary activity. "The basis of the concept of purpose," they wrote, "is the awareness of 'voluntary activity'. . . . When we perform a voluntary action, what we select voluntarily is a specific purpose, not a specific movement." The movements that then accomplish the selected

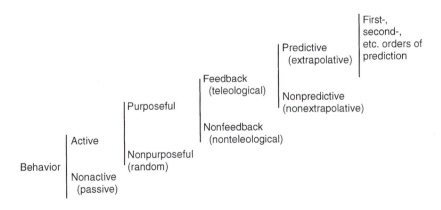

Figure 6.1
A cybernetic classification of behavior. From Arturo Rosenblueth, Norbert Wiener, and Julian Bigelow, "Behavior, Purpose and Teleology," *Philosophy of Science*, Vol. 10 (1943), 21.

purposes are mechanical, directed by feedback of "signals" from the chosen goal. This process parallels the functioning of servo-controlled tracking devices, a prime example being the "torpedo with a target-seeking mechanism," said to exhibit "intrinsic purposeful activity."[32]

"Goal states" explained the organization of behavior and therefore defined meaningful units in advance of experimentation (in contrast to behaviorist dogma). The cybernetic model asserted that goals—including human goals and purposes—could be described within a formalizable, probabilistic, biophysical explanatory space, as direction "to a final condition in which the behaving object reaches a definite correlation in time or in space with respect to another object or event," a concept very different from an Aristotelian *telos*. What is involved in the practice of definition and "selection" of goals, something torpedoes and guided missiles do not do for themselves, was not discussed. But the cybernetic model would ultimately, Rosenblueth et al. believed, explain even these phenomena.[33]

Behaviorist psychologists had constructed the field of the psychological by elevating the unavailability of design-level descriptions to a methodological principle. The emerging cybernetic psychology was revolutionary because the tools it provided could be turned equally to the fully mathematized, behaviorist input/output description and the functionalist, internal-process description. Cybernetic theory gave machine metaphors a new theoretical foundation, creating a new technical terminology and system of quantification—information theory—for describing flexible, self-directed behavior in both machines and minds.

Information theory is part of a general class of mathematical models (including the logical architectures of computers, Turing machines, and the theory of games) that are sometimes called *formal machines*. Such models delimit an ordered set of operations for transforming data; they may, but need not, describe actual machines. "Behavior, Purpose, and Teleology" used formal machines to initiate the development of a new standard of psychological explanation, one I will call *formal/ mechanical modeling*. The terminology is important because cybernetic psychology was *not* simply another form of mathematical psychology. Rather than simply describing observed behavior mathematically, the cyberneticians would treat mathematical models, physical machines, information processing devices, and data from human and animal experiments within a single framework. Unlike "behavior," "association of ideas," "consciousness," and other previous psychological objects, control and communication were *computational* processes susceptible to modeling in terms of devices and formal structures bearing no physical resemblance to the body or brain.

Psycho-Engineering

At the close of their discussion, the proto-cyberneticians examined the structural differences between machines and living organisms and concluded:

> If an engineer were to design a robot [today], roughly similar in behavior to an animal organism,... he would probably build it out of metallic parts, some dielectrics and many vacuum tubes.... In future years, as the knowledge of colloids and proteins increases, future engineers may attempt the design of robots not only with a behavior, but also with a structure similar to that of a mammal. The ultimate model of a cat is of course another cat, whether it be born of still another cat or synthesized in a laboratory.[34]

This striking phrase, paradigmatic of the cybernetic imagination, projects a kind of psycho-engineering as a two-stage formal/mechanical modeling technique: "rough" behavioral simulation via computation and servomechanisms, followed by complete biological replication to capture formal detail in addition to functional effect. Thus cybernetics already conceived biological and psychological replication as engineering problems not qualitatively different from those of the design of electromechanical robots. Bioengineered organisms would differ greatly from machines in structure and materials, but psychologically relevant (i.e., behavioral) categories would remain the same.

A third stage of psycho-engineering lay hidden in the cybernetic dream, implicit in the mixture of apocalypse and elation in Warren McCulloch's words quoted near the beginning of this chapter. This was the moment of technological redesign of the potentially suboptimal organismic system by integrating it with superior materials and mechanisms.[35] Such a goal was avowed most starkly in a passage from Craik, whom McCulloch deeply admired:

> As an element in a control system a man may be regarded as a chain consisting of the following items:
>
> 1. Sensory devices, which transform a misalinement [sic] between sight and target into suitable physiological counterparts, such as patterns of nerve impulses, just as a radar receiver transforms misalinement into an error-voltage.
>
> 2. A computing system which responds to the misalinement-input by giving a neural response calculated... to be appropriate to reduce the misalinement; this process seems to occur in the cortex of the brain.

3. An amplifying system—the motor-nerve endings and the muscles—in which a minute amount of energy (the impulses in the motor nerves) controls the liberation of much greater amounts of energy in the muscles. . . .
4. Mechanical linkages (the pivot and lever systems of the limbs) whereby the muscular work produces externally observable effects, such as laying a gun.

Such considerations serve to bridge the gap between the physiological statement of man as an animal giving reflex and learned responses to sensory stimuli, and the engineering statement in terms of the type of mechanism which would be designed to fulfill the same function in a wholly automatic system. The problem is to discover in detail the characteristics of this human chain, such as its sensory resolving-power, its maximum power-output and optimum loads, its frequency-characteristics and time-lags, its amplitude-distortion and whether or not internal cyclic systems enter into it, its flexibility and self-modifying properties, etc., with a view to showing the various advantages and disadvantages of the human operator as compared with an automatic system.[36]

This passage might have served as the manifesto of cybernetic psychology.

While the discussions of the proto-cyberneticians ranged over many mechanisms, antiaircraft guns, torpedoes, and guided missiles were by far the most central. Concerned to emphasize the universality of cybernetic theory, the cybernetics group also characterized purposeful activity in many other machines. But while these other examples typically varied from work to work, virtually every one of the early articles and books in the emerging field mentioned the self-controlled and servo-guided weapons of World War II.

The centrality of these metaphors reflects a number of features of early cyborg discourse and the emerging cybernetics community. First, these machines embodied shared wartime experiences. Second, semiautomatic weapons systems integrated humans and machines through both mathematical description (formal structure) and embodied practice (mechanism), making them prototypical cyborg devices. Finally, before computers, in terms of information activity such machines were the most advanced devices known to the group. So the war machines were not simply one example among others, but a central, unifying metaphor of early cyborg discourse.

The parallels of Rosenblueth et al.'s vision to the later imagery of the Turing test, described in chapter 1, will by now be obvious. Like Turing, the cyberneticians noticed certain aspects of behavior (target-seeking) that machines (torpedoes, antiaircraft guns) could duplicate. Where the Turing-test metaphor made the content of communication independent of its form, the cybernetic analogy to "purposeful" machines made the form of goal-oriented behavior independent of its content. That is, the analogy to machines rendered irrelevant the specific ways in which human beings generate purposes and goals. Similarly, their discussion described reproduction as modeling or replication, appropriating the language of engineering. Like the Turing machine, the cybernetic principles were putative universals capable of describing any activity at all. "Purposeful machines" became second selves, cyborgs, reflections as much as creations. That metaphorical transformation created a looking-glass discourse in which simulation and reality began to blur into one: "the ultimate model of a cat is of course another cat, whether it be born of still another cat or synthesized in a laboratory."

The Macy Conferences

Three and a half years elapsed between the Cerebral Inhibition Meeting and the first of the postwar conferences organized to take up the cybernetic ideas. Yet Wiener, McCulloch, and some of the others had continued to meet all along, nourishing a dream of interdisciplinarity centered around the "circular causality" idea. Wiener made a special effort to interest his friend and colleague John von Neumann in the servo/organism analogies. Von Neumann, a polymath whose interests spanned virtually every area of science, not only took up the cybernetic analogies but did Wiener one better, adding to them the theory of digital computation and his own experience working on the ENIAC and EDVAC at the Moore School.

The group continued to develop and propagate the machine-organism analogies during the war, with ever-increasing and barely contained self-confidence. By 1944, in a letter, Wiener had "defied" Edwin G. Boring, chair of Harvard's Department of Psychology, "to describe a capacity of the human brain which he could not duplicate with electronic devices." Boring took up the challenge in a 1946 article, "Mind and Mechanism," which also served as that year's Presidential Address to the Eastern Psychological Association. He ended up supporting the Rosenblueth-Wiener-Bigelow pseudo-behaviorist formal/mechanical modeling standard of explanation: all of "the psychological properties of the living human organism ... are to be expressed objectively in terms of stimulus-response relationships, and the way to keep ourselves clear and to avoid vagueness is to think of the organism as a machine or a robot." Boring's response—citing the standard war machine examples—reveals that as early as the end of World War II, the computer metaphor had achieved widespread currency.

We have heard so much during the late war about electronic brains. The [analog] electronic computer on a range-finder figures the range and course and speed of a target, setting the fuses and aiming and firing the gun, all at a speed of which the human brain is incapable. There are now huge electronic mathematicians which will solve mathematical problems with a speed and accuracy and lack of fatigue that puts the mere headwork of the human mathematician out of the running.[37]

While Boring's article was a step toward legitimizing the metaphor within academic psychology, it remained for the cyberneticians to work out its details.

In his mathematical work on digital computer logics, von Neumann used a notation invented by neuropsychiatrist Warren McCulloch and logician Walter Pitts to describe nervous systems. McCulloch and Pitts had applied what was known as Boolean logic to the operation of neurons, analogizing these cells to on-off valves corresponding to the two-state, true-false Boolean system.[38] They had created their notation after reading Turing's groundbreaking paper "On Computable Numbers."[39] Despite their debt to Turing, however, they made no mention of his work in their paper. Nor did they bother to point out that the neural network they described was mathematically equivalent to a Turing machine. Indeed, they worried that their paper's mathematical approach was so marginal it might never be noticed or even published.[40]

Von Neumann learned of the McCulloch-Pitts formalism from Wiener, who had worked with Pitts in 1943 and was greatly excited by correspondences between the formalism and electrical relays. (Claude Shannon's doctoral thesis, just before the war, had made a similar use of Boolean logic in the analysis of telephone relay circuits.) In December 1944 von Neumann, Wiener, McCulloch, Pitts, and Howard Aiken—designer of the Harvard Mark I electromechanical computer—formed the Teleological Society to study "communication engineering, the engineering of control devices, the mathematics of time series in statistics, and the communication and control aspects of the nervous system."[41]

Von Neumann and Wiener planned a meeting with a few colleagues in January 1945 to discuss the cybernetic idea of a "unified mathematical description of engineering devices and the nervous system." At the meeting, Wiener wrote jubilantly,

von Neumann spoke on computing machines and I spoke on communication engineering. The second day [Rafael] Lorente de Nó and McCulloch joined forces for a very convincing presentation of the present status of the problem of the organization of the brain. In the end we were all convinced that the subject embracing both the engineering and neurology aspects is essentially one, and we should go ahead with plans to embody these ideas in a permanent program of research . . .[42]

The group discussed the possibility of establishing a research center after the war at Princeton or MIT. Despite considerable enthusiasm on the part of Wiener and others, this never materialized. Instead, in 1946, McCulloch persuaded the Macy Foundation, via Fremont-Smith, to fund a series of interdisciplinary conferences under the eventual title "Cybernetics: Circular Causal and Feedback Mechanisms in Biological and Social Systems." Ten conferences were held between 1946 and 1953, each involving between twenty and thirty "regular" participants and from two to five invited guests. The group, whose core included most of the original Cerebral Inhibition Meeting participants, comprised engineers, mathematicians, psychologists, neurophysiologists, philosophers, anthropologists, and sociologists.

McCulloch was chairman and primary organizer, though the advice of both Wiener and von Neumann was solicited in selecting participants. The regular members included Wiener's colleagues Rosenblueth and Bigelow, as well as Pitts and Lorente de Nó, a neurophysiologist from the Rockefeller Institute in New York. Gregory Bateson and Margaret Mead were invited as representatives of the social sciences. Bateson, tutored by Mead, had become interested in applying S-R learning theories in cultural anthropology; the two were personal friends of another participant, the psychologist Lawrence K. Frank, who was also a friend of Wiener. Wiener, whose personality has been described as "impish" and self-aggrandizing, and von Neumann were the chief protagonists.[43] This was especially true at the early conferences, when many participants were relatively unfamiliar with information theory and the ENIAC had been operational for just a few months.

The First Meeting: Computers as Brains

The first conference, called the Feedback Mechanisms and Circular Causal Systems in Biology and the Social Sciences Meeting, was held on March 8 and 9, 1946, in New York City, with McCulloch as chair. The very first presentation, by von Neumann, consisted of a description of general-purpose digital computers. According to Steve Heims's interviews with participants, it

included discussion of the greater precision of digital machines as compared to the older analog computers, the use of binary rather than decimal representation of numbers, the stored program concept, various methods available for storing and accessing information, and how in detail arithmetic operations are carried out by these machines. Some methods could not be discussed because they were still classified as military secrets. Von Neumann made semiquantitative comparisons between vacuum tubes and neurons, the overall size of brains and computers, their speed of operation and other characteristics.[44]

Lorente de Nó followed up von Neumann's talk with a complementary discussion of neurons as digital processing units. He described them as binary switches, explaining the physiology behind the McCulloch-Pitts logical model of neural networks.

In the afternoon Wiener and Rosenblueth gave a second team presentation. Like that of von Neumann and Lorente de Nó, it paired a mathematician with a neurophysiologist. Also like theirs, the Wiener-Rosenblueth talk examined machines and organisms in tandem. Wiener spoke about the history of automata, receptors and effectors in machines, and electronic computers as elements of machine systems. Rosenblueth discussed "homeostatic" (self-regulating) neurological mechanisms such as respiration, blood pressure, and body temperature. By design, the effect of the first day's talks was to impress deeply upon the group, in which professional psychologists served as audience to mathematicians and physiologists, the analogies between organisms, psychological concepts, and cybernetic machines, especially computers. The next day von Neumann introduced the group to game theory as applied to economics, going beyond the psychological to the level of the society as a whole.

Like the 1942 meeting, the first cybernetics conference proved an intellectual firestorm, a major event in the lives of many of the participants. The Macy Foundation agreed to continue funding yearly conferences as long as the group remained interested.

Exploring the Metaphor

The cybernetics group's goals were exploratory.[45] The organizing members—McCulloch, Wiener, von Neumann, Pitts, and Rosenblueth—wanted to see how far the analogies between organisms and machines could be pushed. They looked for examples of feedback control processes in every field represented. Some presentations focused directly on functional analogies between computers or servomechanisms, on the one hand, and brains, perceptual and memory mechanisms, or various physical activities, on the other. Others developed information-theoretical approaches to language and communication processes in social systems and groups. A few actual machines, such as Wiener's sensory prosthesis for the deaf and a maze-solving machine built by Claude Shannon, were demonstrated at the conferences.

Verbatim transcripts of the last five meetings were made and published. Since the presentations were informal and the ensuing discussions were also transcribed, these documents provide an invaluable source for analyzing the genesis of a discourse. They show disciplines in the act of merging, sciences in the process of becoming, and the role new machines can take in creating new forms of discourse.

Despite having adopted the McCulloch-Pitts formalism for his computer architectures—and having been among the first to raise the analogy—von Neumann expressed persistent doubts throughout the conferences about whether brains and computers could fruitfully be compared. The conference transcripts contain numerous mentions of "the things that bother von Neumann." These were the limitations on understanding the "information processing" and "storage" capacities of the brain if neurons were really analogous to the dual-state registers of computers. The McCulloch-Pitts formalism suggested that information was processed in the brain by means of two activities of neurons, discharge (also called "firing") and inhibition. Whether or not a neuron fired depended on the ratio of excitatory to inhibitory stimulation it received from other neurons synaptically connected to it. This process could be formally reduced, according to McCulloch and Pitts, to the action of a two-valued, on/off switch. Thus, the entire network of neurons could be modeled as a system of binary-valued digital switches—a formal machine.[46]

Against the enthusiasm of McCulloch, von Neumann counseled despair. Even if the brain were really this simple in organization, theoretical reduction would probably prove completely beyond reach due to its gigantic size (10 to 100 billion cells) and its even more staggering number of possible states (each neuron may send its impulses to hundreds or even thousands of others). At the cellular level, von Neumann wondered whether the system itself, in all its complexity, might not be its own simplest description. Additionally, some evidence pointed to a "mixed" analogical/digital character for the nervous system (e.g., that the continuously variable *rates* of firing were part of the code as well), making the comparison with the purely digital electronic computers untenable. (The psychologists at the first conference raised this objection immediately, but others brushed it aside.) Though he seemed to believe that advances in mathematical logic might one day make possible some sort of description—and certainly saw this as a kind of scientific Holy Grail—he became highly pessimistic about the comparisons conceivable with then-current engineering techniques.[47]

Despite von Neumann's warnings, the cybernetic analogies rapidly grew into much more than suggestive heuristics. The neurophysiologist Ralph Gerard revealed a common discomfort when he opened the seventh conference by lecturing on "Some of the Problems Concerning Digital Notions in the Central Nervous System." Referring to the "tremendous" interest of the scientific community and the general public in the ideas generated by the group as "almost ... a national fad," Gerard complained:

It seems to me ... that we started our discussions and sessions in the "as if" spirit. Everyone was delighted to express any idea that came into his mind, whether it seemed silly or certain or merely a stimulating guess that would affect someone else. We explored possibilities for all sorts of "ifs." Then, rather sharply it seemed to me, we began to talk in an "is" idiom. We were saying much the same things, but now saying them as if they were so.[48]

Comparing the group's "overenthusiasm" to the premature popularity of phrenology in the 1800s, he begged for intellectual "responsibility" in the use of cybernetic terminology.

But neither von Neumann nor Gerard could stop the historical process at work—which Turing by then had articulated as his belief "that at the end of the century the use of words and general educated opinion will have altered so much that one will be able to speak of machines thinking without expecting to be contradicted"—simply by recognizing it.[49] The entailments of computer metaphors were being worked out not only within the cybernetics group, but in other sciences and in popular culture as well. Systems discourses were spreading through the sciences, with new institutions like the Rand Corporation being assembled around them. Systems, information, and communication theories were being reified almost daily in the form of computers, radar-controlled tracking devices, guided missiles, and other machines.

Wiener, von Neumann, and Bigelow, as well as others in the cybernetics group, had spent years working out the design of these very devices. They were committed to the formal machine as a reality and as a standard of explanation, and they would never be satisfied with a psychology that stopped short of the formal/mechanical model, like the black-box behaviorist approach or the vagaries of depth psychology. A cybernetics of the mind would be incomplete until its practitioners had *built one* or at least shown how it could be done—Rosenblueth, Wiener, and Bigelow's "ultimate model." Thus the brain-computer comparisons were not arbitrary, as Gerard's comments implied, but paradigmatic. Most Macy participants agreed, at least tacitly, that servomechanisms and computers afforded the most promising, if imperfect, analogs to human and animal behavior, and they were united in the belief that information theory could render human/animal and organism/machine boundaries permeable to unified explanations.

Challenges to Computational Metaphors

The power of servomechanism and computer metaphors was most readily visible when they were challenged, as when the electrical engineer and conference series editor Heinz von Foerster presented a paper on human memory at the sixth conference. Von Foerster based his argument on an analogy to quantum theory in physics. He proposed a mathematically exact theory of memory in which "impressions," or memory-traces, were preserved as energy states of a hypothetical protein molecule he called a "mem." Memorizing something corresponded to raising the energy level of a mem by a quantum unit; forgetting occurred when the mem discharged a quantum of energy. The theory was supported by experiments in learning nonsense syllables and evidence that forgetting increases with higher body temperatures (which destabilized the mems, von Foerster hypothesized, making them more likely to release energy).

There were superficial parallels to information theory in von Foerster's approach, but the concrete metaphor involved the quantum theory of atomic structure, a subject the Macy group had not discussed (though Wiener and von Neumann, at least, had done theoretical work in quantum mechanics).

In the discussion period, Ralph Gerard and the psychiatrist Henry Brosin expressed "confusion." Gerard did not see "how it could be made to work for any kind of a picture we have of the nervous system." The quantum model did not seem able to account for the organization of the huge numbers of quanta that had to be presupposed in any sensory impression. Though von Foerster could describe his atomic quanta as units of information, this was not his emphasis, nor could his model explain the structuring of that information into patterns in perception or behavior. But another major reason for the psychologists' and neurophysiologists' resistance to von Foerster's ideas seems to have been the foreign character of the metaphor from atomic physics. Von Foerster's audience welcomed the paradigm of servomechanisms and computers, machines they had all seen and worked with, but did not quite know what to do with his model of atomic decay. The quantum mechanical theory was never discussed again by the group.[50]

Another anomalous theory had been presented by the Gestalt psychologist Wolfgang Köhler at the fourth Macy Conference. Köhler's "field" theory of perception was built on the model of electromagnetic fields—another foreign and competing metaphor. His presentation, more influential than von Foerster's, also "created anxiety in the group" but did not significantly affect its theoretical orientation or preferred analogies.[51] But the participants did begin to question whether a rigorous cybernetic approach could go beyond local neuronal mechanisms to account for the large-scale organization of perception as well. Steve Heims has described the relevant differences between the Gestalt and cybernetic theories, and the conflict that resulted, as follows:

To define the relation between psychological facts of perception and events in the brain, Köhler had posited a "psychophysical isomorphism." His starting point had been the empirical psychological facts concerning the phenomena of perception. The cyberneticians ... sought to derive mental processes in the brain by summing or averaging over the interrelated elemental electrical and chemical events (the firing of a synapse) in the brain. Instead of psychophysical isomorphism, [they] posited a relation of "coding" as the paradigm for the connection between perceptions and [brain] events....
[Their] starting point ... was logical mathematical formalism and some facts of ... neurophysiology.[52]

Just as crucial as these theoretical "starting points" were the concrete metaphors used by the cybernetics group to give substance to their formal pictures. Bateson, at the seventh conference, noted that the question of whether the nervous system processes information by analog or digital (continuous or discontinuous) means was the commitment-laden center of the cyberneticians' debate with Köhler.

There is a historic point that perhaps should be brought up; namely, that the continuous/discontinuous variable has appeared in many other places. I spent my childhood in an atmosphere of genetics in which to believe in "continuous" variations was immoral.... There is a loading of affect around this dichotomy which is worth our considering. There was strong feeling in this room the night when Köhler talked to us and we had the battle about whether the central nervous system works discontinuously or, as Köhler maintained, by leakage between axons.[53]

I have been suggesting that while these commitments to the "discontinuous" model had multiple origins, such as those mentioned by Bateson, their immediate sources—the ones unifying the cybernetics group—were digital computers and servomechanisms. The "strong feelings" around the idea of the digital nervous system were generated in part by some of the participants' direct involvement in work on these devices.

The competing theories of von Foerster and Köhler demonstrate by counterexample a number of facts about the cybernetics group that help to explain why alternative analogies failed to capture attention. Some of these can be enumerated as a set of group commitments and innovations, including:

• The centrality of the *mathematical analyses* of Wiener and von Neumann, who continually sought to formalize what they learned about neurobiology and psychology from the group in mathematical terms.

• The requirement of *formal/mechanical modeling* of the object of knowledge (Köhler's field theory had a machine analog, but it was not a formal one, like the electronic logics of computer systems).

• Commitment to *engineering solutions for psychological questions*: analysis of humans as links in control and communications processes with a view to minimizing error.

• Particular *ideas of "information" and "communication,"* derived in part from wartime engineering experience.

The conflicts illustrate how a group whose central concerns were profoundly shaped by some of its members' experiences with military engineering problems continued to think in terms of those problems and created from them a new field of scientific possibilities. The Macy core group of mathematicians, engineers, and neurobiologists had turned its attention to psychological phenomena defined, for them, in terms of the tracking, targeting, and communications tasks of the war. We will now turn to presentations the group received positively, in order to see how these commitments were manifested.

Vision as Tracking and Targeting

John Stroud, research psychologist at the San Diego Naval Electronics Laboratory, was a two-time guest at the conferences. He was trained in electrical engineering and X-ray physics. During the war Stroud became interested in the nature of psychological time, which he thought might be connected to the rhythms of brain waves. He took a degree in psychology at Stanford "as a kind of renegade physicist."[54] At the sixth conference he began his lecture on "The Psychological Moment in Perception" by acknowledging his intellectual debt to the McCulloch-Pitts theory of neural nets and, especially, to the work of Craik on servo-controlled antiaircraft guns, which he then introduced to the group. Stroud had served in the RAF during the war, learning of Craik's work while in England.[55]

Stroud described experimental observations of the low-frequency periodicities by which test subjects corrected the path of tracking devices, a rate of two or three corrections per second. Stroud deduced from this that "experience is quantal in nature": experience occurs in quantum units he called "psychological moments."[56] He conjectured that the central nervous system acts like a computer using a periodic scanning mechanism, possibly based on the brain's alpha rhythm. Note that while Stroud also invoked the term "quantum," he did so on the basis of a concrete metaphor different from von Foerster's: the now-standard servomechanisms and digital computer analogy, rather than atomic physics. For him the idea of a quantum implied a kind of digital/analog distinction rather than an atomic model.

Following Craik's model, Stroud posed the psychological problem in now-familiar terms:

In the firing of guns we have to use human operators to make certain decisions but today we have to fire them very rapidly. We have tried our level best to reduce what the man has to do to an absolute minimum.... We know as much as possible about how all the associated gear which brings the information to the tracker operates and how all the gear from the tracker to the gun operates. So we have the human operator surrounded on both sides by very precisely known mechanisms and the question comes up, "What kind of a machine have we placed in the middle?"[57]

Stroud went on to present a detailed analysis of the human as servomechanism, via a set of experiments on subjects' accuracy and response rates in visually "tracking" objects moving on a variety of more or less predictable paths. To Stroud, these experiments suggested the characteristics of what might be termed the tracking-targeting cyborg, that is, the human as an element of a tracking system. First, "operators" could "track" accurately by using pointers controlled by any combination of displacement, velocity (first derivative of displacement), and acceleration (second derivative of displacement). They adapted readily to the possibilities presented by the experimental setup, rather than being locked into a single mode of prediction. Second, Stroud concluded, "man is a predictor and says 'I shall continue to do whatever my last solution predicted will be right so long as no detectable difference arises.'"[58] In other words, human subjects (unlike the simplest servomechanisms) depended on feedback control only until a suitable solution could be found, which they then employed until changed conditions required a new one. Third, the fact that subjects continued to make corrections for a short period when pointer and target were suddenly obscured suggested that operators used information gained in advance of the corrective action itself in making their corrective movements: their corrections were not made in "real time," but in a predictive, psychological time.

Finally, Stroud found that tracking performance was quantal or discontinuous, and he spoke of the operator as an information processor:

Our operator receives his information by way of his eyes.... This is where the information goes into the human organism, and it goes in, to all intents and purposes, continuously. When we analyze what comes out of the organism, every set of records of sufficient sensitivity... has shown low frequency periodicities.[59]

Input-output—as opposed to stimulus-response—analysis began to yield a formal/mechanical theory of the "machine... in the middle."

The group's response reveals a struggle over key metaphors

between the mathematicians and engineers, on the one hand, and the psychologists on the other. The latter were unimpressed with Stroud's information-processing analysis. Neurologist and psychoanalyst Lawrence Kubie viewed the "moment" theory as merely another instance of the well-documented phenomenon of "reaction time" and pointed out that it varied among individuals and according to emotional and situational conditions. In the ensuing general discussion of the information retention and processing capacity of the nervous system, experimental psychologist Heinrich Klüver mentioned the "ten thousands of pages of Wundtian psychology" concerning attention span and instantaneous perception. Like Kubie, Klüver downplayed the significance of the new metaphors by reading them as disguised or rephrased versions of traditional problems in psychological theory. Invoking the Gestalt theory of wholes and fields, he insisted that the group's quantitative, mathematical emphasis would remain "meaningless" as long as the context and organization of psychological "information" was not considered.

Wiener, McCulloch, and Pitts, by contrast, were pleased and stimulated by Stroud's method of analysis. With no stakes in the disciplinary history of psychology, they responded well to the mathematical regularities and mechanical comparisons Stroud used. Wiener's excitement was palpable; his exuberant comments amounted to a running dialogue with Stroud. He discussed a series of machine analogs to Stroud's ideas, all devices Wiener had helped to design. During a discussion at the Hixon Symposium six months previously, McCulloch had invited Stroud to present his theory as reinforcement for McCulloch's own version of the scanning computer analogy. Stroud spoke the engineering language the cyberneticians understood, and they were unconcerned by what the psychologists heard as his loose usage of technical psychological terms (such as "phi" for "apparent motion") and his reformulation of traditional psychological problems in formal/mechanical terms.

The different responses of these two factions within the Macy group point again to formal/mechanical modeling's revolutionary character. Those trained as psychologists started with the phenomena of conscious experience and observations about human responses, and they looked for relations between these and stimuli in the physical world. But the cyberneticians began with the formal machines they had made and sought to build a machine that would model its builder.

Soon afterward, Stroud himself went to work as a scientist for the U.S. Navy, where he spent the rest of his working life, mostly on classified research. The cybernetic metaphors became permanent features of Stroud's thinking; in his only published article (in 1950) Stroud asserted that "man is the most generally available general purpose computing device." In 1972 he told Steve Heims, raising a Turing-like claim, that "a system is what it does" and asserted that conceptual boundaries between living and nonliving systems should be drawn, or erased, along functional lines. "If the actions [of a system] are calculating, analyzing information, measuring, deductive reasoning, guiding and controlling a ship or missile, then humans and machines—although different in the details of their actions and capabilities—are comparable living systems."[60]

Psychological tracking and targeting research became an ergonomic discipline aimed at the construction of integrated human-machine cyborgs.[61] Among its most fertile offshoots were military experimental training simulators similar to modern video games, complete with joystick controls. The eyes themselves were militarized through the tracking and targeting metaphor. "The human eye," proclaimed the NRC's widely distributed 1943 Army pamphlet *Psychology for the Fighting Man*, "is one of the most important military instruments that the armed forces possess."[62] The distinguished British visual psychologist R. L. Gregory, in 1966, called the retinal edge "an early-warning device, used to rotate the eyes to aim the object-recognition part of the system onto objects likely to be friend or foe rather than neutral." He described the two systems for visual identification of movement:

we name them (a) the image/retina system, and (b) the eye/head system. (These names follow those used in gunnery, where similar considerations apply when guns are aimed from the moving platform of a ship. The gun turret may be stationary or following, but the movement of the target can be detected in either case.)[63]

Thus we find, again, the legacy of world war in the sciences of mind and brain: the human capacities of perception and thought as the design features of technological soldiers.

Project X: Noise, Communication, and Code
Another guest at the Macy Conferences was Claude E. Shannon of Bell Laboratories. A mathematician of nearly the same rank as Wiener and von Neumann, Shannon had taken his Ph.D. at MIT at

the age of 24. In 1937, while still a doctoral student, he had held a summer job at Bell Labs, during which he had demonstrated the application of Boolean algebra to relay circuit analysis. The mathematical model he created bore direct parallels to both Turing's universal machines and the McCulloch-Pitts neural logic. His achievement was credited with changing relay circuit design from an "esoteric art" to an engineering discipline.[64] Shannon's systematization had significant impacts on early electromechanical digital computer design, since the latter employed telephone relays as registers. It was also related to von Neumann's later studies of logical architectures.

Shannon worked with the NDRC in 1940–41 before rejoining Bell, where he immediately became active in research on cryptanalysis. In many ways Shannon was the American counterpart to Turing: a mathematician, interested in what would soon be called digital logic, whose wartime contributions involved him in cryptology.

One of the Bell inventions to which he contributed was the top-secret "Project X" speech encipherment system. This was the first digital speech transmission system. Previous voice coding systems had depended on analog techniques; they were simply more or less radical distortions of ordinary speech waves, whose receivers reversed whatever distorting process was used in the transmitters. All of these methods produced sound that, while distorted and irritating, could be fairly easily understood by a determined listener without special equipment. By contrast, the X system quantized the speech waveform and added it to a digital "key" before transmission. Careful listening was no longer sufficient: to decipher this signal, an eavesdropper would have to possess not only the equipment for converting the digitized information into sound waves, but also the key pattern.

Shannon's wartime work on secrecy systems was "intimately tied up together" with his most important theoretical contribution, "A Mathematical Theory of Communication." An essential precursor to that paper was his 1945 Bell memorandum, "Communication Theory of Secrecy Systems."[65] In these papers Shannon outlined a mathematical analysis of general communications systems of the form shown in figure 6.2. Like Wiener, Shannon achieved mathematical generality by defining "information" and "communication" in technically specific, quantitative terms:

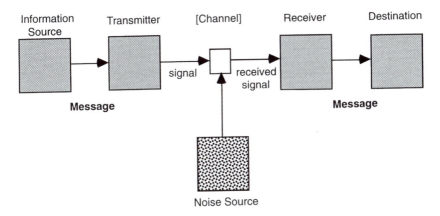

Figure 6.2
Generalized communications system. After Claude Shannon and Warren Weaver, *The Mathematical Theory of Communication* (Urbana: University of Illinois Press, 1949), 5.

The fundamental problem of communication is that of reproducing at one point either exactly or approximately a message selected at another point. Frequently the messages have meaning; that is, they refer to or are correlated according to some system with certain physical or conceptual entities. These semantic aspects of communication are irrelevant to the engineering problem. The significant aspect is that the actual message is one selected from a set of possible messages. The system must be designed to operate for each possible selection, not just the one which will actually be chosen since this is unknown at the time of design.[66]

For Shannon, communication theory stopped short of semantics, finding its true purposes in the design and analysis of X systems and telephones. Nevertheless, the possible application to human language, as a code bearing information in the technical sense, intrigued him.

Shannon's presentation on "The Redundancy of English" at the seventh Macy Conference concerned written language as a discrete (discontinuous, or digital) information source. Shannon used a slightly altered version of figure 6.2, renaming the transmitter and receiver "encoder" and "decoder," respectively, and thus explicitly identifying communication with cryptanalysis.

He considered the twenty-seven characters of the English alphabet (twenty-six letters and the space) as elements in a stochastic coding process. Letter sequences differed from the "ideal" information source, where all messages are equally likely, since the probability of any character's occurrence depends to some degree on the characters preceding it. He estimated the redundancy of typical English prose to be about 50 percent, interpreting this to mean that "when we write English half of what we write is determined by the structure of the language and half is chosen freely."[67] To compile statistics on letter and word frequency and interdependency, Shannon simply opened books at random, located the last letter or word in the sequence he was recording, and wrote down the letter or word that followed it. This method tacitly accepted a crucial premise of behaviorist theory of language, namely that

> sentences are produced in a left-to-right sequence conceived as a series of probabilistic events in which each stimulus word in the sequence has a specifiable probability of eliciting the next word response which, in turn, acts as a stimulus for the succeeding word response.[68]

Like the behaviorists, Shannon tried scrupulously to keep semantic questions at arm's length. But his treatment of language still seemed to imply that the theory might be general enough to cover, as Warren Weaver pointed out in his *Scientific American* review of Shannon's work,

> all of the procedures by which one mind may affect another. This ... involves not only written and oral speech, but also music, the pictorial arts, the theater, the ballet, and in fact all human behavior. In some connections it may be desirable to use a still broader definition of communication ... [including] the procedures by means of which one mechanism (say automatic equipment to track an airplane and to compute its probable future positions) affects another mechanism (say a guided missile chasing this airplane).[69]

The one-way communications system of figure 6.2 becomes a feedback loop when the receiver transmits control information back to the transmitter. The communications device also makes a picture of neurological phenomena such as the reflex arc (with the transmitter as afferent and the receiver as efferent nerve) or the control of movement by the brain via the kinesthetic sense. Shannon later worked with Noam Chomsky, for whom linguistics became the study of the human as X system, performing complex transformations on "kernels" of information, decoded by reverse transformations in the listener. Shannon also built a maze-solving machine, maintained a serious interest in chess-playing programs, and mentored the young John McCarthy, who went on to coin the phrase "artificial intelligence."[70]

Warren Weaver was Director for Natural Sciences of the Rockefeller Foundation and thus an immensely influential figure in the construction of postwar scientific research programs. He felt that Shannon was wrong to exclude semantic considerations from the theory. In his own view, communication could be analyzed as a hierarchy of three problems:

Level A. How accurately can the symbols of communication be transmitted? (The technical problem.)

Level B. How precisely do the transmitted symbols convey the desired meaning? (The semantic problem.)

Level C. How effectively does the received meaning affect conduct in the desired way? (The effectiveness problem.)[71]

Where Shannon limited himself conservatively to the technical, engineering question (Level A), Weaver thought "analysis at Level A discloses that this level overlaps the other levels more than one could possibly naively suspect.... Level A is, at least to a significant degree, also a theory of levels B and C."[72] Weaver, then, saw in the mathematical theory of communication far more than a technical aid for designing telephone systems. It was the beacon of a complete statistical analysis of human social activity. Twin cyborgs—X system, tracking-and-targeting servomechanism—became concrete symbols of intelligence as a system of control and communication. Finding the "key patterns" coding perceptual "input" into digitized neural representations, transformations, and "output" as speech or human-machine interaction became the ultimate goal of psychological and social science.

This sort of optimism provoked strenuous debate at Shannon's Macy Conference presentation. The social psychologist Alex Bavelas wanted to use Shannon's ideas to describe the communication of "second-order information," or those messages about status, relationships, desires, and so on implied but not directly expressed in conversation or by other means. He thought, for example, that "a change in emotional security could be defined as a change in the individual's subjective probabilities that he is or is not a certain kind of person or that he is or is not 'loved.'"[73] As a selection of one of a set of possible messages about the individual's relationship to the group, this would qualify as information in the strict sense. Walter Pitts and the skeptical mathematician Leonard Savage objected to this picture of emotional contact as purely "informative."

Shannon pointed out that using communication-theoretic terms "at the psychological level" created difficulties of definition, especially with concepts like "signal" and "noise": "if a person receives something over a telephone, part of which is useful to him and part of which is not, and you want to call the useful part the signal, that is hardly a mathematical problem."[74] In the X system, the generation of "information" by the source and its use by the destination—its interpretation—are "semantic" and fall outside the boundaries of the formal machine. The theory is essentially concerned only with defining the accuracy of transmission of already-constituted messages. Messages generated by a source must be elements of some specifiable, predefined set, such as sound waves or lettering. Similarly, the criterion of accurate transmission is a comparison of the message at the source with the message at the destination; the "effectiveness" of the message on the addressee forms no part of the theory.

The metaphorical extension Bavelas hazarded at Shannon's presentation was exactly the kind of transition from "as if" to "is" against which Ralph Gerard had railed in his talk opening the same session. The exchange over the idea of "second-order information" reveals the process of metaphorical elaboration at work. Bavelas was trying to bend the technical term "information" to the purposes of a wider, less restricted conversation. Whether or not there was theoretical justification for such an application of the term, Bavelas perceived an available metaphorical entailment.

Shannon's theory described the substance of communication as information. But Shannon's theory applied directly only to observables, externals—preconstituted messages without semantic content. It was a theory of communication between and through machines. Communication between human beings, as Bavelas understood, has semantic values at its heart. His description of emotional security as a "change in subjective probabilities" transferred the notions of "message" and "information" onto the internal experience of the individual. What Weaver called the "semantic problem" and the "effectiveness problem" were areas where the extension of theory and the growth of metaphor blurred together. Understanding the one without the other, given the self-elaborating tendencies of discourse systems, had already become a doomed rear-guard action.

What these dead ends in metaphorical elaboration revealed was that a *behaviorist*-mechanist analysis would, finally, fail. Internal processes—not just the transmission, but the constitution and selection of "messages"—would have to be included in the model. For that purpose computers, not communications systems, would provide the key.

The Chain of Command
In engineering, however, the semantic and effectiveness problems could be ignored as long as, with Shannon, source and destination were conceived as the beginning and end of the communication process. Even complex systems may be analyzed without considering the semantic dimension. In the case of feedback-controlled machines, for example, the destination doubles as source for a transmission back to the original source, which in turn becomes the destination for the control message. Or the process can continue sequentially, as when orders are relayed along a human chain of command, with each destination becoming a selective source for the following link. For processes like these, the message need not be "understood" or "interpreted" at the destination in order to be processed and be effective.

Computer development had been funded by the military with an eye to the potential for automation of command, control, and communications in the SAGE project. In the same way, the enhancement of command-control processes motivated military investment in communications technology and psychology over the long term. It was again the problem of the "machine in the middle," whose slow speeds and tendencies to failure made the human an unreliable part of the military machine. J. C. R. Licklider of the Harvard Psycho-Acoustic Laboratory, whose guest Macy presentation immediately preceded Shannon's, made the following comment in discussion:

During the war, especially when the military was picking people for fairly crucial jobs involving listening—the military had the notion, as you probably all know, that everything must go through many links of a chainlike communication system before it becomes really official—they were interested in the problem of training listeners.[75]

As links in this chain, human listeners in communications systems added unacceptably high levels of noise. For the purposes of the Psycho-Acoustic Lab's wartime research (as we will see in chapter 7) the "job" of "listening" had been defined as accurate decoding. The listener as destination became an essentially mechanical linkage in the command circuit, carrying out the orders propagated by the electronic components and mediating among them. Human listeners in military roles were themselves conceived as X systems, natural-technical devices for decoding signals. The soldier-listener was trained as well as operated by the techniques of feedback control: "the way to teach a person to listen is to provide high motivation..., lots of speech for him to listen to, and ... knowledge of results. (The listener must know whether he heard it correctly or not.)"[76] Licklider's laboratory tested and trained the human elements of the chain of command as component parts whose accuracy and error rates would affect the performance of the system as a whole. The criterion of that performance was the undistorted flow of command and control signals.

Under the emerging electronic communications regime of the war and the postwar military world, command-control-communications systems operated as a nested hierarchy of cybernetic devices. Airplanes, communications systems, computers, and antiaircraft guns occupied the micro levels of this hierarchy. Higher-level "devices," each of which could be considered a cyborg or cybernetic system, included aircraft carriers, the WWMCCS, and NORAD early warning systems. At a still higher level stood military units such as battalions and the Army, Navy, and Air Force themselves. Each was conceptualized as an integrated combination of human and electronic components, operating according to formalized rules of action. Each level followed directives taken from the next highest unit and returned information on its performance to that unit. Each carried out its own functions with relative autonomy, issuing its own commands to systems under its control and evaluating the results using feedback from them.

This, at least, was the formal picture of the operation of military hierarchies, the picture drawn by the emerging class of science-guided, control-oriented military managers. Transforming institutions with a deep historical command tradition to correspond with this formal picture became a kind of Holy Grail for military technologists and the sciences behind them in the postwar years. The human elements of cybernetically organized systems question, negotiate, make mistakes, delay, and distort messages; this made them problematic from the military formalist point of view. As a "battle-scarred and ribbon-covered Admiral" introduced the crew of the USS Missouri to NDRC scientists during World War II: "Twenty-five hundred officers and men: gentlemen, twenty-five hundred sources of error."[77]

Prior to the advent of cybernetic machines, this "problem" had been addressed by techniques of discipline and normalization, focused on incorporating the soldier into the human chain of command.[78] With cybernetic devices, however, it became possible to minimize or entirely eliminate some human roles. The transition to cybernetically automated military organizations was abrupt. As Jay Forrester put it,

One could probably not have found [in 1947] five military officers who would have acknowledged the possibility of a machine's being able to analyze the available information sources, the proper assignment of weapons, the generation of command instructions, and the coordination of adjacent areas of military operations.... During the following decade the speed of military operations increased until it became clear that, regardless of the assumed advantages of human judgment decisions, the internal communication speed of the human organization simply was not able to cope with the pace of modern air warfare. This inability to act provided the incentive [to begin replacing the 'human organization' with computers].[79]

Such automation required a formal/mechanical model of the "human organization."

Thus cybernetic psychology, as both theory and practice, both mirrored and transformed the chain of command. Like the hierarchy of behavior pictured by Rosenblueth, Wiener, and Bigelow in figure 6.1, the chain of command does not set goals or define values, but only carries out orders. Though individuals inside a military system certainly make decisions and set goals, as links in the chain of command they are allowed no choices regarding the ultimate purposes and values of the system. Their "choices" are, to paraphrase Shannon, always the permutations and combinations of a predefined set. The "military machine" was a metaphor built from a palpable cybernetic device, an interlocking assemblage of human and electromechanical parts, its coherence and continual refinement activated by the theory of information.

The Combat Information Center of the USS Lexington, January 1945. Courtesy National Archives and the Naval Historical Foundation.

7
Noise, Communication, and Cognition

The cybernetics community's leaders were primarily mathematicians and engineers. They succeeded in altering psychological approaches precisely because they were *not* psychologists. They lacked commitment to the behaviorist paradigm, and they focused on practical design problems rather than pure theory-building and on brain models rather than black-boxed reflex systems. But for these same reasons, the cyberneticians' effects on academic psychology were muted and far from immediate. Instead, cybernetic ideas filtered slowly into psychological theory, channeled through interpreters within the discipline rather than taking it by storm.

What were cognitivism's origins within psychology itself? Who were these interpreters, and how did they learn of the cybernetic ideas and computer metaphors? Why did they find them so appealing? How did they turn them into an extraordinarily productive research program in experimental psychology?

As a partial answer to these questions, this chapter investigates the work of the largest World War II institution for experimental psychology, Harvard's twin Psycho-Acoustic and Electro-Acoustic Laboratories (which we shall refer to by their acronyms, PAL and EAL). These laboratories studied problems of communication in the context of noise as simultaneously technological and psychological issues. PAL veterans such as George A. Miller went on to become early proponents of information theory in psychology; to create research programs in psycholinguistics; and eventually to found institutions (such as the Harvard Center for Cognitive Studies) that established the dominance of the cognitive paradigm. Miller's postwar research bears the marks of the PAL's wartime concerns and of social networks constructed during that period. He would eventually work with Allen Newell and Herbert Simon, founders of artificial intelligence, and co-author *Plans and the Structure of Behavior*, arguably the single most significant work in the early history of cognitive psychology. The PAL's legacy thus exemplifies one of the major historical trajectories of cognitivism, from wartime work on human-machine integration to postwar concerns with information theory to the computer as metaphor for the human mind.

The Psycho-Acoustic Laboratory and the Problems of Noise

World War II nearly drowned in the noise of its own technology. Added to the older sounds of artillery and gunfire were the newer ones of rockets, airborne bombs, and various combat vehicles—tanks, airplanes, submarines, half-tracks, trucks, jeeps, and a host of others—that had gradually joined military arsenals during and after World War I. The tremendous din of mechanized battle created two unprecedented military problems.

The first problem came from the fact that complex machines like tanks and airplanes had become critical to military success. The battle-worthiness of these machines depended crucially on the intercommunication of the soldiers inside them. Such technologies were not simply machines "used" by men, as railroads were used to transport armies that then continued toward the front on foot in World War I. They were a new cavalry, a new kind of integrated fighting unit. The airplane, the tank, and the submarine were primitive examples of what would eventually be labeled "cyborgs": biomechanical organisms made up of humans and machinery.[1] Their internal and external linkages took the form of electronic feedback circuits. These included not only interphones and radiophones for communicating with other humans but the dials, controls, and bombsights through which humans communicated with the machines. Should any of these information linkages fail, a cybernetic weapon could be totally disabled. The limits of communication under noisy conditions formed, therefore, ultimate limits to their effectiveness. The noise the machines made and endured had to be checked, or else, somehow, their communications systems had to be made to function despite the noise.

The second new problem came from the radical increases in the speed of motorized military units and air warfare, which necessitated matching increases in the rapidity of communication. This meant new technologies, both human and electronic, for transmitting information. As one psychophysicist posed the problem,

> The waist-gunner with a new "bandit" looming in his gunsight has no time to code his message to the rest of the crew in dots and dashes. "Tally-ho, three o'clock," fills his quota of seconds.... Even speech was often too slow when the kamikaze came boring in and the radar lookout had to relay his data on the "contact" through the ship's information center to the gun directors. On too many of our ships the tearing boom of a suicide strike preceded the bark of the gun that should have dropped the attacker into the sea.[2]

Techniques for extremely rapid communication, and even for taking certain information processing tasks out of human hands altogether, therefore became an important war research goal.

These two design problems—the "human engineering" of cyborgs to counter the problem of noise, and the engineering of communications for maximum speed and efficiency—involved academic and industrial psychologists in the problems of war. Interpreting military requirements in light of their existing research programs and theoretical commitments—largely psychometric, psychophysical, and behaviorist in character—experimental psychologists directed their efforts toward specifying the design parameters of the human organism, in order to insert that organism into electromechanical military systems.

At the war's beginning the noise made by the machines themselves, especially airplanes, overwhelmed existing communications technologies created by peacetime engineering in quiet laboratories. In 1942, a study reported that

> all earphones now used in U.S. planes must be classed as unsatisfactory. They ... blast the ears at some frequencies and fail to respond effectively at others. Their response is so sharply peaked at about 1000 cps that when they are operated at a level permitting the listener to hear the other speech frequencies the sounds near the resonant frequency are loud enough to be painful.[3]

In communication between aircraft carriers and their fighter planes, and between ships, "fifty percent of the words transmitted ... must be repeated," according to shipboard communications officers.[4] Long-range bombers were so loud they endangered their crews: "As far back as 1940, the British reported that airplanes returning from long flights were crashing short or to one side of landing fields for no other reason than the stultifying reaction of engine roar upon the pilot's perception."[5]

During initial mobilization for the war, these problems were addressed by a Committee on Sound Proofing of Aircraft under the National Research Council, chaired by the physicist Philip M. Morse of MIT.[6] By late 1940, however, the real extent of the noise problem had become clear, and Morse organized a group at Harvard to work on soundproofing and communications in all kinds of vehicles. The work was divided between the Electro-Acoustic Laboratory, under Leo L. Beranek, where physical and electronic elements of the problem were studied, and the Psycho-Acoustic Laboratory, under S. S. Stevens, which focused its efforts on "those problems arising from the fact that a human being is part of the total circuit."[7] The two laboratories eventually took complete control of this research as Section 17.3 of the NDRC. They also explored problems around designing and testing acoustic equipment and selecting and training communications personnel.

Ultimately, the two labs employed over seventy-five scientists. Roughly half of these were psychologists. The PAL was the "largest university-based program of wartime psychological research" in the country. Over the course of the war it received almost $2 million from the NDRC—a vast amount, by then-current standards, for psychological research.[8]

The PAL played a crucial role in the genesis of postwar information processing psychologies. A large number of those who worked at the lab either during or after the war—when it continued its work under military, NSF, and NIMH contracts—helped to develop computer models and metaphors and to introduce information theory into human experimental psychology. These included Wendell R. Garner, George A. Miller, J. C. R. Licklider, E. B. Newman, Leo J. Postman, Walter Rosenblith, Karl Pribram, and Eugene Galanter. Of some sixty-one graduate and "other" research assistants employed by the PAL between 1945 and 1961 on Office of Naval Research contracts, a substantial number went on to become important cognitive psychologists, including Richard Held and Ulric Neisser.[9]

Why was this laboratory such a seedbed for cognitive psychology? What was its role in the establishment of the cognitive research paradigm, and how did that role develop from the lab's beginnings in military research? In the rest of this chapter I want to show how the view of the human as an information processing system, in a precise, quantifiable sense, emerged for the psychologists of the PAL under three kinds of influences: military needs, contact with communications engineers, and experience with communications hardware and concepts.

Many of the lab's tasks were the mundane and fairly straightforward chores of developing and/or testing earplugs, microphones, headsets, and acoustic insulators for use in intense noise.[10] Some of these projects produced dramatic results. The EAL's research on sound control in airplanes led to the invention of Fiberglas, installed as an acoustic insulator in most long-range military aircraft built after 1941.[11] The V-51R Ear Wardens—painstakingly engineered vinyl earplugs—probably had the most widespread impact of any PAL project. Demand from all branches of the military outstripped supply within months of their introduction. Tens of thousands of pairs were distributed over the course of the war. The Flint, Michigan, *Journal* reported under the headline "Navy Ear Plugs Cut Noise, Yet Let Sailors Hear Commands" that "with the ear warden ... loudness of both battle noises and the voice is brought down within the limits of hearing.... The command can be heard."[12] Noise caused chaos by breaking the links of the chain of command. Ear Wardens restored order.

The Chain of Communication

Several PAL projects reached beyond ordinary engineering problems to analyze humans as elements of communications networks, susceptible to the same criteria of performance and general methods of study as their electronic and mechanical counterparts. The military nature of these projects had a significant influence on the lab's work, in two major respects.

First, the conditions of extreme noise were specifically and intimately related to war. Much of the noise "problem" was as essential to war as communication itself: explosions, electronic jamming of enemy signals, giant machines laboring at their limits of performance. This narrowed the range of possible solutions dramatically. Furthermore, because of the "transient nature and violence" of gunfire, explosions, and other battle noise, the PAL found it more expedient to concentrate only on combat vehicles, systems that could be easily obtained, studied, and simulated.[13] The PAL's task could not be to eliminate the sources of noise, but only to reduce noise locally, to ensure the functioning of the human components of cybernetic weapons.

Second, nowhere else were solutions as urgent. Communication was an absolute prerequisite of effective combat. The PAL's job was to analyze and optimize the physical and psychophysical basis of the chain of command. The lab—mirroring Shannon's conception of the communications chain—studied the technologies that passed orders from commanders to soldiers and returned information in the opposite direction. It made no distinction between the technology of hardware and the technology of language and listening.

War noise thus helped to constitute communication as a psychological and psychophysical problem. The first task in approaching this problem was to bring noise into the laboratory by measuring and then simulating it.[14] According to the final report of Division 17.3, "the study of this problem formed a sizable part of the total effort of the two laboratories until the close of World War II."[15] Noise levels and spectra in combat vehicles were carefully measured and recorded. The EAL then built generating equipment to duplicate these sounds, and the PAL used this equipment to test men and communications devices for operating efficiency in noise. (Noise from the huge sound generators carried for several blocks around the lab, and they made for impressive public relations displays to visiting military officers.) The EAL also constructed the world's largest and most advanced "anechoic chamber" (a name invented by Beranek) by covering the walls, floor, and ceiling of a large room with wedges of sound-absorbing material. This acoustically "dead" room, built to simulate open-air conditions at the high altitudes where airplane crews operated, was the inverse of the noise generator. The eerie quiet of the chamber stood, opposite the cacophony of battle, for the silent but ubiquitous presence of scientific laboratories at the heart of high-technology war.

One of the first experiments by the PAL established that the effects of noise were militarily relevant only at the level of the command-control-communications system, that is, at the level of information processes. Conscientious-objector volunteers were tested on a series of motor coordination tasks, such as visual tracking and marksmanship, under simulated airplane noise as loud as 115 decibels. Even after spending eight hours a day, four days a week for four weeks under these conditions, no significant difference could be detected between their performance in noise and in quiet conditions. The only measurable effects were marked temporary deafness and a "subjective" feeling of fatigue.[16] *Environmental* noise did not affect the motor skills of soldiers; it did not alter their bodies or their behavior. But it did interfere with the transmission of messages.

After these experiments the PAL devoted all its efforts to reducing noise in communications systems. Noise was not a physical but an abstract threat: a threat to the mind, not the body—a threat to "information" itself.

Thus psychoacoustics began to theorize the chain of command as a chain of communication, and the person as one of its units. With the EAL, the PAL projects covered every element of the command-control-communications system, from radio components to training in speaking and listening to engineering military languages. Table 7.1,

Table 7.1
Some Considerations Affecting the Intelligibility of Speech Communication

Speaker
1. Quality and intensity of voice.
2. Correctness of pronunciation.
3. Manner of holding microphone, etc. (These factors call for selection and training of operators.)

Speech
1. Phonetic composition of oral codes, Alphabetic Equivalents, etc. (If properly chosen, these can add greatly to intelligibility.)

Microphone
1. Frequency-response characteristics.
2. Non-linear distortion.
3. Efficiency and impedance.
4. Behavior at different altitudes.

Amplifier
1. Frequency-response characteristics.
— etc. —

Radio Link
1. Over-all fidelity (response characteristic).
2. Signal/noise ratio.
3. Loudness of side-tone channel heard by speaker.
— etc. —

Listener
1. State of hearing (deafness).
2. Masking of speech by noise entering ear.
3. Basic ability to understand speech when distorted and masked. (There are great individual differences in this ability—calling for a program of selection and training.)

Source: From OSRD Report No. 901, "The Performance of Communication Equipment in Noise," Psycho-Acoustic Laboratory, Harvard University (1942). Harvard University Archives, 4.

from the research prospectus for a PAL study of radio-telephone systems, illustrates the comprehensive character of the PAL's research on communication systems.

The comprehensive system here conceived—much like Kenneth Craik's analysis of radar tracking systems, cited in chapter 6—includes the bodies of operators, the electrical and acoustic properties of equipment, the physical properties of spoken sound, the information content of speech, and the mental abilities of listeners.

The PAL's primary investigative tool was the "articulation test." In this test selected syllables, words, or phrases are spoken over a communications system. The percentage of them correctly heard by the listeners is the *articulation score* and identifies the efficiency of the system's components. This technique had emerged from research on telephone equipment at the Bell Telephone Laboratories in Murray Hill, New Jersey. Though the PAL psychologists, who were professionally interested in acoustic testing techniques, had of course known about articulation testing before the war, it was by collaborating with Bell Labs, where Division 17.3 of the NDRC was headquartered, that they learned the full range of its possibilities. M. H. Abrams and J. Miller of the PAL, involved in a speech training project, returned from a visit to Bell in late 1943 to report they were "amazed at the scope of the experimentation in the way of articulation testing, etc., which the Bell Labs had done."[17]

PAL scientists collaborated closely with Bell Labs, through correspondence, mutual sharing of reports, and occasional exchange of scientists between Cambridge and Murray Hill. As the war drew to a close, Bell indicated strong interest in hiring a substantial number of PAL psychologists, including J. C. R. Licklider, George A. Miller, W. R. Garner, and James P. Egan. (The PAL's John Karlin was eventually hired.)

The PAL also consulted with the Massachusetts Institute of Technology's Radiation Laboratory, especially in the later years of the war when both labs were working on "blind" flight guidance systems, and the two labs tested equipment for each other. Several PAL scientists held teaching positions at MIT during the decade after the war's end, including Licklider and Miller.

Language Engineering

Articulation testing was primarily used to determine the fidelity of communications equipment; but in terms of the eventual construction of "human information processing" as a new research paradigm, its most significant applications were to the analysis and engineering of language itself. The PAL investigated the relative intelligibility of words and phrases used as alphabetic equivalents, telephone directory names, and tactical call signs in noisy radio and telephone links. This project produced a general list of 1,000 "highly intelligible" common words, handed on to each of the armed services for use in battlefield communication. These lists, still employed today, are familiar from war movies in which radiomen shout hoarsely into their microphones such phrases as "Charlie Baker Zebra uh-One, do you read me?"

At the same time, the PAL developed methods for selecting and training talkers and listeners for military communications posts. Wide variation was found in the ability of individuals to understand and to make themselves understood through communications systems under noisy conditions. The PAL also discovered that brief training in articulation and listening improved performance measurably. The lab found that tests conducted in conditions of quiet did not predict performance in noise well; war, that is, gave speech new parameters of intelligibility. Finally, the PAL researched the jamming of voice communications signals, developing methods for "masking" speech with noise and annoying enemy listeners.

Language engineering had two important meanings for the future of psychological theory. First, it was the cornerstone of an emerging science of psycholinguistics. The PAL had to focus not only on human communicators and their equipment, but on the very substance of their communication: speech. Psychological theories of language of course predated the war, but their primitive state is indicated by the fact that in 1946, when George Miller was to teach a course on the psychology of speech and communication, he could find no suitable textbook and ended up writing his own (largely based on the lab's work).[18]

Second, with the engineered vocabulary of the call signs, language itself had become a field for scientific intervention. Like noise control, this became both necessary and possible because of the war, where the performance of information systems stretched to their outermost limits became a matter of life and death. Communications engineering, previously concerned with technologies and their interface with people, now took over the articulation of the messages themselves as well. "Natural" language was converted into a technology, a code or cipher device. Conceiving the communication relation as a kind of cryptography replaced the question of meaning with the problems of information, channel capacity, and accurate transmission as foci of the theory of language.

The Systems Research Laboratory
As an early PAL report put it, anticipating systems theory: "ultimate perfection of communication demands that each link in the total system be designed to complement and work effectively with every other link."[19] The Harvard group carefully detailed the questions to be asked of every element of the communication system. The most comprehensive research project of this nature began in mid-1943. At that time the Navy Coordinator of Research and Development requested that the PAL study and recommend improvements in the combat information centers (CICs) aboard Navy carriers and battleships. CICs were known as the "nerve centers" of the complex cybernetic devices that battleships were becoming. They coordinated battle activity in all parts of the ship by means of the sound-powered interphone system, communicated with other ships in the battle group, directed carrier-based fighter aircraft, and watched for incoming sea and air attacks on the ship's radar. Under pressure of Japanese kamikaze attacks late in the war, maximal efficiency and rapidity of response in the CIC became critical to the ship's survival.

By 1945, this problem had become acute enough to justify establishment of a third Division 17.3 lab at Jamestown, Rhode Island. The Systems Research Laboratory (SRL) created complete working models of the CICs. It added a new ingredient to the disciplinary mix of the PAL and EAL: "The research staff included about twenty research men, mainly psychologists and physicists, but also a sprinkling of specialists in the field of 'time and motion' engineering who were charged with the diagnosis and remedy of inefficiency in the activities and movements of the operating personnel at battle stations."[20] Essentially, "systems research" meant "human engineering." As the postwar proposal to continue SRL research put it,

the primary concept of basic Systems Research on Informational Devices and Centers is the scientific study of personnel and equipment as they must operate together under conditions of actual use. Systems Research utilizes the techniques of many fields of which engineering, physics and psychology are, perhaps, the most important. Systems Research is not primarily concerned with the development of individual equipments for specific applications, but rather with determining that these equipments when operated by average human beings are assigned appropriate tasks and contribute their correct share to the over-all performance of an organization.[21]

This approach was similar to that of the operations research concepts developed during the war (see chapter 4). The focus was on optimizing man-machine interaction, rather than on perfecting machine performance alone.

The Postwar Era
Both the Systems Research Lab and the Psycho-Acoustic Lab managed, after minor bureaucratic struggles, to obtain continued funding under combined military and academic sponsorship after the war. The SRL moved to Johns Hopkins under psychology professor Clifford T. Morgan. The PAL's wartime success was something of a coup for S. S. Stevens, its director, who "received tenure in 1944 and became the second-ranking member of the Department of Psychology after [Edwin G.] Boring." This was remarkable for a man of less-than-perfect intellectual pedigree whose academic skills had once been in question. "Only a few years before President Conant had told [Stevens] to expect to remain an assistant professor indefinitely."[22]

The PAL benefited, somewhat serendipitously, from tensions within the Harvard Psychology Department between social psychologists such as Gordon Allport and experimental psychologists like Stevens. In 1946, the Psychology Department split, with the social and clinical psychologists joining the new Department of Social Relations under the empire-building sociologist Talcott Parsons. The remaining members of the Psychology Department, primarily experimentalists, moved to the Psycho-Acoustic Lab's location in the basement of Memorial Hall. With Boring approaching retirement, Stevens effectively became the department's leader. In 1948, B. F. Skinner joined the department, moving into an office at the opposite end of the corridor from Stevens, a position interpreted by many colleagues as symbolic of a polarity between Skinner's strict behaviorism and Stevens's "operationism."[23]

Stevens established the postwar PAL under contract to the Office of Naval Research. Its last military contracts concluded at the end of 1961, though the lab continued its work under National Science Foundation and National Institute of Health grant support. In the sixteen years between the end of the war and the termination of its ONR contracts, the lab employed thirty-one psychologists, physicists, and physiologists, sixty-one research assistants, and some two dozen technical support staff. Lab scientists published 267 articles and books during this period, including a number of landmark studies in audition, Miller's textbook *Language and Communication*, and Stevens's monumental *Handbook of Experimental Psychology*, which replaced Murchison's 1934 text as the bible of the field.[24] PAL associates wrote six of the handbook's 36 chapters.

The PAL also maintained its connections with industry after the war, providing an avenue for exchange between academic and industrial psychology. Research on psychoacoustics was, naturally, relevant to the industrial production of communications equipment. The relationship with Bell Labs, in particular, flourished. John Karlin, who had been hired from the PAL by Bell, wrote to Stevens in 1952 to "clarify . . . possible future relations between Bell and psychology." He suggested that "good psychologists here at Bell would not only lead to better solutions of our problems, but would also develop psychologists and a body of psychological knowledge and thinking which would feed back effectively into psychology in general."[25] There was considerable revolving-door activity in the postwar years, with PAL scientists moving between Bell, other industrial posts, military laboratories, universities, and engineering academies like MIT.

Thus in less than four years, the Harvard Psycho-Acoustic Laboratory created a number of innovations that proved vital for the future of cognitive theory. First, its psychoacoustic studies created a background for an emerging psycholinguistics. Second, it helped to construct visions of human-machine integration as problems of psychology. Third, it established the psychological laboratory as what Bruno Latour has called an "obligatory passage point" in the study of human-machine systems. That is, any study of such systems would henceforth have to proceed by way of the tools, techniques, and concepts established by psychology and controlled by professional psychologists.[26] Since human-machine integration was a key interest of postwar military agencies, this effectively linked the PAL to the success of these agencies' research programs. Collectively, these accomplishments made the PAL the central locus for the nascent cognitive approach to psychology.

These achievements were, I have been arguing, tightly interwoven. The lab built upon an existing base and program of psychological research, and many of the scientific, technological, and social developments leading to information theory's emergence after World War II as the dominant life and social science paradigm would probably have occurred anyway. But the war accelerated them, lent them foci they might not otherwise have had (such as the problems of noise and hierarchical communications systems), and provided an institutional and social nexus for their crystallization.

As was true of research at many American university laboratories, the PAL's work in the postwar era bore an ambiguous relationship to its military sources of funding. Most of its research was only indirectly linked with specific military technologies. Yet the ONR never ignored the military relevance of the work it supported; its deep interest in operations research concepts led it to fund many kinds of research in human-machine systems, for which the PAL was perfectly positioned and toward which its research remained directed.[27] As Terry Winograd has pointed out, military funding for basic research can strongly influence the content and direction of research without intervening directly in the research process, simply by making vast amounts of money available for projects in a given area.[28] This was especially true in the postwar research climate, when military agencies entirely dominated U.S. R&D funding and the National Science Foundation had not yet emerged to provide a source of funds controlled by scientists themselves.

The PAL's request for postwar government funding stressed that

the experience of the present war has demonstrated that, in the field of communications, both the requirements and the stresses against their satisfaction are far greater in the midst of battle than in the realm of the commercial art. The intense noises of machinery and explosives handicap the ordinary processes of speech and hearing. Furthermore, the ears are called upon to perform many functions peculiar only to the military situation.[29]

This analysis suggests the special relationship between military structures and needs and the research pursued by the lab after the war. Compare the similar assessment of wartime acoustic technology requirements by Bell Laboratories historians:

Military requirements were quite different from those of commercial systems, since the former called for operation under conditions of use, noise, and power output bearing little resemblance to those of the ordinary telephone environment. Thus, military acoustical devices were not just copies or minor physical modifications of existing instruments but rather basically new designs.[30]

The connection of PAL research to specific military needs was diffuse. But like the Whirlwind group at MIT, the PAL participated in a process of mutual orientation with its military sponsors, each guiding the other's conception of research problems and potential solutions.

The war effort helped show psychologists how measurements at the outer limits of human and electronic capacities could reveal truths about ordinary situations. It emphasized to them the importance of such concepts as noise and information (even before formal information theory had spread); psychoacoustics helped join information theory and information technology to psychophysics. Finally, the research context of transmissions through a command chain became a paradigm for future work on communication. These factors help to explain why ex-PAL scientists like Miller, Pribram, and Neisser would later become primary exponents of the view of the mind as a hierarchically structured information processor incorporating multiply redundant systems as hedges against stressed, noisy communications channels. The PAL's work thus stands as an important precursor to the postwar restructuring of human experimental psychology around its new object, human information processing.

Psychoacoustics and Cognition: George A. Miller

Perhaps the best way to see how the work of the Psycho-Acoustic Laboratory blended with the other trends I have been discussing is to examine the intellectual trajectory of George Armitage Miller, one of the lab's most influential scientists.

Miller's academic career began at the PAL. Early work on problems of psychoacoustics led him to the psychology of language. Later, contact with information theory and computers through work and colleagues at MIT, Rand, and Lincoln Laboratories enabled him to become an important early exponent of the computer and information-processing models within academic psychology. In 1960, Miller co-founded Harvard's Center for Cognitive Studies, marking the beginning of cognitive psychology as a paradigm. Miller was, in many ways, at the center of the set of events I have been describing.

Born in 1920, George Miller took B.A. and M.A. degrees in Speech Communication from the University of Alabama in 1940 and 1941, remaining there afterwards to teach an introductory course in psychology. He moved to Harvard during the war to pursue his Ph.D. There Miller was trained in the behaviorist methodology and theory typical of the times. His first published papers, on simple learning problems, used rats as subjects and couched their findings in the standard language of stimulus and response.[31] Though he maintained his interest in speech and communication, such training was an unavoidable rite of passage in every major psychology department of the 1940s. As Miller recalled later, behaviorism

was perceived as the point of origin for scientific psychology in the United States. The chairmen of all the important departments would tell you that they were behaviorists. Membership in the elite Society of Experimental Psychology was limited to people of behavioristic persuasion; the election to the National Academy of Science was limited either to behaviorists or to physiological psychologists, who were respectable on other grounds. The power, the honors, the authority, the textbooks, the money, everything in psychology was owned by the behaviorist school.... [T]hose of us who wanted to be scientific psychologists couldn't really oppose it.[32]

In 1944, however, Miller became a fellow of the Psycho-Acoustic Laboratory, and he remained there for the next seven years.

At the PAL Miller worked on ways to jam enemy voice communications. He experimented with many different kinds of sound, such as static, bad music, and obnoxious voices, trying to determine which would produce the greatest interference with a listener's understanding of the primary signal in radio voice communications. At the PAL he was exposed to communication theory, language engineering, articulation testing, and technologies of communication. Out of this experience came both a doctoral thesis, *The Design of Jamming Signals for Voice Communications*, and a book, *Transmission and Reception of Sounds under Combat Conditions*, the NDRC summary technical report describing the PAL's history and wartime research (co-authored with F. M. Wiener of Bell Labs and PAL director S. S. Stevens).[33] As first author of this book, Miller learned details of every wartime PAL study.

Language and Communication
Throughout the postwar decade Miller continued to publish on technical problems in psychoacoustics,[34] but he also developed two other major areas of expertise: the psychology of language, and the application of information-theoretical techniques to psychological problems. Both interests were heavily marked, in their origin and in their development, by his experiences at the PAL. Miller's second

book, *Language and Communication*, was the first major American text on the psychology of language and the first textbook in the field now known as psycholinguistics. A significant proportion of its bibliography and contents came from PAL studies, which provided much of the experimental basis for theories of speech perception.

The introduction to *Language and Communication* shows Miller's behaviorist training beginning to give way to a submerged but nevertheless real cognitivism. On its first page, Miller wrote that the book's "bias" was behavioristic and that this was equivalent to taking a "scientific" approach to psychology. Miller worried that theories of organized information would seem insufficiently "objective," and he wrote defensively (in a dreadful mixed metaphor):

> It is necessary to be explicit about this behavioristic bias, for there is much talk in the pages that follow about patterns and organizations. Psychological interest in patterning is traditionally subjective, but not necessarily so. Discussions of the patterning of symbols and the influences of context run through the manuscript like clotheslines on which the variegated laundry of language and communication is hung out to dry. It is not pleasant to think that these clotheslines must be made from the sand of subjectivity.[35]

In fact, Miller's "clotheslines" were made not from sand but from mathematics, specifically the mathematical theory of communication and information.

Miller embraced a behaviorist theory of verbal learning:

> In general we learn to repeat those acts which are rewarded. If bumping into a chair is never rewarded, we soon stop behaving that way and start walking around it. In such cases the nature of the physical situation ensures that our responses develop in a certain way. Our response to the word "chair," however, develops differently. In order that we learn to respond correctly to the word "chair," it is necessary for another organism to intervene and reward us each time we respond correctly.[36]

But a few paragraphs later we find him discussing what looks like a quasi-Chomskyan picture of language as rule-governed: "Our choice of symbol sequences for the purposes of communication is restricted by rules. The job is to discover what the rules are and what advantages or disadvantages they create."[37] Immediately below, Miller gives a detailed description of communications systems in Shannon's formal-mechanical terms—source, destination, channel, and code—and compares the "human speech machinery" to a transmitter and the ear to a receiver, establishing the "vocal communication system" as another in a general class of such systems. (Shannon's communication theory, which Miller was able to read immediately since the PAL received Bell Labs technical reports, was perhaps the single most important influence on Miller's early theoretical development.)[38] Ten years later, with Chomsky's polemical review of Skinner's *Verbal Behavior*, the picture of language as an innate X system generating grammatical symbol sequences out of unstructured kernel sentences would devastate the behaviorist model.[39] In 1950, however, formal-mechanical modeling techniques had not yet spread to learning theory, and the two could therefore still coexist—if uncomfortably—in a single text.

Bringing Information Theory to Psychology

While he was writing *Language and Communication*, Miller was also trying to apply the statistical techniques of information and communication theory to the S-R model. In a series of articles, Miller and Frederick Frick analyzed rat behavior sequences and simple verbal learning using finite Markov processes, the statistical technique Shannon had employed to illustrate how messages can be built up stochastically on the basis of known letter or word frequencies.[40] Wiener's cybernetics and von Neumann's game theory were also mentioned as roots of what Miller called "statistical behavioristics."

The immense number of possible sequences made the Markov-chain approach of analyzing the dependencies of one behavioral "unit" on the unit or units preceding it more an interesting curiosity than a useful tool. But it represented another step toward cognitivism, since it too involved the relatively radical understanding of behavior as ordered sequences and patterns rather than simple stimulus-response pairs. The Miller-Frick articles were the first in academic psychology to employ concepts from information theory; another major article using communications-theory concepts was published by Miller's close PAL colleague and former roommate Wendell R. Garner in 1951.[41] Garner, analyzing the growth of cognitivism, argued later that the 1949 Miller-Frick article "could well be considered the birth date of cognitive psychology."[42]

By this point Miller was not only fully aware of developments in computers and cybernetics, but deeply involved in a number of ways. He knew of the work of the cyberneticians through Stevens (who attended the Neurological Supper Club with Wiener, Karl Lashley, and others starting in the late 1930s) and was especially impressed with the Rosenblueth-Wiener-Bigelow article, Wiener's cybernetics, and the McCulloch-Pitts theory of neural nets. He spent the summer of 1950 at the Institute for Advanced Study, studying mathematics and computers with John von Neumann. There, he said, "I learned that anything you can do with an equation you can do with a computer, plus a lot more. So computer programs began to look like the language in which you should formulate your theory."[43]

The trend toward information and communication theory in Miller's work took a sharp upward turn around 1951. In that year Miller moved from Harvard, where he had served as assistant professor of psychology since 1948, to MIT, where former PAL researcher and Macy conference veteran J. C. R. Licklider was building a new psychology program.

In fact, Licklider, Miller, and PAL veteran Walter Rosenblith not only taught at MIT but all worked on the SAGE project at Lincoln Laboratories. MIT's interest in building a psychology department stemmed less from intellectual imperatives than from a need to incorporate psychologists into work on the human-machine interface problems of the SAGE systems. Miller, Licklider, Rosenblith, and others were part of the "presentation group" that helped develop the audiovisual systems used by SAGE. With military money, Miller and Patricia Nicely built a laboratory at MIT for the study of speech perception. From 1953 to 1955 Miller worked at Lincoln Laboratories, leading a group carrying out studies of perception and its role in human-computer systems. At MIT Miller also had more extensive contact with the cyberneticians and their ideas.[44] Thus Miller, long before most other psychologists, came into contact with computers operating as part of a vast command-control system—a gigantic version of the communications systems the PAL had studied during World War II. Under these influences, Miller began to publish a steady stream of articles on the uses of information theory in psychology, including the important "What Is Information Measurement?" (1953), frequently noted in the psychological literature as the first general introduction to information theory for psychologists.[45]

Miller's early work was part of the broader sweep of human-machine integration studies known as "human factors" and "human engineering" research, whose major base lay in the military services. By 1946 the Navy Special Devices Center, which was also responsible for Whirlwind, had established a Human Engineering Section with at least six university contracts, including the SRL at Johns Hopkins.[46] By 1953 two of the ONR Psychology Division's four branches were Human Engineering ("the successful design of equipment for human use") and Physiological Psychology ("concerned with research on man's sensory, neurophysiological, and motor potential [and also] psychophysiological factors in complex work situations").[47] The Psychology Division controlled grant budgets in the $1.5 to $2 million range in the late 1940s and early 1950s; it and other military agencies supported most of Miller's postwar work. In 1950 the Air Force hired Fred Frick, Miller's chief collaborator in the statistical behavioristics studies, as a division head at its Human Factors Operations Research Laboratory. As Newell and Simon observed in their historical essay on the information processing approach in psychology, the human factors work created conceptual, personal, and communicative linkages between the emerging fields of computer science and information theory and of research on concept formation within psychology.[48]

In 1955, Miller returned to Harvard as an associate professor and once again worked in the Psycho-Acoustic Laboratory. In that year, with PAL/ONR support, he wrote perhaps his most important article, "The Magical Number Seven, Plus or Minus Two: Some Limits on Our Capacity for Processing Information."[49] Here Miller showed his colleagues how information theory could transform psychological research in perception and memory. Miller first reported how new experiments, which "would not have been done without the appearance of information theory on the psychological scene," had led to a new interpretation of traditional experiments in "absolute judgment" (judgments about the place of some stimulus, presented alone, on an absolute scale, such as the loudness of a pitch or the salinity of a solution). Considering experimental subjects as communications channels, the psychologist could test their "channel capacity" for absolute judgments.

For absolute perceptual judgments the relevant measure of channel capacity is the bit (the minimal unit of information corresponding to a binary choice). Channel capacities for many different sense dimensions fall in the neighborhood of three bits (five to eight alternative possibilities). However, in immediate (short-term) memory people can store much larger numbers of bits by "chunking" information into familiar units such as words, letters, or decimal digits.

Miller named verbal behavior as one such chunking or "recoding" process:

> The most customary kind of recoding that we do all the time is to translate into a verbal code. When there is a story or an argument or an idea that we want to remember, we usually try to rephrase it "in our own words." ... Our language is tremendously useful for repackaging material into a few chunks rich in information. I suspect that imagery is a form of recoding, too, but images seem much harder to get at operationally and to study experimentally.[50]

Recoding allows us to defeat the "information bottleneck" imposed by the stringent biological limitations on channel capacity: the span of absolute judgment and the span of immediate memory. The human, read as a communications channel, was again under scrutiny as a potential element of a cyborg for which information flow and rates of error were crucial variables.

Miller's experience with computers provided him with a major clue to this process. The chunking process was

> vividly illustrated for me the first time I saw one of those digital computing machines that have small neon lights to show which relays are closed. There were 20 lights in a row, and I did not see how the men who ran the machine could grasp and remember a pattern involving so many elements. I quickly discovered that they did not try to deal with each light as an individual item of information. Instead, they ... grouped the lights into successive triplets and gave each possible triplet pattern a number as its name, or symbol.[51]

"The Magical Number Seven" ended with a vision of the bright future of information-processing psychology: "A lot of questions that seemed fruitless twenty or thirty years ago may now be worth another look. In fact, I feel that my story here must stop just as it begins to get really interesting."[52]

A Cognitive Paradigm

Miller himself marks the year 1956, when he returned to Harvard, as the great transition. In that year his studies of language, information theory, and behavior crystallized into a new research paradigm. In an unpublished essay, Miller recounts his experience of the second Symposium on Information Theory, held at MIT on September 10–12, 1956. There he had his first realization, "more intuitive than rational, that human experimental psychology, theoretical linguistics, and the computer simulation of cognitive processes were all pieces from a larger whole."

At the symposium, Shannon and others gave papers on coding theory in engineering. Newell and Simon, fresh from the 1956 Dartmouth Summer Seminar on artificial intelligence organized by John McCarthy (see chapter 8), presented their Logic Theory Machine. Noam Chomsky discussed an early version of transformational-generative (TG) grammar in his "Three Models of Language." (TG grammars would rapidly replace the Markov-chain model of Shannon and Miller with their more powerful formal-mechanical model. Miller recalls that Newell told him Chomsky "was developing exactly the same kind of ideas for language that he and Herb Simon were developing for theorem proving.")[53] Miller himself presented his theory of "chunking" from the "Magical Number Seven" paper. Miller notes in his memoir that "what was remarkable was the coming together of [computer modeling of cognitive processes] with the other psychological and linguistic papers at the Symposium. Nearly every aspect of what we now call cognitive science was represented on that day."[54]

Two years later Miller, with Simon and Carl Hovland of Yale (who studied concept formation using information theory), organized a summer Research Training Institute on the Simulation of Cognitive Processes at the Rand Corporation. There he worked with Newell, Simon, and Shaw and their early artificial intelligence programs, the Logic Theorist and the General Problem Solver. Marvin Minsky, the MIT computer scientist whose influential paper "Steps toward Artificial Intelligence" had by then been widely circulated, was the other major leader of the institute.[55]

By 1958, then, Miller had worked in all of the following areas:

• study of rat behavior along standard S-R lines,

• psychoacoustic phenomena related to voice transmission technologies,

• mathematical theories of information, communication, and language,

• probability theory and its application to behavioral description,

• computer interface design, and

• computer modeling of cognitive processes.

That fall Miller arrived at Stanford's Center for Advanced Study in the Behavioral Sciences for a fellowship year carrying a sheaf of still-unpublished documents on computers and cognition by Newell, Simon, and Shaw.[56] There he discussed this material, among other things, with two other fellows: Eugene Galanter, another psychologist, and Karl Pribram, a neurophysiologist with interests in psychology.

Plans and the Structure of Behavior
This triad embodied the complex personal, institutional, and intellectual interconnections of postwar cybernetic psychology. Galanter had worked at the PAL in 1955–56. Pribram had been a visiting lecturer at MIT in 1954 and at Harvard in 1956, and he had worked at the PAL in 1951–52. Also, from 1946 to 1948 Pribram had worked at the Yerkes Laboratories of Primate Biology, then directed by the brain scientist Karl Lashley. (Lashley, also connected with Harvard, was an important early critic of behaviorism and delivered a major paper at the 1948 Hixon Symposium.[57] Lashley's prewar discussions with Norbert Wiener at the Neurological Supper Club had been central to the formation of cybernetics.[58] His Hixon paper argued that complexly patterned, rapidly executed behaviors such as the playing of arpeggios proceeded far too quickly for neurological execution of the chains of conditioned reflex arcs postulated by behaviorists.) Pribram's later work in neurology continued to develop Lashley's arguments for complicated internal structures.

Miller's discussions with Galanter and Pribram rapidly became a collaboration that produced the immensely influential book *Plans and the Structure of Behavior* (hereafter *PSB*). This book used a detailed computer metaphor to describe human behavior. It adopted the terms "Plan" and "Metaplan" (roughly, program and system program) to describe the structure of purposive activity. In postulating a complex internal logical structure, Miller and his colleagues abandoned the notion that observation alone could support the construction of psychological theory.

The central theoretical concept in *PSB* was that of the TOTE unit, an acronym for "test-operate-test-exit." This is a basic feedback loop in which a system performs an operation, then tests the results for congruity to a goal until the goal is reached. It is the same feedback principle upon which Wiener, Rosenblueth, and Bigelow had based

"Behavior, Purpose, and Teleology"; indeed, *PSB*'s announced goal was to "discover whether the cybernetic ideas have any relevance for psychology."[59]

The TOTE unit's psychological interest lay in its ability to incorporate transformations of information and control as well as energy within the same theoretical structure—a thoroughly cybernetic conception. As a flow of energy, the TOTE unit could describe any number of physical processes, from the neural reflex arc to thermostats and servomechanisms. When defined in terms of information, the test process could be made more abstract than a physical threshold, and the unit could be redescribed as a channel that correlates input information with output in a predictable way. But the most useful level of discussion for Miller, Pribram, and Galanter was also the most abstract: feedback as a flow of "control."

This concept appears most frequently in the discussion of computing machines, where the control of the machine's operations passes from one instruction to another, successively, as the machine proceeds to execute the list of instructions that comprise the program it has been given.[60]

The authors described looking up references in an index as an instance of a "control" TOTE. Behavior in searching for one page after another is under the control of successive index entries. "Here we are not concerned with a flow of energy or transmission of information from one page number to the next but merely with the order in which the 'instructions' are executed."[61] Just as a computer uses the same system to control its own programs as it does to process information, so a person can use the same number either as a "control" instruction (e.g., as a page number in an index) or as an item of information (e.g., as a price).

The remainder of *PSB* was devoted to the analysis of various forms of behavior, including "cognitive behavior," as the execution of instruction sets in the form of TOTE units. The key notions were organization, hierarchy, and feedback in the passage of control from one instruction to the next. This is the same structure I have been calling the "chain of command," relocated within the individual.

The influence of the work of Newell, Simon, and Shaw pervaded *PSB*.[62] Miller, who added most of the documentation after the book was written, referenced their papers more often than any others, and he devoted ten pages to their studies of heuristic processes in constructing computer programs to play chess and solve logic and trigonometry problems. Miller's group followed Newell, Simon, and Shaw in adopting the engineering/programming approach to understanding behavior: analysis of system design and formal-mechanical modeling. They spoke of "man viewed as a system for processing information"[63] and noted that with the TOTE concept, that system could be modeled and certain features of behavior explained in terms of design:

- The hierarchical structure underlying behavior is taken into account in a way that can be simply described with the computer language developed by Newell, Shaw, and Simon for processing lists.
- Planning can be thought of as constructing a list of tests to perform. When we have a clear Image of a desired outcome, we can use it to provide the conditions for which we must test, and those tests, when arranged in sequence, provide a crude strategy for a possible Plan. . . .
- The operational phase can contain both tests and operations. Therefore the execution of a Plan of any complexity must involve many more tests than actions. This design feature would account for the general degradation of information that occurs whenever a human being is used as a communication channel.[64]

In other words, behavior was the execution of a program; planning was programming; and defects in the design of the human cybernetic device explained communications and information processing failures.

Discussing computer models of problem-solving, Miller and his colleagues wrote:

It is impressive to see, and to experience, the increase in confidence that comes from the concrete actualization of an abstract idea—the kind of confidence a reflex theorist must have felt in the 1930s when he saw a machine that could be conditioned like a dog. . . . That confidence is no longer reserved exclusively for reflex theorists. Perhaps some . . . conjectures of the "mentalists" should now be seriously reconsidered. Psychologists have been issued a new license to conjecture.[65]

The passage indicates the crucial significance of the experience of interacting with computers to *PSB*'s claims.

Instead of simply rejecting the claims of behaviorism, the information-processing theorists, like Wiener and the cyberneticians before them, *subsumed* the S-R model within their own as a special case. Simultaneously, they discounted simple S-R models:

Reinforcements are... a special kind of feedback that should not be identified with the feedback involved in a TOTE unit. That is to say: (1) a reinforcing feedback must strengthen something, whereas feedback in a TOTE is for the purpose of comparison and testing; (2) a reinforcing feedback is considered to be a stimulus (e.g., pellet of food), whereas feedback in a TOTE may be a stimulus, or information (e.g., knowledge of results), or control (e.g., instructions); and (3) a reinforcing feedback is frequently considered to be valuable, or "drive reducing," to the organism, whereas feedback in a TOTE has no such value.[66]

Plans and the Structure of Behavior, using the tools of cybernetic theory, computer modeling, communications engineering, and linguistic logic, was the first major work within academic psychology to construct a comprehensive, coherent picture of cognition as information processing. It was built, as we have seen, around computer models and concepts of hierarchically organized control systems. These, in turn, relied in part on the metaphor of the chain of command Miller had first studied at the Psycho-Acoustic Laboratory during the war.

Though nontechnical and speculative, *Plans and the Structure of Behavior* became a manifesto for cognitivism. It was the first text to examine virtually every aspect of human psychology, including instincts, motor skills, memory, speech, values, personality, and problem solving, through the lens of the computer analogy. It also made explicit the methodological and theoretical approach of formal-mechanical modeling, including computer simulation. This approach became the core of the cognitive paradigm.

PSB excited a great deal of interest. It received long reviews in at least six major psychological journals, including reviews by the influential D. O. Hebb and by Macy Conference participant David McKenzie Rioch. Between 1966 and 1975 alone, other psychologists cited *PSB* well over 500 times (more than almost any other book).[67] Moreover, Miller's work with Galanter and Pribram was a key factor in his founding of the Center for Cognitive Studies with Jerome Bruner in 1960, after his return to Harvard. *PSB* thus propagated cyborg discourse throughout the psychological disciplines.

The Center for Cognitive Studies
The Center for Cognitive Studies (CCS) was the first major institutional embodiment of cybernetic psychology. The word "cognitive" was used "defiantly" in its title, Miller later wrote, since "most

respectable psychologists at the time still thought cognition was too mentalistic for objective scientists."[68] One CCS fellow recalled that upon his arrival at Harvard, he was summarily informed that "the Center was in the business of demolishing behavioristic doctrine and replacing it by a mentalistic approach."[69]

The CCS was an interdisciplinary group devoted to the study of "questions concerning the nature, organization, and transformation of human knowledge ... questions of perception, memory, thinking, learning, and decision making."[70] Over the years it became a gathering place for those scientists who were most active in the blending of psychology, linguistics, computer modeling, philosophy, and information theory that Miller and Bruner were backing. Noam Chomsky, Nelson Goodman, Benoit Mandelbrot, Donald Norman, Jerrold Katz, Thomas Bever, Eric Lenneberg, and Joseph Weizenbaum were only a few of the dozens of visitors who spent a year or more at the Center between 1960 and 1966. More than $2 million in grant money flowed into the Center's coffers, mostly from the Carnegie Corporation, the National Institutes of Health, and ARPA, the Advanced Research Projects Agency of the Department of Defense.[71]

The "General Description" of the Center written by Miller and Bruner at its founding notes the importance of computers and other mechanical analogs to the new field of study. Cognitive theory was to be a theory not simply of human cognition but of information devices in general. Today, they wrote, questions about the nature of thought and knowledge

have application not only to the study of man but also to the devices man uses to amplify his cognitive control over his environment. Consider an example: the study of memory systems and devices now extends far beyond any philosophical or psychological formulations. Librarians, geneticists, educators, computer engineers, geologists, and historians share the psychologist's desire for a more general theory of memory.

The Center for Cognitive Studies is concerned with how information is stored, processed, and communicated—both by human beings and by the devices human beings invent in order to cope with information.[72]

Simulation of cognitive processes by computer was one of the main components of the Center's program. One of the advantages of cognitive simulation was that "human information processing is placed in its proper frame of reference as one of many alternative ways an intellectual system might function, thus adding perspective to our understanding of mental functions."[73] Thus Miller and his colleagues translated the wartime program of the "machine in the middle" into the primary object of knowledge for cognitive theory's first academic institution.

Since the Center's founding Miller has been one of the foremost exponents of cognitive psychology, writing and editing numerous volumes on psychological theories of language, communication, and perception. Interestingly, like his former student Ulric Neisser he later came to believe that "how computers work seems to have no real relevance to how the mind works, any more than a wheel shows how people walk."[74]

Miller's career embodies the weaving together of the several strands of an emerging cognitive science. Beginning with psychoacoustics and the wartime problem of turning meaning into noise (jamming), Miller moved through information theory, linguistics, and stochastic statistical behaviorism. Working on SAGE in the Lincoln Laboratories presentation group, he returned to the problem of human-machine integration, this time with digital computers; the air defense chain of command and control linked "cognition" to roots in the wartime ergonomic studies of antiaircraft fire control and the defeat of noise in the cyborg fighters. Through the Rand Corporation Miller learned about artificial intelligence and cognitive simulation. Over the course of his early career, he gathered up the object now studied as cognition from its several sources: language, perception, memory, problem-solving, and computer simulation. With *Plans and the Structure of Behavior* and the Center for Cognitive Studies, Miller helped establish the "cognitive revolution" in psychology as an accomplished fact.

Conclusion

In this chapter we have explored the roles of one institution, and one of its most illustrious alumni, in creating the research program and theoretical perspective of cognitive psychology. The PAL faced problems of human-machine integration; it responded by probing the nature of perception and language, on the one hand, and of electronic communication systems, on the other. George Miller, focusing on problems of language, sought concepts and tools for analyzing internal processes too complex to fit neatly into the black boxes of behaviorism. These projects oriented Miller and the PAL toward the ideas and metaphors of cybernetics. These led, in turn, to the cyborg psychology of symbolic information processing.

In one way, this chapter has been about the origins of cognitivism in a standard historical sense: who had the first ideas, when, where, and why. In another way, however, my goal has been to illustrate three more theoretical points. First, I emphasized the close relationship between the PAL's research and wartime needs as a way of pointing to the *socially constructed* character of scientific theory. Social networks developed to solve the war's practical problems helped convey new ideas, such as those of cybernetics and information theory, among disciplines. Engineering projects founded in wartime technological strategies channeled experimentation in particular directions, forced psychologists to face the implications of their theories for design, and led them into finely detailed studies of relationships between people and machines. Second, I drew attention to *the key role of information-machine metaphors*, drawn in large part from the wartime work in human-machine integration, in constituting cognitive theories. The computer eventually became the most important of these. But for psychologists experienced with electronically linked communications webs in combat machines—real war cyborgs—it was not the first. Finally, I have chosen these examples to illustrate the intricate *connections among levels* typical of science as a sociopolitical activity. In describing Miller's work, for instance, I stressed the many strands of a career, intertwining personal experiences and intellectual acts with formal training, institutional locations like the PAL and Lincoln Laboratories, collaborations with other individuals, access to advanced computer equipment, military financial support, and so on. Miller's *ideas* were only one part of all this. Yet all of it was significant to the particular way in which he carried cybernetics to psychology—not just as abstract concepts, but as concrete examples, experimental projects, techniques, technologies, and metaphors.

The social construction of theory, the centrality of metaphor, and the interconnections among social levels point us once more toward the concept of discourse. Miller and the PAL did much more than run experiments and prove theories. They helped set the terms of cyborg discourse, the understanding of minds as information processors and information machines as potential minds. Placing psychology at a central leverage point in the analysis of human-machine systems, they primed the field to participate in the construction of real cyborgs. Through a mutual orientation process much like the one described in chapter 3, their military supporters came to seek out psychological theories of the "machine in the middle" for their nascent electronic battlefield. Building the scaffolding of cognitive theory from computer metaphors and information processing, Miller and the PAL produced a psychology of humans as natural cyborgs. In so doing, they helped create the cyborg subject, for whom experience and knowledge are built bit by bit, as it were, from pure information.

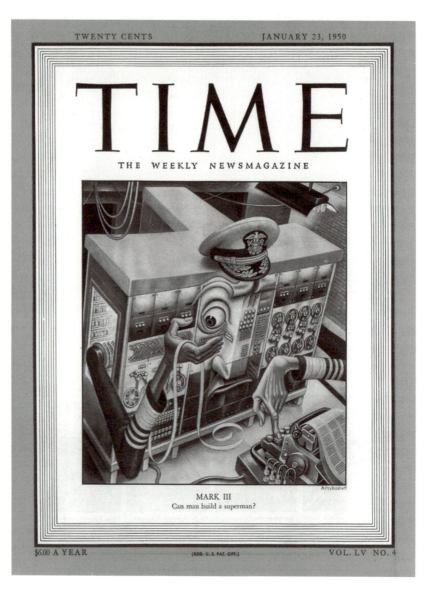

Courtesy Time/Life Syndication.

8
Constructing Artificial Intelligence

Where cognitive psychology analyzed human intelligence as an information process—human minds as cybernetic machines—artificial intelligence (AI) sought information processes that could exhibit intelligent behavior—cybernetic machines as thinking minds. AI established a fully symmetrical relation between biological and artificial minds through its concept of "physical symbol systems." It thus laid the final cornerstone of cyborg discourse.

Even more than that of cognitive psychology, AI's story has been written as a pure history of ideas.[1] AI inherited the ambitions of cybernetics, and like the cyberneticians, the founders of AI have been much concerned to document their own history, which they (quite naturally) view mainly in intellectual terms. Yet AI, too, has a prehistory, tightly linked to that of cybernetics, that dates to World War II and before. This chapter, like the last two, sets the birth of a new science against a wider background of postwar practical needs, political discourses, and social networks. Rather than rehearse the standard stories, I focus here on what has been obscured by the canonical accounts.

Instead of modeling brains in computer hardware—the central goal of cybernetics—AI sought to mimic minds in software. This move from biological to symbolic models has usually been interpreted as an abrupt intellectual break, a sudden shift in orientation from process to function. I see, on the contrary, a more gradual split. With the possible exception of John McCarthy, all of AI's founders were significantly influenced by the biological models of cybernetics. Even as late as the 1956 Dartmouth conference, the birthplace of AI as a systematic research program, the rift between brain modeling and symbolic processing remained incomplete.

I am going to argue that this shift occurred for reasons that had as much to do with the practical problems and subjective environments of computer use as with purely theoretical concerns. Links, almost fortuitous in nature, between AI and time-sharing systems brought the budding discipline into close connection with military projects for human-machine integration in command-control systems. Because it tied AI to a realizable development program with a wide range of practical uses, this connection led to AI's most important institutional support: the Information Processing Techniques Office (IPTO) of the Advanced Research Projects Agency (ARPA). IPTO's founder, PAL and SAGE veteran J. C. R. Licklider, aggressively promoted a vision of computerized military command and control that helped to shape the AI research agenda for the next twenty-five years.

From Cybernetics to AI: Symbolic Processing

The founders of cybernetics were mostly mathematicians, like Wiener and Pitts, working with brain scientists, like Rosenblueth and McCulloch. While their theories of information and communication succeeded in erasing the boundary between humans and machines, allowing the construction of cyborgs, the cyberneticians themselves remained largely within the perspective of the mechanical. From W. Ross Ashby's *Design for a Brain* and Grey Walter's mechanical tortoises to von Neumann's self-reproducing factories and Shannon's maze-solving mechanical mice, the cyberneticians always ultimately sought to build *brains*. For them the formal machine itself, the logical model, was first and foremost a *machine*, an actual device with fixed properties and hard-wired channels for inputs and outputs. Even "the things that bothered von Neumann" at the Macy conferences had to do with whether a computer, despite its physical differences, could fruitfully be compared to the human brain; he spent his last days worrying over that very question.[2] The principal methods of cybernetics as practice, like those of its behaviorist antecedents, built complex structures out of simple, low-level, physical units such as neurons or reflex arcs. Frequently, based on then-current beliefs about how the brain worked, cyberneticians tried to design self-organizing machines that would achieve complex behavior through encounters with their environments. The subject, in both senses, of cybernetics was always the *embodied* mind.[3]

The next intellectual generation, the students of McCulloch, Pitts, Shannon, and Wiener, took one further step. They placed the emphasis of formal-mechanical modeling on the side of the formal, the disembodied, the abstract—on the side of the mind rather than that of the brain. Some of them were too young to have been shaped by the wartime experience of combining academic science with military engineering; for this generation, computers appeared not primarily as tools for solving practical problems, but as automated mathematical models with a powerful intellectual appeal. The key to a truly comprehensive theory of intelligence lay, for them, in the notion of "symbolic processing." This idea, in turn, acquired concrete standing and directionality through the craft activity of computer programming. This direction in computer evolution had been foreshadowed by Alan Turing's earliest work.

The Turing Machine

Turing's paper "On Computable Numbers, with an application to the *Entscheidungsproblem*," was published in 1937. It proved an ingenious solution to the last unsolved prong of mathematician David Hilbert's three-part challenge concerning the ultimate foundation of mathematical reasoning. Hilbert had asked whether mathematics was complete (whether every mathematical statement was either provable or disprovable), whether it was consistent (whether two contradictory mathematical statements could never be reached by a valid sequence of steps), and whether it was decidable (whether some mechanically applied method, which is to say some formal machine, could guarantee a correct analysis of the truth of any mathematical assertion). Kurt Gödel's famous theorem of 1930 had answered the first two questions, demonstrating that arithmetic and all other mathematical systems of comparable richness are incomplete, and that they cannot be proved consistent. Turing's paper answered the third, proving additionally that mathematics is not decidable—that some mathematical problems are not susceptible of any algorithmic solution.

Of greatest practical importance in Turing's work, however, was not the mathematical result but the unique method he devised to reach it. Turing proposed an imaginary machine something like a typewriter, operating on a paper tape of infinite length divided into unit squares. Each square would either be blank or contain a single mark. The machine's operations would be limited to "reading" the tape (recognizing blanks and marks); erasing marks; writing marks in blank squares; advancing to the right or returning to the left on the tape; and stopping.

Using only these basic operations, the machine could be "configured" in an infinite number of ways. For example, it could be designed to move to the right upon encountering a marked square, fill in the first blank square it came to, and continue to the right until it encountered a second blank, at which point it would move one square to the left, erase that square, and stop. This operation would make a single string of marked squares from two adjacent strings, preserving the total number of marks—the equivalent of adding two numbers. Such a Turing machine could function as an adding machine. An infinite number of other configurations are possible, and ultimately, any rule-based symbolic operation could be modeled by the machine. Many of the key elements of the electronic digital computer were already present in this purely hypothetical device: binary logic (squares are either blank or marked), programs ("configurations"), physical "memory" (the tape), and basic reading and writing operations.

Turing proved that this extremely simple machine, given the appropriate finite sequence of "configurations" or instructions, could solve almost any precisely specified symbolic problem that had a solution.[4] In other words, we can simulate almost any mathematical operation, of any degree of complexity, with a Turing machine by reducing it to a series of simple steps. Thus the Turing machine can also simulate any other logic machine. Hence most actual digital computers, while quite different in design from the primitive tape reader of Turing's paper, are in fact specimen Turing machines. The Turing machine is thus a conceptual bridge between the universalist ideas of logical automata (developing since Babbage and Boole), the metamathematics and metalogic of Hilbert, Gödel, and the logical positivists, and the practical problem of numerical computing machinery.[5]

Even in this early work, Turing had considered the relationship between the infinite set of "configurations" of his simple machine and the mental states of human beings. A human "computer" (the word was not applied to machines until the mid-1940s) performing the operations of a Turing machine by hand would necessarily, on his view, proceed through a sequence of discrete "states of mind" directly parallel to the states of the machine. "The operation actually performed is determined ... by the state of mind of the [human] computer and the observed symbols. In particular, they determine the state of mind of the computer after the operation is carried out.... We may now construct a machine to do the work of this computer."[6]

Elsewhere in the paper Turing made it clear that the essential move in his analogy was to reduce each state of mind of the (human) computer to its atomic units. This could be achieved by breaking down any complex operation the person performed into a series of definite steps—precisely the basic operating principle of the Turing machine. A computer programmed to carry out the same steps would thus be doing exactly the same thing as the person. Hence its activity would be at least directly analogous, and perhaps even identical, to the series of mental states experienced by its human counterpart in performing the same operation.

Despite its cogent formulation of the basic principles of machine computation, however, "On Computable Numbers" had very little effect on the invention of actual computers in America. Most of the early computer pioneers read Turing's paper, if they read it at all, only after their own work was done.[7] The sole exception was John von Neumann, who met Turing in Cambridge, England, and who also became the only one of the early computer pioneers to explore the computer-brain analogy in a serious way. Instead, Turing's first strong influence on American thought occurred through the cybernetics work of McCulloch and Pitts, whose theory of the brain as a logical automaton was partly inspired by "On Computable Numbers."

Turing's explorations of the mind-machine analogy were a minor, if significant, element of his original paper. But over the next twelve years these analogies grew more full-blooded, coming eventually to occupy a central place in Turing's thought. In World War II Turing found an opportunity to build real versions of his hypothetical machine at Bletchley Park (see chapter 1). By 1948 he had composed a technical report for the British National Physical Laboratory entitled "Intelligent Machinery." This essay relied heavily on the same behaviorist picture of the brain used by McCulloch, Pitts, and the cyberneticians. Namely, the brain was a self-organizing machine that required extensive experience and training (analogous, for Turing, to programming) in order to realize its potential for intelligence. At this stage, Turing's focus on learning gave his ideas about intelligent machines a quasi-biological cast, while his method of symbolic programming more closely resembled what AI would eventually become. Thus this work occupied a transitional place between the brain models of cybernetics and the symbolic processors of artificial intelligence.

For the future of psychology and AI, Turing's most important work was unquestionably his 1950 article in *Mind*, "Computing Machinery and Intelligence." This paper, by proposing the Turing test, effectively focused most of the future debate about machine analogs to human thought processes. The Turing test's hypothetical computer could interpret and compose not only numbers and logical notation, but also *written language*. The computer became far more than a calculator—it was a symbol processor par excellence, a universal information machine.

No machine then in existence could produce anything remotely resembling the performance demanded by the Turing test. Nor did Turing provide even a hint of how such abilities might be created. Yet the concept of communicating with computers in ordinary language foreshadowed the evolution of computer software over the decade 1945–1955. During that period computer languages grew increasingly sophisticated and increasingly removed from the level of digital computation, albeit for practical reasons and not primarily because of Turing-like goals. Not until high-level languages began to become available did Turing's vision of a computerized mind become something more than a speculation or a dream. To understand how and why this happened, we must briefly review how computers work.

Symbolic Computing: Levels of Description
The operation of computing machinery can be described at a number of levels. The "lowest" of these is the electronics of hardware: electronic switches, resistors, magnetic storage devices, and so on. Descriptions are causal and physical: electrons, currents, voltages. The second level is digital logic. The hardware electronics are designed to represent logical or mathematical operations, such as "AND" or the addition of binary digits or "bits"; the hardware itself is thus described as a set of logic "gates." Descriptions are logical, not physical, but they are still tied to the hardware itself, whose structure determines how each operation affects its successors. At a third level lies the "machine language" of the programs that "run" on a particular machine. Machine language consists of the binary representation of program instructions—the language the machine itself "speaks." Machine language also remains tied to a particular machine; programs written in one machine's language generally will not run on a different machine. However, machine-language programs are not a fixed part of the hardware (like its logic gates), determined by its physical structure; they are user-determined instructions. Assembly languages, consisting

of alphanumeric mnemonics for machine language instructions, constitute yet another hardware-dependent tier.

At an even higher level of description are compiler programs. Compilers are metaprograms, originally written in machine language, that translate programs written in "user-oriented" or "high-level" languages (now usually just called "computer languages," such as Pascal, Ada, and C) *into* machine language. Compilers make high-level languages "machine-independent"; given a compiler, the same program in a high-level language may run on any machine, irrespective of its particular electronics, digital logic, or machine language. (High-level languages cannot, in fact, function without compilers, since only a machine-language program can actually run a computer.) By translating statements in high-level languages into the calculations of machine language, compilers also create the possibility of easy symbolic manipulation.

High-level languages *may* be mathematical, like the algebraic languages FORTRAN or ALGOL, but they may also be symbolic—that is, they may take the form of language or logic, as in the business language COBOL or the AI language LISP. At the highest level of all lie another set of metaprograms—user interfaces and operating systems—usually themselves written in some high-level computer language. These metaprograms control how users interact with the machine, managing its processors, memory, disk, and other resources. Examples include the Unix operating system and the Windows or Macintosh graphical user interfaces.

These nested levels are relatively opaque to one another. In general, those concerned with one level need not know anything about levels "below" or above the ones they use. Programmers need not know how the interface works or concern themselves with the particular hardware of the machines on which their programs will run. Designers of user interfaces need only understand the high-level computer language in which they compose their programs, not those of compilers, machine languages, digital logic, or chip hardware. Chip designers need not understand or even consider compilers or high-level languages. Someone writing a program to manipulate symbols, such as a word processor, need not comprehend how the compiler translates each instruction into a series of calculations in binary arithmetic.

Each level is *conceptually* independent of the ones below and above, while remaining *practically* dependent on the lower levels.[8] Higher levels are "reducible" to lower levels in the brute sense that any higher-level operation may be described as a definite series of lower-level operations. But while the cliché that computers understand nothing but ones and zeroes is true in this sense, it is false in the more important sense that "what the computer is doing" *for its users* is defined by the higher levels of description—for example, by the computer language and the user's goals and methods. This insight into the possibility of a symbolic, machine-independent level of description in computing was the conceptual foundation of artificial intelligence.

As with other issues, such as the advantages of digital over analog computing, the practical conditions of computer development obscured the possibility of symbolic computing foreseen by Turing and others. To the majority of computer designers and users in the first decade of electronic computing, the conceptual relationships I have just described were scarcely visible. An intellectual history might say, anachronistically, that those practical conditions simply delayed the genesis of symbolic computing, a development that merely played out an inevitable conceptual logic. But in fact symbolic computation did *not* emerge mainly from theoretical concerns. Instead, its immediate sources lay in the practice of the programming craft, the concrete conditions of hardware, computer use, and institutional context, and the metaphors of "language," "brain," and "mind": in other words, the discourse of the cyborg.

Writing Programs, Building Levels: Programming and Symbolic Computing

The (extremely limited) memory capacity and other hardware features of early digital computers placed a premium on efficient algorithms and compact program code. In the ENIAC, programming still involved the mechanical operation of physically interconnecting the machine's registers by means of plugboards. From the EDVAC on, programs themselves were coded as binary numbers and stored in memory along with the data. In their case the numbers stood for instructions (e.g., ADD a TO b) or addresses in memory (i.e., the location where the values of a or b were currently stored), thus incorporating the fundamental insight of Gödel and Turing that algorithms could be written in the same language as the data upon which they would operate. Until the late 1940s all program code was written in machine language; any numerical data had to be converted from decimal to binary form before the machine could process it. Because it consisted entirely of concatenations of only two symbols (ones and zeroes), machine language was extremely difficult to use and even more difficult to debug.

In 1949 Eckert and Mauchly introduced Short Code (for the ill-fated BINAC). Employing alphanumeric equivalents for binary instructions, Short Code constituted an "assembly language" that allowed programmers to write their instructions in a form somewhat more congenial to human understanding. In Short Code, to represent the equation "$a = b + c$" one would write the instruction "S0 03 S1 07 S2," where S0, S1, and S2 stood for a, b, and c respectively; 03 meant "equal to"; and 07 signified "add."[9] A separate machine-language program called an "interpreter" translated Short Code programs into machine language, one line at a time. Interpreters were quick and efficient, and they elevated the symbolic level of programming a notch upwards toward algebraic language. But instructions still had to be entered in the exact form and order in which the machine would execute them, which frequently was not the conventional form or order for composing algebraic statements. It also required breaking down every complex operation into simple components.

The earliest programmers were primarily mathematicians and engineers who not only programmed but also designed logic structures and/or built hardware; for this tiny group, interpreters often seemed adequate to the job. Among them, the mathematical aesthetic of brevity-as-elegance, also motivated by the economics of memory and computer time, prevailed. Short, efficient algorithms and highly mathematical code ruled this culture's values. (As late as the middle 1960s at MIT, such an aesthetic reigned among the hackers who wrote much of the operating system software and many important utility subroutines for MIT's computer systems, even though by then higher-level languages had become the norm.)[10] Somewhat ironically, perhaps, programming a computer became a kind of art form.

But in order for nonexperts to write computer programs, some representation was required that looked much more like ordinary mathematical language. Higher-level languages, in turn, required compilers able not merely to translate statements one-for-one into machine code, as interpreters did, but to organize memory addressing, numerical representation (such as fixed- or floating-point), and the order of execution of instructions. (Frequently a single instruction in a higher-level language will be compiled into a dozen or more machine-language instructions.) Such programs required what then amounted to exorbitant quantities of memory and machine time.

Philip Morse (whose NDRC Committee on Sound Control in Aircraft had spawned the Psycho-Acoustic Laboratory during the war) chaired MIT's Committee on Machine Methods of Computation in the early 1950s and was named director of its Computation Center in 1955. He recalled that bitter struggles over higher-level languages sprang from the aesthetics of programming and the economics of machine time and memory.

> The argument over the desirability of compiler programs was intense and acrimonious. The hardware experts were horrified. It was inefficient, they protested; every computation would have to run twice through the machine, once using the compiler to prepare a program in machine language, then running this program to direct the machine to do the actual calculation. In addition, the compiler itself... would have to be a program in machine language, and it would have to be the great-grandfather of all programs, because it would have to foresee all the different operations used in any sort of computation and would have to guard against all the logical errors that might occur. It would take dozens of man-years of the best programmers' time to write one, and, when written, it would be so big there would not be room for anything else in the machine's memory.[11]

At first, these predictions proved accurate enough. Compilers *did* produce inefficient machine code. The first compilers, written between 1950 and 1953, slowed even huge, superfast machines such as Whirlwind to a veritable crawl. Furthermore, debugging a compiled program often required sorting through the compiler-generated machine code itself, a task rendered more difficult by the fact that the code followed the machine's logic rather than that of a human author.

As sales of commercial computers mounted rapidly, so did the number of programmers. In 1950, with only a few computers in existence, there were only a few hundred programmers. Most of these people would not have identified themselves as such, since they also played other roles in machine design or construction. By 1955 the one thousand or so extant general-purpose computers required the services of perhaps 10,000 programmers. Five years later, in the midst of a booming commercial computer market, programming had suddenly become a profession in its own right, with about 60,000 practitioners servicing some five thousand machines. Programming began to emerge as a craft, a specialized practice; already the amount of mathematical skill it required had begun to diminish. In the words of Michael Mahoney, it was "technical rather than technological, mathematical only in appearance."[12]

As we saw in chapter 3, programming for the SAGE system contributed its massive momentum to this process through the Systems Development Corporation and its many spin-offs. SDC hired and trained thousands of new programmers beginning in the mid-1950s. Music teachers and women without specialized backgrounds were among the most successful recruits.[13] Such groups could not be expected to learn machine code or produce mathematically elegant algorithms; to make this new work force effective required symbolic languages easily learned by nonspecialists. Businesses, too, wanted to write their own software without hiring expensive experts. Not only did they want to do this, they had to if they were going to use computers at all, since before the 1960s "off-the-shelf" software was virtually unknown. Software had to be written specially for each type of machine and, in many cases, for each *individual* machine. The exponentially increasing demand for software, and thus for nonspecialist programmers, helped drive the movement toward higher-level languages.

The first true algebraic-language compiler was written in 1953 for Whirlwind, but its slow speed and the unavailability of other equally capable machines prevented its widespread use. In an attempt to kick-start the field, the Office of Naval Research sponsored symposia on "automatic programming" in 1954 and 1956. The name "automatic programming" for compilers is itself revealing. "Programming" still referred to the composition of the machine-language instruction list. Algorithms written in a higher-level language were not yet seen as actual programs, but rather as directions to the compiler to compose a program; compilers thus performed "automatic programming." The independence of symbolic levels in computing had not yet achieved the axiomatic status it later acquired.

The first commercially viable higher-level language was FORTRAN, an algebraic, scientific programming language. IBM researchers completed the FORTRAN compiler in 1956–57. In 1959, at the request of a consortium of universities, computer users, and computer manufacturers, the Defense Department convened a Conference on Data Systems Languages (CODASYL). CODASYL soon produced specifications for the English-like, data-processing-oriented high-level language COBOL (for COmmon Business-Oriented Language).

Among the most durable of all computer languages, COBOL still finds widespread use today. Though COBOL was designed and used primarily for business data processing, the DoD's leadership in the CODASYL standard-setting reflected its continuing status as the major customer and primary supporter of computer research.[14]

Though programmers developed user-oriented computer languages for purely practical ends, the process of creating these languages generated awareness of the potential of computers as manipulators, not just of numbers, but of symbols of *any* type. The compilers turned English-like commands such as PRINT, ADD, and GOTO into binary code. Thus while the *theoretical* possibility of computers conversing in natural language had been raised by Turing and others in the 1940s, it was in fact the *practical* work of composing software that led to the first symbolic languages. Even for many of AI's founders, the idea of the computer as embodying a universal representational system, a "language of thought" in a later phrase,[15] emerged not abstractly but *in their experiences with actual machines*, as both users and developers of higher-level languages.

Intelligence as Software

Allen Newell and Herbert Simon, as we saw in chapter 4, both recalled their first experiences with programming computers to simulate live radar screens as moments of epiphany in which they perceived computers as symbol systems. Another epiphany for Newell was a November 1954 Rand presentation by Oliver Selfridge, who was then working at Lincoln Laboratories. Selfridge had been Norbert Wiener's assistant, in which capacity he had proofread the manuscript of *Cybernetics* and attended some of the Macy cybernetics conferences. At Lincoln, Selfridge was working on the computerized pattern-recognizing system that later evolved into the highly influential "Pandemonium,"[16] a device that recognized letter forms and simple shapes. To accomplish this, a number of subprocesses analyzed various features of a figure; they then, in effect, "voted" for the result by computing a value and comparing it with a set of norms. In their use of a large number of simple processors that combined values to yield an outcome, and especially in their ability to "learn" by adjusting their own functions, Pandemonium and its predecessors resembled the neural nets of McCulloch and Pitts. At the same time, they were symbolic models in the sense that each subprocess used logic to analyze and categorize features. Thus, like Turing's speculations, Selfridge's programs occupied the transitional space between brain models and symbolic information processing.

Newell recalled the Selfridge presentation as utterly galvanizing: "I can remember sort of thinking to myself, you know, we're there.... And [Selfridge's program] turned my life. I mean that was a point at which I started working on artificial intelligence. Very clear—it all happened one afternoon."[17] Despite its cybernetic flavor, what Newell took from Pandemonium was the concept of symbolic processing. He began working on chess programs. By the time of their Logic Theorist experiments, Newell, Simon, and Shaw were also developing their own symbolic computer language, the IPL (Information Processing Language) series of list processing languages.

Edward Feigenbaum, one of Simon's students and later the major exponent of the "expert systems" approach to AI, remembered that when Simon handed out IBM 701 manuals to his course in early 1956, he went home and "read it straight through, like a good novel."[18] Feigenbaum went on to intern at IBM that summer, down the hall from the team working on the still-unreleased FORTRAN, and to help develop IPL-IV in the summer of 1957 at Rand. John McCarthy, as we shall see, also developed a list processing language. For all of these AI founders, the experience of symbolic programming constituted a significant step toward their vision of intelligent symbolic machines.

In symbolic processing the AI theorists believed they had found the key to understanding knowledge and intelligence. Now they could study these phenomena and construct truly *formal-mechanical* models, achieving the kind of overarching vantage point on both machines and organisms that cybernetics had attempted to occupy. Newell and Simon, first of the post-cybernetic AI theorists, rapidly raised to axiomatic status the idea that intelligence was a *symbol manipulation process* capable of being modeled by a computer.

Following on the methods used in Newell's experimental work on air defense simulation, Newell and Simon began around 1955 to work with "protocols" in studies of human problem-solving behavior. To determine a protocol, the experimenters would pose a complex logic problem to a subject and ask him to talk while he attempted to solve it, describing every one of his thoughts and perceptions as best he could. A collection of individual protocols would often reveal common strategies and techniques. Following Turing's plan of reducing each "state of mind" to a sequence of atomic procedures, these could be specified as lists of instructions, programmed on the computer, tested, and refined. One result of this research program was the 1957 General Problem Solver (GPS), a program that could solve certain formal puzzles such as cryptarithmetic equations and the missionary-cannibal problem. In Newell and Simon's view, programs like the GPS simultaneously constituted both a psychological theory and a rudimentary form of artificial intelligence.

To write effective protocols, Newell and Simon did not need access to lower-level processes such as neural nets, just as with the IPLs they no longer required access to the computer's machine language to compose programs. As long as the program produced results similar to those of the human problem-solver, its purely symbolic level of description was all that mattered. In their first article directed at academic psychologists, they stressed the point: "our theory is a theory of the information processes involved in problem-solving and *not a theory of neural or electronic mechanisms* for information processing."[19] Ultimately Newell and Simon established this principle as an axiom of both AI and cognitive science: their "physical symbol system hypothesis," under which people and all intelligent entities were essentially active symbol systems, physically instantiated.[20]

Newell and Simon thus began to displace the cybernetic computer-brain analogy with the even more comprehensive and abstract computer-*mind* metaphor of artificial intelligence. The two metaphors shared concepts of coding and information. But where the cyberneticians' ideal systems were weakly structured, self-organizing, and engaged with the environment, AI systems were highly structured, manipulating pre-encoded and pre-organized knowledge rather than building it through sensory encounters. Instead of feedback, reflex, and neural networks, the AI theorists thought in terms of instructions, languages, goals, and logical operations. The physical machine became little more than an arbitrary vehicle for the interactions of pure information.[21]

The Dartmouth Conference

In 1956, the Summer Research Project on Artificial Intelligence met for two months at Dartmouth College. The Dartmouth conference is generally recognized as the conceptual birthplace of AI (though Newell and Simon's work in fact predates it). The second MIT Symposium on Information Theory, from which George Miller dates the inception of cognitive psychology, occurred a few months later, and its protagonists included many of the same people.

The Dartmouth institute's chief organizer was John McCarthy, who coined the term for the grant proposal. As a twenty-year-old undergraduate mathematician, McCarthy had literally wandered into the 1948 Hixon Symposium at the California Institute of Technology, where he heard von Neumann talk on automata theory. He claims to have conceived the idea of artificial intelligence then and there (though not the phrase itself). He spoke to von Neumann about the idea, receiving the curt reply, "Write it up."[22] Other organizers included Claude Shannon, Marvin Minsky, and Nathaniel Rochester of IBM. Both McCarthy and Minsky had worked for Shannon at Bell Laboratories in the summer of 1952, and Shannon and McCarthy had already edited a volume on *Automata Studies* together.[23] Newell and Simon attended the conference as well, bringing with them their already-operating Logic Theorist and their computer language IPL.[24]

In the proposals for work at the conference, the still-nascent split between the computer-brain and computer-mind metaphors already appears clearly. For example, Shannon, representing the cyberneticians, planned to do brain modeling (using information theory). But McCarthy, in his work on the *Automata Studies* volume, had increasingly come to see Shannon's agnostic approach toward meaning as conservative and restricting. He felt that automata studies would never lead to true artificial intelligence since it deliberately avoided what was, for him, precisely the central issue. (This was in large part why he had insisted on the more radical and provocative term "artificial intelligence," over Shannon's objections, when organizing the conference.)

McCarthy's own goal for the summer was

to attempt to construct an artificial language which a computer can be programmed to use on problems requiring conjecture and self-reference. It should correspond to English in the sense that short English statements about the given subject matter should have short correspondents in the language and so should short arguments or conjectural arguments. I hope to try to formulate a language having these properties and in addition to contain the notions of physical object, event, etc.[25]

At the time of the conference, Marvin Minsky had just made a transition from mathematics and neurology in the McCulloch-Pitts line to symbolic models. With Dean Edmonds, Minsky's first major project, as a Harvard undergraduate, had been a working simulation of a neural network based on the neurological learning theory of the influential psychologist D. O. Hebb.[26] The machine, essentially an analog computer, used three hundred vacuum tubes, motors, potentiometers, and automatic clutches to simulate a highly interconnected set of neuron-like elements with "learning" abilities. Following on this experiment, Minsky had written a dissertation on *Neural Nets and the Brain-Model Problem*.[27] By the time of the Dartmouth conference, however, he had become disenchanted with the neural-network approach, in part under Shannon's influence and in part because it had become "boring" since "it didn't work."[28] The appeal of symbolic processing, for him, had a practical flavor: computer programs, unlike his analog brain models, could *do interesting things* with symbols, whether or not they resembled how brains worked.

Minsky wanted to use the Dartmouth institute to develop a geometry theorem-proving program something like the Newell-Simon-Shaw Logic Theorist. He circulated drafts of what later became his influential paper "Steps Toward Artificial Intelligence," essentially an essay-review of significant work on symbolic processing, including search, pattern recognition, learning systems, problem-solving, and heuristic programming. The paper noted but chose not to discuss brain modeling research as less significant to AI than symbolic processing.

McCarthy soon went on to develop the primary artificial intelligence programming language LISP (LISt Processing), known for its high level of symbolic abstraction. LISP manipulates lists (of symbols of any sort) and lists of lists; programs written in LISP are themselves lists. The language is highly recursive, meaning that definitions of terms may include those terms and that program processes may invoke themselves.[29] Among LISP's chief advantages is its ability to manipulate terms with many complex interrelations, like the words of a natural language.

By 1957, the year after the Dartmouth conference, Newell and Simon had completed their General Problem Solver and firmly believed they were on the track of a general, *hardware-independent* model of intelligence. Symbolic processing was the key. Where the PAL psychologists had engineered human language to render it compatible with electrical communications systems, the AI theorists engineered computer languages to make them compatible with human thought processes.

Newell and Simon were confident enough of the basic vitality of their approach to make four very strong predictions about their information-processing psychology. Within ten years, they claimed,

- a computer would be world chess champion,
- a computer would compose aesthetically valuable music,
- a computer would discover and prove an important unknown mathematical theorem, and
- most psychological theories would take the form of computer programs.[30]

An ultimate rapprochement between human and computer thought loomed on the horizon. Human mental software would be decoded and recompiled in machine language.

By 1958 McCarthy and Minsky had teamed up to establish an Artificial Intelligence Group at MIT's Research Laboratory of Electronics. That year McCarthy "proposed that all human knowledge be given a formal, homogeneous representation, the first-order predicate calculus."[31] Similar optimism reigned throughout the field. George Miller and Yale psychologist Carl Hovland had worked with Simon, Newell, and Minsky that summer at the Rand Research Training Institute on the Simulation of Cognitive Processes. Miller, Pribram, and Galanter were composing what would become *Plans and the Structure of Behavior*. Psychology, cognitive simulation, and artificial intelligence seemed increasingly to be parts of a single whole, united through the abstraction of symbolic processing. Cyborg discourse crystallized around its Foucaultian "support": the computer-centered research programs of AI and information-processing psychology.

In rejecting the cybernetic brain models and learning machines, AI also rejected a model of mind as inherently embodied. The brain-model approach relied intrinsically on interaction with the world; repeated experience, not formal analysis, was supposed to shape the weighted connections of neural elements into a functional system. Symbolic AI instead sought first to formalize knowledge of the world, injecting it into computer systems predefined and predigested. Logic, not experience, would determine its conclusions. (Yet this logical method reflected the AI founders' own experiences of the power of symbolic programming.) In its Enlightenment-like proposals for an encyclopedic machine, AI sought to enclose and reproduce the world within the horizon and the language of systems and information.[32] Disembodied AI, cyborg intelligence as formal model, thus constructed minds as miniature closed worlds, nested within and abstractly mirroring the larger world outside.

Time-Sharing: Linking AI to Command and Control

At this point, through a rather convoluted series of historical connections, AI intersected with another practical concern of programmers: the availability of computer time. The linkages that coalesced around this tie proved extremely significant to AI's success as a research program. Without them, AI might have had a very different future.

Until the late 1950s all computers (with the sole, partial exception of the SAGE machines) were "batch processors." This meant that each program, with its data, was run on the machine as a unit or "batch"; while that program was running, the computer could do nothing else. However, the speed of input and output (I/O) devices was far lower than that of the computer's central processing unit (CPU). This meant that the CPU, the computer's heart and (then) its most expensive element, actually sat idle most of the time, while it waited for I/O units to do their jobs. The use of compiler programs further slowed "run" times.

Furthermore, programs usually had to be run many times before all errors were found and fixed. Since the debugging process was slower than CPU or I/O times by yet further orders of magnitude, after receiving their output and fixing their programs programmers would have to wait, frustrated, in a queue until the machine was again free. Finally, the relatively small number of available computers (especially in universities) meant intense competition for computer time.

Jay Forrester had foreseen the bottleneck created by the disparity between CPU, I/O, and human time scales as early as 1948. His group designed Whirlwind to mitigate the problem with a technique known as "multiprocessing," which allowed the CPU to perform a number of predetermined tasks simultaneously.[33] With a later technique known as "multiprogramming," several programs could be run at once under the direction of an "executive" program. But the multiprogramming systems, sometimes also called "time-sharing" systems, were still designed with only a single user-operator in mind. More programs could be run faster under multiprogramming, but the essential approach remained batch processing.

To allow many users at once to take advantage of the computer's speed, John McCarthy conceived the modern concept of time-sharing around 1958. In time-sharing the computer (under one typical technique) partitions its memory into sections, each containing some program and its data. The CPU then cycles among these programs, performing one operation from each on every cycle. Many users connect to the same machine via remote terminal displays. Since the computer operates so rapidly, these users typically experience little or no delay between starting a program and receiving its results. Dozens of terminals may be connected to a single computer, but each user experiences the computer as a private domain.

Time-sharing (in McCarthy's sense) would produce two major changes in how computers were used. First, it would take full advantage of the CPU's speed, allowing one computer to perform work previously requiring many. Second, it would permit individual users to operate the computer "interactively"—privately, personally engaged with the machine, without the need for queues and delays between program runs. This, in turn, would create the possibilities of on-line debugging (fixing programs while they were running, with the effects of each change instantly visible to the operator), use of graphic displays rather than paper output, and a myriad of other "interactive" features.

At MIT in the late 1950s, time bottlenecks on computer systems had reached major proportions. MIT's Computation Center used the Whirlwind I until 1956, when IBM donated one of its 704 machines. This computer was shared by three users: MIT, IBM, and a consortium of New England area colleges and universities. IBM used the machine ten hours a day, while MIT and the university consortium each got a seven-hour shift. The 704 was replaced by the more powerful 709 in 1960. By this point MIT had also acquired several other computers, including an IBM 650 at its Instrumentation Laboratory and a Bendix G15 at the Naval Supersonic Laboratory. Lincoln Laboratory had donated its experimental TX-0 to the Research Laboratory of Electronics. Some, but not all, of these machines were available for general use, but demands for computer time increasingly outstripped this very limited supply.

By 1960, courses in computing, thesis work, data reduction, and numerical processing devoured most of the available computer time. Yet symbolic processing—such as virtually all of the work McCarthy, Minsky, and their AI group wanted to do—required an increasingly large proportion (in 1960, 28 percent of total research computing time).[34] AI research, computer-aided design, machine translation, and library information retrieval systems all required ever-larger programs in symbolic languages. By early 1961 an MIT report on computer capacity, composed by a committee that included McCarthy, Minsky, and V. H. Yngve (working on machine translation of natural languages), foresaw "needs for extreme capacities in the way of memory sizes and operating speeds" in order to carry out research in "language translation and analysis...heuristic problem solving [i.e., AI]...and computer-aided design." The latter would require "real-time operation...since the system will use pictorial and written language forms through a real-time console." For AI, the report concluded, "to tackle problems of more than trivial interest, very large memory and high processing speeds are needed." By this point McCarthy's work on time-sharing was well along under an NSF grant. IBM had provided hardware modifications to MIT's 704 allowing his plans to be put into action, but this machine's capacities remained too limited to accommodate McCarthy's vision. The report accordingly recommended acquiring an "extremely fast" computer with a "very large core memory" and a time-sharing operating system as the solution to the time bottleneck.[35]

McCarthy's time-sharing idea bore no inherent connection to AI. It was simply an efficient kind of operating system, an alternative way of constructing the basic hardware and software that controls how the computer processes programs. Yet AI work influenced his invention of time-sharing in two major ways.

First, AI programs consumed vast quantities of memory and machine time. Time-sharing would allow AI workers more access to the essential tools of their trade. Second, and more importantly in this connection, McCarthy needed time-sharing to *provide the right subjective environment* for AI work. He promoted time-sharing "as something for artificial intelligence, for I'd designed LISP in such a way that working with it interactively—giving it a command, then seeing what happened, then giving it another command—was the best way to work with it."[36] McCarthy and many of his coworkers wanted not simply to employ but to *interact* with computers, to use them as "thinking

aids" (in their phrase), cyborg partners, second selves. They wanted a
new subjective space that included the machine.
Through McCarthy's work, AI became linked with a major change
not only in computer equipment but in the basic social structure and
the subjective environment of computer work. Under time-shared
systems, rather than submit their programs to what many referred to
as the "priesthood" of computer operators, users would operate computers themselves from terminals in other rooms, other buildings—in
their offices and, eventually, even at home. They would see programs
operate before their eyes, rather than find their results hours later in
a lifeless printout. They could engage with the machine, create new
kinds of interaction, experience *unmediated* communication between
human and computer.

Human-machine integration, driver of computer development
from the beginning, would now be applied to computers themselves.
In the process AI researchers would acquire the interactive access to
machines McCarthy's vision of AI demanded. This connection between AI and time-sharing eventually led, as we shall now see, to AI's
most important institutional relationship.

The Advanced Research Projects Agency

By the mid-1950s military research agencies were already not only
aware of, but actively seeking out, work in the emerging field of computer simulation of cognitive processes. Though the Dartmouth institute's primary sponsor was the Rockefeller Foundation, the Office of
Naval Research was also involved.

The ONR had long since decided that future military forces would
require computer technologies for "decision support." As Marvin
Denicoff, one of its leaders, later recalled, "In the early fifties, it occurred to some of us [at the ONR] that we ought to begin supporting
decision makers where that decision maker could be an inventory
specialist, a command and control specialist, a tactical officer, a battlefield officer, a pilot—any of many categories. Also, we should begin to
go out to universities that had strong programs or emerging
programs in computation and fund them."[37]

Denicoff's brief memoir stresses the similarity between the operations research (OR) methods familiar to the ONR's leaders and the
emerging programming techniques of AI. Like OR, AI sought close
fits, best approximations, and heuristic rules rather than the more

rigid full-solution methods of other techniques. When McCarthy planned the 1956 Dartmouth conference, the ONR gladly footed part of the bill. By then it had already sought out Newell, Simon, and others at Carnegie Tech and the Graduate School of Industrial Administration (where Simon taught): "We wrote a long-term contract essentially to explore new approaches to decision making with all of those people [from Carnegie] involved."[38]

Yet AI was only one of a vast number of human-machine relationships military sponsors sought to explore. It might have remained a minor part of this wider program had it not been for the Advanced Research Projects Agency (ARPA, renamed the Defense Advanced Research Projects Agency, DARPA, in 1972) and in particular for one person within ARPA: J. C. R. Licklider. In Licklider and his Information Processing Techniques Office, the closed-world military goals of decision support and computerized command and control found a unique relationship to the cyborg discourses of cognitive psychology and artificial intelligence.

The Eisenhower administration founded ARPA hastily early in 1958, in the aftermath of the 1957 Sputnik launch. It served as an interim space agency, a kind of holding tank for civilian and military space programs while the government devised a more permanent institution to lead the space race. In an attempt to neutralize bitter competition among the three services over which should control what kinds of missiles and space technology, the administration chose an institutional location under the central Office of the Secretary of Defense. When space programs were transferred to the newly founded National Aeronautics and Space Administration (NASA) in 1960, other development-oriented military programs (such as the Intercontinental Ballistic Missile) reverted to the services. This left ARPA, at the start of the Kennedy administration, in charge of a widely varied roster of experimental scientific and technological research projects.

ARPA's post-NASA mandate was broad and also unique. According to a recent history of the agency, its new role would be

to advance defense technology in many critical areas and to help the DoD create military capabilities of a character that the Military Services and Departments were not able or willing to develop for any of several reasons: because the risks could not be accepted within the limits of the Service R&D and procurement budgets; because those budgets did not allow timely enough response to newly appearing needs; because the feasibility or military values of the new capabilities were not apparent at the beginning, so that the Services declined to invest in them; or because the capabilities did not fall obviously into the mission structure of any one Service.[39]

As an early official chronicle notes, ARPA was "spawned in an atmosphere that equated basic research with military security."[40]

Because of this military association, ARPA-sponsored research was subject neither to peer review, like that of the National Science Foundation, nor to other traditional equalizing principles designed to distribute research money widely. In effect, ARPA reincarnated the World War II OSRD. The agency's small directorate of scientists and engineers chose research directions for the military based on their own professional judgment, with minimal oversight. ARPA concentrated its funding in a small number of elite "centers of excellence," primarily in universities. Lacking both heavy congressional oversight and the development orientation of the services, the agency could support far-sighted, high-risk projects that might help avoid "technological surprises" (in the phraseology of a later era) from other countries.[41] ARPA's tacit funding agreements could also span much longer terms than those of other grantor agencies. These features of ARPA sponsorship had the effect of permanently "addicting" ARPA-supported laboratories to Defense Department funding.[42]

The newly inaugurated president, John F. Kennedy, in a March 1961 message to Congress, called for stepped-up research on command and control.[43] In June, the Director of Defense Research and Engineering (DDR&E) assigned command and control research to ARPA under the heading of the ballistic missile defense project DEFENDER. (DEFENDER also included research on phased-array radar tracking, high-energy lasers, and a number of other technologies picked up again in the 1980s by the Strategic Defense Initiative.)[44]

Computers were to be the focal technology of the ARPA command-control effort, in part for reasons of pure expedience:

DDR&E's problem appears to have been the existence of a rather expensive computer (the [transistorized] AN/FSQ-32 XD1A), built as a backup for the SAGE air defense program, which the Air Force had determined was no longer needed and hence was available for other purposes. There was also considerable interest within DDR&E in computer applications to war gaming, command systems studies and information processing related to command and control, as well as concern about the continued utilization of the System Development Corporation (the major software contractor for the Air Force), which apparently was experiencing some cutbacks in support due to the stage of development of Air Force programs [i.e., the imminent completion of SAGE]. DDR&E thus had a major piece of computer hardware begging for use, strong interest in computer applications to command and control, and an available contractor asset with appropriate credentials.[45]

DDR&E assembled the pieces of this puzzle by giving ARPA the FSQ-32. ARPA then initiated a command-control project at SDC for something called the "Super Combat Center" (a kind of super-SAGE), simultaneously rescuing SDC and providing a use for the extra computer.[46] This new direction led to the creation of ARPA's Information Processing Techniques Office (IPTO) in 1962.

Thus in the aftermath of SAGE, by this rather circuitous route, cutting-edge computer research once again found support under the aegis of air defense, part of the increasingly desperate, impossible task of enclosing the United States within an impenetrable bubble of high technology.

J. C. R. Licklider

IPTO's first director was none other than Psycho-Acoustic Laboratory and Macy Conference veteran J. C. R. Licklider.

In the postwar years Licklider's interest in computing had grown into a career. As we saw in chapter 7, in 1950 he had organized the psychological side of the Lincoln Laboratory "presentation group" responsible for SAGE interface design, which also included PAL veterans and information-processing psychologists George Miller and Walter Rosenblith. Like his PAL colleagues, Licklider's concern with human-machine integration dated to his wartime work in psychoacoustics; his particular specialties had been high-altitude communication and speech compression for radio transmission. At MIT he joined the series of summer study groups, beginning with 1950's Project Hartwell on undersea warfare and overseas transport and continuing with the 1951–52 Project Charles air defense studies that led to SAGE. As he recalled later, the summer studies

> brought together all these people—physicists, mathematicians. You would go one day and there would be John von Neumann, and the next day there would be Jay Forrester having the diagram of a core memory in his pocket and stuff—it was fantastically exciting. Project Charles was two summer studies, with a whole year in between, on air defense. At that time, some of the more impressionable ones of us were expecting there would be 50,000 Soviet bombers coming in over here.[47]

In his first years at MIT Licklider split his time between the Lincoln presentation group, MIT's new psychology program, and the Acoustics Laboratory.

The MIT Acoustics Laboratory, originally formed by Philip Morse, was then being directed by Leo Beranek, wartime head of the Harvard Electro-Acoustic Lab (the PAL's engineering arm) and Licklider's former colleague. The Acoustics Lab, within the Electrical Engineering (EE) Department, functioned as a "sister laboratory" of the Research Laboratory of Electronics (RLE, successor to the famed wartime Radiation Laboratory), whose programs had doubled in size "under strong military support" with the outbreak of the Korean War in 1950.[48] Licklider soon became deeply involved.

The early 1950s were a period of profound intellectual excitement in these laboratories. In the years following Licklider's arrival, the list of his RLE and EE Department colleagues reads like an almost deliberate merger of cyberneticians, Macy Conference veterans, Psycho-Acoustic Laboratory staff, and AI researchers. Norbert Wiener, Alex Bavelas, Claude Shannon, Walter Rosenblith, Warren McCulloch, Walter Pitts, Jerome Lettvin (a key McCulloch-Pitts collaborator), George Miller, John McCarthy, and Marvin Minsky were all part of the lab, as were the linguists Noam Chomsky, Morris Halle, and Roman Jakobson. As Jerome Wiesner later recalled, "the two decades of RLE [were] like an instantaneous explosion of knowledge." Wiener "was the catalyst," making daily rounds of the laboratory to investigate everyone else's research and to hold forth on whatever new idea had happened to seize his restless mind.[49] Licklider counted himself among the members of the cyberneticians' inner circle, attending Wiener's weekly Cambridge salons. He himself "exercised a profound influence on the growth of the EE Department," according to its official chronicles.[50]

In the Lincoln presentation group, the RLE, and the Acoustics Laboratory, Licklider learned about the most advanced digital machines and the problems of human-computer interaction (though his own work at the time involved building *analog* computer models of hearing). When Miller went to Lincoln to become a full-time group leader in 1953, Licklider remembered, "I stayed [at MIT]. Then I found out I really had to learn digital computing, because I couldn't do this stuff with analog computers, and there was no way, as a psychologist over in the Sloan Building, for me to get a digital computer." He eventually "struck this deal with Bolt Beranek and Newman" (BBN) to use the firm's Digital Equipment Corporation PDP-1.[51]

BBN was a research and consulting organization established by Leo Beranek and two other members of the MIT Acoustics Laboratory. It was among the most successful of the MIT/Lincoln spin-off companies and is still a major force in advanced computing research. In 1957 Licklider left MIT to work at BBN full-time on "man-machine" research sponsored by the Air Force Office of Scientific Research.[52] He led BBN's Psychoacoustics, Engineering Psychology, and Information Systems Research Departments, and eventually he became the firm's vice-president. Around the time Licklider arrived, BBN had begun a large project on time-sharing systems in which both Minsky and McCarthy were involved. Licklider audited one of McCarthy's MIT courses and underwent, in his own words, a "religious conversion to interactive computing."[53]

The groundwork for this "conversion" had been long prepared by Licklider's experience with the SAGE system's interactive computers. In a different interview, he recalled: "I was one of the very few people, at that time, who had been sitting at a computer console four or five hours a day."[54]

"Man-Computer Symbiosis"

Licklider's paper on "Man-Computer Symbiosis," published in 1960 and bearing the visible stamp of McCarthy's ideas, argued that batch-processing computers failed to take full advantage of the computer's power. Computerized systems, he wrote, generally served an automation paradigm, replacing people; "the men who remain are there more to help than to be helped."[55] Licklider proposed a more integrated arrangement to which computers would contribute speed and accuracy, while men (sic) would provide flexibility and intuition, "programming themselves contingently," as he put it in a perfect cyborg metaphor.

By this point Licklider was clearly an AI partisan, though his expectations—echoing the 1943 robot cat example of Rosenblueth, Wiener, and Bigelow—reflected uncertainty about whether biocybernetic or computer methods would first reach the goal. "It seems entirely possible that, in due course, electronic *or chemical* 'machines' will outdo the human brain in most of the functions we now consider exclusively within its province," he wrote, referencing the work of Newell, Simon, Shannon, Wesley Clark (designer of Lincoln Laboratories' TX-0 and TX-2 computers), and others.[56] Citing an Air Force study that had "estimated that it would be 1980 before developments in artificial intelligence make it possible for machines alone to do much thinking or problem-solving of military significance," he contended that while awaiting AI's maturity a regime of "man-computer symbiosis" could be attempted. The key to such a goal was time-sharing for real-time, interactive computing.

Licklider's example of the problems with batch processing shows the influence of his military concerns:

Imagine trying... to direct a battle with the aid of a computer on such a schedule as this. You formulate your problem today. Tomorrow you spend with a programmer. Next week the computer devotes 5 minutes to assembling your program and 47 seconds to calculating the answer to your problem. You get a sheet of paper 20 feet long, full of numbers that, instead of providing a final solution, only suggest a tactic that should be explored by simulation. Obviously, the battle would be over before the second step in its planning was begun.

Calling upon both early AI and the SAGE system as precedents, he speculated that men and computers might become so tightly coupled that

in many operations... it will be difficult to separate them neatly in analysis. That would be the case if,... for example, both the man and the computer came up with relevant precedents from experience and if the computer then suggested a course of action that agreed with the man's intuitive judgment. (In theorem-proving programs, computers find precedents in experience, and in the SAGE system, they suggest courses of action. The foregoing is not a far-fetched example.)

High-level symbolic computer languages were another key to man-computer symbiosis. Procedural languages such as FORTRAN and ALGOL should be supplanted by goal-statement languages more similar to those of human beings; here Licklider mentioned the work on "problem-solving, hill-climbing, self-organizing programs," referring to the work of both AI theorists and cyberneticians. Another issue in the "language problem," and the point at which Licklider's psychoacoustics research linked with AI, was speech recognition and production. Computers should eventually be able to converse directly with humans in ordinary spoken language. Licklider noted that business computing, with its relatively slow time scales, might not require the rapid interactive capability of computer speech. But, in a direct reference to military command-and-control issues, he wrote:

> The military commander... faces a greater probability of having to make critical decisions in short intervals of time. It is easy to overdramatize the notion of the 10 minute war, but it would be dangerous to count on having more than 10 minutes in which to make a critical decision. As military system ground environments and control centers grow in capability and complexity, therefore, a real requirement for automatic speech production and recognition in computers seems likely to develop.

"Man-Computer Symbiosis" rapidly achieved the kind of status as a unifying reference point in computer science (and especially in AI) that *Plans and the Structure of Behavior*, published in the same year, would attain in psychology. It became the universally cited founding articulation of the movement to establish a time-sharing, interactive computing regime. The following year, for example, when MIT's Long Range Computation Study Group recommended purchasing a very fast, large-memory computer for a time-sharing system, it pointed "the reader interested in a discussion of the idea of a time-shared machine used as a 'thinking center'" to Licklider's paper. The same report noted that AI research *required* a symbiotic method: "We will have to use man-machine interaction for such research; we do not yet know enough to set the machine up completely on its own to solve complex analytic problems."[57]

Licklider's metaphor of "man-computer symbiosis" crystallizes many facets of cyborg discourse in a single phrase. Recalling chapter 1's summary, we can see that the phrase and its history combine

• *techniques* of automation and integration of humans into computerized systems;

• *experiences* of computers as symbol processors and of intimate interaction with computers (and the deep desire for more of this);

• the computer as a *technology* with linguistic, interactional, and heuristic problem-solving capacities;

• particular *practices* of computer use, especially high-level languages, heuristic programming, and the interactive computing developed for SAGE and other military systems;

• *fictions* and *fantasies* about cyborgs, robots, and intelligent machines, like those of the cyberneticians, military futurists, and emergent AI, as well as *ideologies* of AI and interactive computing as military weapons, "thinking aids," and scientific research goals; and

- computer *languages* for formal representation of natural language and logic, such as LISP, whose optimal use requires interactive computing.

The idea of "symbiosis," a biological metaphor applied to human-machine interaction, perfectly captures a moment in the history of this discourse. The computer is not yet, in Licklider's metaphor, a fully realized artificial intelligence or cognitive simulation. It remains a *second* self, in Sherry Turkle's phrase, a partner in thought.[58] "Programming themselves contingently," men will work alongside computers, passing out of the age of "mechanically extended man" into a world of human-machine collaboration, "to think in interaction with a computer in the same way that you think with a colleague whose competence supplements your own."[59]

The Information Processing Techniques Office

Before his appointment to IPTO, Licklider met with then-ARPA director Jack Ruina to discuss computers for command and control. He was accompanied by Fred Frick, George Miller's PAL collaborator from the late-1940s statistical behavioristics studies. (Frick had served as a division head at the Air Force Human Factors Operations Research Group from 1950 to 1954. After that he had worked in various posts at the Air Force Cambridge Research Center, a locus of frequent contacts between MIT, Lincoln Laboratories, and Air Force scientists since World War II.) Licklider recalled their meeting:

I had this little picture in my mind of how we were going to get people and computers really thinking together. Ruina was thinking of this in terms of command and control, and it didn't take really very much to see how this would work... so I had the notion [that] "command and control essentially depends on interactive computing and there isn't any interactive computing so the military really needs this." I was one of the few people who, I think, had this positive feeling toward the military. It wasn't just to fund our stuff, but they really needed it and they were good guys. So I set out to build this program.[60]

Ruina assigned Licklider responsibility not only for IPTO, but also for ARPA's Behavioral Sciences program. Licklider interpreted his behavioral-science mandate to encompass the interactive computing work, reasoning that cognitive psychology, too, would be best served by investments in advanced computer technology. Thus a range of existing connections both broad and deep carried Licklider into ARPA as an advocate of interactive computing, time-sharing systems, cognitive simulation, and artificial intelligence.

Licklider's career path, like George Miller's, demonstrates how military research problems, and a location within institutions dedicated to their solution, could shape scientists' intellectual interests and visions of the future. As a psychologist, Licklider worked on the human side of the wartime human-machine integration problems of the PAL. At Lincoln he confronted the earliest issues of computer interface design for SAGE. In both laboratories the overarching context was the problem of command and control in electronically mediated, partially automated, eventually computerized systems. At BBN Licklider encountered time-shared computer systems and artificial intelligence. Throughout, as a psychoacoustician and a close colleague of George Miller, he maintained strong interests in language as both a formal system and a human-machine interface. The ideas Licklider enunciated in "Man-Computer Symbiosis" reflected all of these concerns. So Licklider "came to ARPA in 1962 because he interpreted fundamental advances in command and control to be heavily dependent on fundamental advances in computer technology, and [he] was committed to seeking advances in that field, particularly interactive computing."[61]

Histories of ARPA and AI frequently refer to the role of Licklider's "vision" in establishing IPTO as the major backer of AI, time-sharing, and computer networking. This version of the story undoubtedly holds some truth. But our exploration of Licklider's intellectual biography reveals that the "vision" was much more than a personal ideal. It was at least as much a product of the wider discourses of the closed world and the cyborg, technological approaches to politico-military problems and cybernetic metaphors of computers as minds and brains. Licklider's vision emerged from interactions with institutions and other individuals working on specifically military research problems. While it also, of course, encompassed more general concerns, its roots in these interactions played a major role in its construction *as a problem of command and control*.

Furthermore, Licklider did not carry this "vision" to a benighted military. A number of military agencies, including the ONR, the Air Force Office of Scientific Research, Lincoln Laboratories, the Defense Director of Research and Engineering, and Robert McNamara's Office of the Secretary of Defense itself had more or less definite plans of their own for increasing the use of computers in command and control and "decision support." What Licklider did contribute, through his contacts with AI, was an understanding of time-sharing systems as the specific vehicle of high-technology command-control solutions. Thus AI piggybacked its way into ARPA, riding along with interactive computing.

The goals articulated in "Man-Computer Symbiosis" became, almost without revision, the agenda of ARPA's IPTO under Licklider: time-sharing, interactive computing, and artificial intelligence. Upon his arrival he immediately initiated support for two major time-sharing projects.

The first of these projects took advantage of the recent cancellation of the Systems Development Corporation's Super Combat Center project. Once again rescuing SDC from the loss of military funding, Licklider altered the company's contract, refocusing it on the new goal of a time-sharing system. In December 1962, Licklider challenged SDC to produce a functioning system in six months. SDC met the deadline. Its system, known simply as the Time-Sharing System (TSS), incorporated a General Purpose Display System (GPDS) that included a primitive graphical user interface—the first of its kind. With graphical symbols for input and output functions, the GPDS thus foreshadowed modern icon-based graphical user interfaces such as Windows or the Apple operating system.[62]

The second, far better known project was MIT's Project MAC, begun in 1963. MAC stood variously for Man and Computer, Machine-Aided Cognition, or Multi-Access Computing. With a mandate broader than that of the SDC, Project MAC explored a wide range of interactive computing technologies. MIT had its first time-sharing system up and running by November 1963, and Project MAC ultimately produced a wide range of highly influential time-sharing systems (including CTSS and MULTICS), high-level computer languages (such as MACSYMA), interactive graphic display technologies (such as the KLUDGE computer-aided design system), and programming utilities (compilers and interactive debugging systems). ARPA spent about $25 million (current) on Project MAC between 1963 and 1970.

MAC was part of ARPA's centers-of-excellence strategy, and one of its effects was to establish MIT as the country's premier computer research laboratory. AI research benefited enormously from Project

MAC's support of Marvin Minsky and his many students. The program also had major impacts on commercial computing. IBM lost a competition for MAC hardware contracts and collaborative development to General Electric. This loss "resulted in a considerable reaction by IBM," which instituted a large time-sharing development program of its own as a direct consequence.[63] Military purchases of MAC-based equipment represent yet a third major impact: "The WWMCCS had spent, by 1979, about $700 million [current] on Honeywell 6000-type computers, peripherals and software" developed from the GE equipment originally built for MULTICS.[64]

In 1962, John McCarthy had left MIT to join the growing computer science group at Stanford University. The following year, soon after Project MAC began, he founded the Stanford AI Laboratory, which immediately became another major locus of AI research. Funding from ARPA was virtually automatic; Licklider simply asked McCarthy what he wanted and then gave it to him, a procedure unthinkable for most other government agencies. Licklider remembered that "it seemed obvious to me that he should have a laboratory supported by ARPA. . . . So I wrote him a contract at that time."[65]

Artificial intelligence per se was only one of many kinds of computer research IPTO backed; ARPA budgets did not even include AI as a separate line item until 1968. Numerous other IPTO-funded projects reaped major advances. Perhaps most significant of these was the ARPANET computer network, which eventually spawned the MILNET military network and the modern worldwide network of networks known as the Internet. Other ARPA-supported work has included supercomputing (as in the ILLIAC IV) and advanced microprocessor research (including work on gallium arsenide semiconductors and very-large-scale integrated circuits). IPTO supported advanced computing and AI not only at MIT and Stanford but at Rand, Carnegie-Mellon, SRI, SDC, BBN (Licklider's own company), and seven other "centers of excellence."

As the project with the least immediate utility and the farthest-reaching ambitions, AI came to rely unusually heavily on ARPA funding. As a result, ARPA became the primary patron for the first twenty years of AI research. Former director Robert Sproull proudly concluded that "a whole generation of computer experts got their start from DARPA funding" and that "all the ideas that are going into the fifth-generation [advanced computing] project [of the mid-1980s]—artificial intelligence, parallel computing, speech understanding, natural-languages programming—ultimately started from DARPA-funded research."[66] In the late 1980s, DARPA remained the largest single funding source within the military for computer and behavioral sciences.[67] Since its founding, IPTO has typically provided between 50 and 80 percent of the federal government's share, which is usually by far the largest share, of AI research budgets in the academic centers it funds.[68]

Conclusion: The Closed World and the Cyborg

When Vannevar Bush submitted his report on postwar prospects for government-sponsored scientific research to President Truman in July 1945, he stressed the necessity for national security of a continuing OSRD-style scientific research program in peacetime. Military strength was no longer simply a matter of competent, well-equipped armies. War, he noted, "is increasingly total war, in which the armed services must be supplemented by active participation of every element of the civilian population."[69] "Every element" of that population could be researched, rationalized, and reorganized and its efficiency improved. With computers, in Licklider's vision, a new "population" could "actively participate" in the preparations, thinking alongside human beings as if "with a colleague whose competence supplements your own." Both could be integrated into combat machines through an analysis of two complementary problems in high-technology war.

One was a kind of automation: how to "get man out of the loop" of precision-critical machines, to duplicate and then improve on human prediction and control functions by artificial means. The other was integration: how to incorporate men more smoothly and efficiently into those "loops" where their presence remained necessary—into the chain of command—by analyzing them as mechanisms of the same type and knowable through the same kinds of formalisms as the machines themselves. As we have seen, such a program constituted a central agenda of postwar research. Computers promised *general* solutions to problems of this nature. Thus they made rigorous theoretical analysis on the basis of the erasure of human/machine boundaries both practical and necessary for the first time.

Military sponsors, staffed by veterans of World War II agencies and laboratories concerned with operations research and human-machine integration, could easily perceive, in broad terms, the military utility of research programs aimed at integrating humans and machines.

Such programs took many forms, from directly military engineering (as in SAGE, its "Big L" C^3I offspring, or the McNamara Line of Operation Igloo White) and "human factors" research to experimental studies of cognitive processes and artificial intelligence software to demonstrate the empirical validity of information-processing psychology. Yet all reached toward the same general ends. Thus, however basic and benign it may have seemed to the researchers working on individual projects, the process of producing knowledge of (and for) cyborgs was militarized by the practical goals and concrete conditions of its formation.

Academic psychologists and computer scientists generally did not understand the major role they played in orienting the military toward automation, "symbiosis," and artificial intelligence as practical solutions to military problems. Some caught a glimpse of the overall pattern—such as Wiener, who renounced all military associations in 1946, and Licklider, who deeply desired to contribute to new military technologies from his areas of expertise. But many others simply pursued their own interests, oblivious or indifferent to their place in a larger scheme. They could do so precisely because for the most part there *was* no scheme, in the sense of some deliberate plan or overarching vision. Instead, this larger pattern took the form of a discourse, a heterogeneous, variously linked ensemble of metaphors, practices, institutions, and technologies, elaborated over time according to an internal logic and organized around the Foucaultian support of the electronic digital computer.

Supported by computers, the military-industrial drive to engineer semiautomatic command-control systems intersected with the postwar political hegemony of the United States in closed-world discourse. The central metaphor of "containment" combined the closures of Cold War ideology and military global reach with computerized systems for total central defense. Likewise supported by computers, the academic-intellectual drive to create artificial intelligence, cognitive psychology, and "man-computer symbiosis" intersected with the psychology of information processing in cyborg discourse. The central metaphor of the computer as a brain or mind combined information theory with concepts of complex mental processes that could ultimately be modeled with computers.

The closed world and the cyborg have in common not only their history and their technological support, the computer, but also their institutional backing by the armed forces, an ideology of formal-mechanical modeling, a metaphorical system, and a cybernetic language of natural-technical objects. Together, the two discourses constituted an organized, coordinated collection of institutions, technologies, and ideas for integrating human beings and information machines.

Ultimately, closed-world discourse represents the political logic of the cyborg. Seen against its backdrop, military support for cognitive research and artificial intelligence is part of the practical future of military power. The closed world, with its mathematical models, tactical simulations, and electronic battlefields, represents the form of politics and war for brains seen as computers and minds conceived as information processors.

Cyborg discourse is the political subject position, the "psycho-logic," of the closed world's inhabitants. Artificial intelligence, man-computer symbiosis, and human information processing represent the reductions necessary to integrate humans fully into command and control. The cybernetic organism, with its General Problem Solvers, expert systems, and interactive real-time computing, is the form of a mind in symbiotic relationship with the information machines of the electronic battlefield.

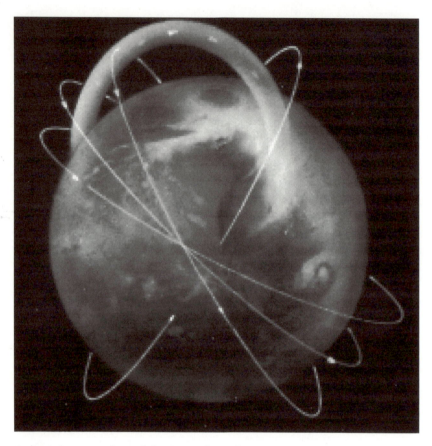

An artist's conception of the Strategic Defense Initiative space-based ballistic missile defense, circa 1985. Courtesy Lawrence Livermore National Laboratory.

9
Computers and Politics in Cold War II

This chapter and the next one carry the themes of the book into the era of Ronald Reagan, personal computers, and Star Wars, when the role of computers in renewed Cold War politics bore striking links with the projects of the 1950s.

In the early 1980s, discourses of the closed world and the cyborg found their apotheosis. As the generation that came of age during the 1950s Cold War reached the pinnacle of its political power, Cold War rhetoric, high-technology defense solutions, and containment politics all saw revivals. The most controversial military program of the period, the Strategic Defense Initiative, relied to an unprecedented degree on centralized computer control, while its rhetoric employed extraordinary closed-world iconography. The related Strategic Computing Initiative sought to fulfill the promise of military artificial intelligence in autonomous weapons and battle management systems, putting cyborg theory into practice. Both programs represented the culmination of long-term research programs whose essential aims had not changed since their initiation in the 1950s. Both also represented a return to military-sponsored research and development on a scale unknown since the first Cold War.

In the Reagan era, closed-world and cyborg discourses regained political and cultural salience. The wave of mass anti-technological sentiment inaugurated by the Vietnam War passed into a renewed optimism as new technologies—most visibly computers themselves—led an economic upswing. Apple and IBM brought personal computers to a broad consumer class. At the same time, the first commercial AI products entered the marketplace. Technological advances created smaller, more powerful, more "user-friendly" machines. The iconography of computers and AI achieved a new cultural centrality, while public understanding of their nature and importance dramatically increased. Popular cultural forms brought forth increasingly sophisticated articulations of the relationships between closed-world politics and the subjectivity of mental machines. Themes whose origins lay in the 1940s and 1950s thus achieved their richest and most thoroughly articulated forms three decades later.

The leap to the 1980s, then, will lead us to the final subject of this book: the forms of subjectivity inside the closed world. In this chapter, I explore the computer-based military systems of the 1980s and their meaning for political culture in what Fred Halliday has called Cold War II.[1] In chapter 10, which concludes the book, I study the staging of closed-world politics in science fiction and science fiction film from the entire Cold War era. These fictional constructions captured the political and conceptual connections among information tools, war machines, and artificial minds within a single cultural gestalt. In displaying the relation between the closed-world stage and its subjective spaces, science fiction enacted the subjectivity of cyborg minds.

The Era of Détente

The election of Richard Nixon in 1968 marked a notable change in the tone of closed-world discourse, as the United States stepped back from the brinksmanship and confrontations of Cold War I.

President Nixon and his secretary of state, Henry Kissinger, embarked upon a policy of détente (relaxed relations) toward the Soviet Union. Nixon signed the nuclear nonproliferation treaty, negotiated the first Strategic Arms Limitations (SALT I), and established diplomatic relations with the People's Republic of China. Under Kissinger's tutelage, Nixon focused new attention on the Middle East, especially Israeli-Egyptian relations—a conflict that lay outside the Cold War frame of capitalist-communist struggle. Nixon and Kissinger strove to create an image of themselves as stern but cooperative negotiators ready for peaceful coexistence with the Russians. Journalists frequently described Kissinger's policies as "Realpolitik," that is, as being formulated without idealism inside a sober assessment of what was and was not politically possible. Threats of military intervention generally dropped out of the political vocabulary.

Yet the closed-world discourse that structured the Cold War lived on in Kissinger's concept of "linkage." Under this policy, the United States attempted to manipulate Soviet behavior by making various favors conditional on Soviet actions, such as ceasing its arms shipments to Hanoi. As Stephen Ambrose describes it,

Linkage assumed that world politics revolved around the constant struggle for supremacy between the great powers. Like Dulles and Acheson and Rusk, Kissinger regarded North Vietnam, South Vietnam, Cambodia, and Laos as pawns to be moved around the board by the great powers. He insisted on viewing the war as a highly complex game in which the moves were made from Washington, Moscow, and Peking.... Everything was linked—the industrial nations' oil shortage, the Vietnam War, wheat sales to Russia, China's military capacity, etc.[2]

Direct interventions were out, but the Nixon-Kissinger strategy employed the CIA and other covert-action units to secure the overthrow of leftist regimes (as in Chile) while maintaining a conciliatory public posture. Thus, even as they repaired public relations with the communist world, Nixon and Kissinger sought to destroy it. As much as any of their predecessors, they viewed the world as a closed, "linked" system subject to American control.

The rhetoric and gesture of détente served as a veneer concealing the continued pursuit of American control. The new relations with China, for example, constituted (in Kissinger's grand scheme) the final "playing of the China card," an attempt to divide the communist world against itself to the ultimate benefit of the United States. As for nuclear weapons, even as the Nixon administration negotiated for arms control, it assembled ever larger nuclear arsenals. SALT I limited the number of intercontinental missiles each side could possess, but not the number of warheads each missile could launch. New MIRV (Multiple Independently targeted Re-entry Vehicle) systems gave each ICBM as many as fourteen warheads. By 1973, the United States had six thousand warheads, well over twice the Soviet count. By 1977, the U.S. force "limited" by SALT I had swollen to ten thousand warheads, while the Russian arsenal counted four thousand.[3] Nevertheless, under Nixon and Kissinger relations with the communist world did undergo a real and sustained thaw. Nixon prolonged the Vietnam War until 1973, when the end of his presidency loomed under the Watergate scandal. But at last he did withdraw all American troops, ending America's final attempt to "contain" communism through direct, frontal military action.

In the years following Nixon's disgrace and resignation, the traumas of Vietnam and Watergate deeply shaped both foreign policy and domestic politics. Leaders sought alternatives to confrontation with the Soviets and shied away from frontal interventions in the Third World. Gerald Ford (1974–76) essentially continued the policies of his predecessor. With Jimmy Carter's election in 1976, the swing away from containment doctrines reached its zenith.

The first half of Carter's administration (1977–78) was marked by an idealist retreat from Cold War rhetoric. Carter completed the Middle East peace mission begun by Kissinger and promised to withdraw U.S. forces from South Korea. The nuclear weapons buildup on both sides continued virtually unchecked during this period. But in rhetoric, at least, and in his attempts to negotiate SALT II, Carter promised to reorient America toward a genuine policy of peaceful coexistence. Cyrus Vance, Carter's liberal secretary of state, supported détente and ridiculed the idea that "the United States can dominate the Soviet Union" or in any other way "order the world just the way we want it to be."[4] Another discourse seemed poised to emerge from the aftermath of containment's failure in Vietnam.

Cold War Redux

Carter's policy appointees, drawn from a wide ideological spectrum, also included a number of certified Cold Warriors, especially his national security advisor Zbigniew Brzezinski. From the beginning this conservative element asserted itself, for example in declaring challenges to the Soviet position in the Third World and supporting destabilizing new weapons such as the neutron bomb and Stealth radar-evading aircraft. Carter's inexperience, the repeated failure of the Soviets to respond in kind to his conciliatory initiatives, and bruising conflicts among his appointees created a crisis of direction by the end of his second year in office. In addition, new Soviet initiatives on the nuclear weapons front, especially the deployment late in 1977 of SS-20 medium-range missiles aimed at Western Europe, alarmed American conservatives and produced new calls for a tougher military posture. The liberal elements gradually gave way, starting with the resignation in 1978 of Paul Warnke, head of the Arms Control and Disarmament Agency. Brzezinski's views increasingly came to control the administration's foreign policy. Vance finally quit in 1980.

The outcome was a purge of Carter's mixed liberalism in favor of a far tougher, more militarized Cold War stance. Carter embarked upon the major defense buildup that would be completed by (and generally named after) Ronald Reagan. He authorized deployment of major new weapons systems, including the Cruise and Pershing II intermediate-range European-theater missiles and the extremely accurate MX counterforce ICBM. He established a Rapid Deployment Force designed to carry out swift, large-scale interventions anywhere on the globe. Carter also made official, in 1980, the shift from "countervalue" (city-busting) to counterforce in nuclear strategy, opening the way for new waves of ballistic and cruise missile technologies capable of destroying hardened military targets and thus starting another round of arms racing.[5]

Late in 1979 the Soviet Union invaded Afghanistan, and this event shook away the final vestiges of Carter's liberalism. He retaliated with policies that went beyond even Kissinger's "linkage," restricting sales of grain, computers, and other high-technology equipment to the USSR and boycotting the 1980 Moscow Olympic Games. He took up the ready resources of closed-world discourse: "Aggression unopposed," he declared (echoing Lyndon Johnson's domino theory and Dean Acheson's apple-barrel metaphor), "becomes a contagious disease," and he called the Afghanistan invasion "a stepping stone to [the Soviets'] possible control of the world's oil supplies."[6] These strong reactions were soon complemented by the Carter Doctrine, which defined the Persian Gulf oil-producing region as an area of vital U.S. interests and bluntly proclaimed U.S. readiness to defend that region by force.

Thus despite the cracks that had appeared in the façade of closed-world politics following the Vietnam War, by 1979 it had regained its status as the dominant American politico-military discourse. Fred Halliday notes that just as in the Cold War of the 1940s and 1950s, Cold War II

> involved a concerted and sustained attempt by the USA to subordinate the various dimensions of its foreign policy, and that of its allies, to confrontation with the USSR. The image of a "Soviet Threat" was used not merely to elicit increased vigilance against the Soviet Union, but also to create a strategic framework within which other issues should be seen and given their due proportion and to mobilise the European allies and Japan for economic pressure on the USSR.[7]

Cold War II seized once again on containment's formula for global politics as the bipolar, apocalyptic struggle of a radically bounded closed world, radically divided against itself.

Ronald Reagan's 1980 electoral platform stressed alleged U.S. weakness. The immediate reference was the humiliating Iran hostage crisis, but Reagan's campaign promise to "make America strong again" ultimately referred directly to the supposedly inadequate U.S. military forces. In a direct echo of the 1950s bomber and missile "gaps," Brzezinski had talked of a nuclear "window of vulnerability," open while U.S. force modernization lagged behind that of the Soviets. This idea finally played against Carter. Despite its heavy retrenchment in 1978–80, his administration was widely viewed as weak-willed and insufficiently responsive to the Soviet threat.

Reagan's first administration went beyond traditional conservatism to embrace a New Right ideology that included a fervent, quasi-Biblical anticommunism. "Supply-side" economics, enormous tax cuts for the wealthy, and fundamentalist Christian moralism were blended with a hammering Cold War rhetoric. Reagan's infamous remarks about the Soviet Union as the "Evil Empire" and the "focus of evil in the modern world" captured this renewal of apocalyptic themes, accompanied in practice by loud saber-rattling. The Cruise and Pershing II theater nuclear missiles authorized by Carter were actually installed in Europe under Reagan, in 1983. He ordered toy-war skirmishes in Grenada (1983) and Libya (1986) designed to flex American muscle on the global stage.

By 1985 Reagan had distilled his foreign policy into a formal doctrine of renewed interventionism, including both direct and proxy methods, which again defined the world situation under the overarching scheme of superpower struggle. "Our mission," his State of the Union address proclaimed, "is to nourish and defend freedom and democracy, and to communicate these ideals everywhere we can.... We must stand by all our democratic allies. And we must not break faith with those who are risking their lives—on every continent, from Afghanistan to Nicaragua—to defy Soviet-supported aggression and secure rights which have been ours from birth.... Support for freedom fighters is self-defense."[8] Reagan backed this policy with massive support for counter-revolutionary civil wars in Afghanistan and throughout Central America. The Reagan Doctrine, thus defined, bore a striking similarity to the 1940s idea of "rollback."

Carter had already raised defense expenditures by a small percentage, but Reagan's defense budgets reached truly awesome proportions. Spending rose from FY 1980's $197 billion to $296 billion in FY 1985, a 51 percent increase in just five years. Much of the increase was technology-related. "Modernization" of strategic nuclear forces, particularly the B-1 bomber program, the MX missile, and the

Trident submarine, accounted for more than a quarter of this amount. "General-purpose force programs," including such technology-intensive areas as communications, logistics, and intelligence, comprised the rest.[9] By the middle years of his administration, President Reagan's defense budgets routinely exceeded $300 billion.

Like his World War II–era predecessors, Reagan saw the Cold War as a conflict that encompassed the entire society. His Pentagon relied increasingly on the notion of a "defense industrial base" (DIB): the segment of the national economy that would have to remain intact and self-sufficient in order for the United States to fight a protracted central war. It included not only directly military industries like aerospace and weapons producers, but upstream suppliers such as the electronics, computer, "strategic" minerals, and petroleum industries. Given this sweeping vision of its purview—which recalled Vannevar Bush's 1945 pronouncement that "war is increasingly total war, in which the armed services must be supplemented by active participation of every element of the civilian population"[10]—the Pentagon developed programs for broad support of DIB sectors, largely through R&D funding. At the same time, it attempted to impose heavy restrictions on trade in both goods (especially high technology) and information, including classification of academic research and limits on the rights of foreign nationals to attend scientific conferences and to work in American laboratories. Such restrictions, which were intended to cover *all* R&D the Pentagon saw as defense-sensitive and not just work actually funded by the DoD, further illustrated the Reagan administration's all-encompassing militarism. The Reagan vision of the globe as a closed world was thus matched at home by a picture of America itself as a gigantic, integrated, self-sufficient military machine.

The Reagan administration's policies thus recapitulated all of the essential Cold War themes described in chapter 1: the transference of all political conflict onto the mythic, apocalyptic struggle with the USSR, heavy reliance on new technology and faith in its power, and the view of the world as a closed system manipulable by American force. Reagan's was also the most popular peacetime presidency in history.

Computers and War in the 1980s

In chapters 2–4 we explored the role of the Cold War and its institutions in the development of computer equipment and applications, ending with an early version of the computerized "electronic battlefield" in Operation Igloo White. How did this role evolve during the era of détente? What was the effect of Cold War II on military sponsorship of computer research?

The Return of Military-Led R&D
By 1967, the height of the Vietnam War, the military-led computer R&D of the 1940s and 1950s had given way to a mature commercial computer industry capable of gigantic R&D investments of its own. In the late 1950s IBM, soon to become the world's largest computer manufacturer, instituted a strategy of heavy investment in R&D. Within a few years, the company was spending almost half its net earnings on internal research projects.[11] This pattern soon typified IBM's competitors as well; indeed, such investment became paradigmatic of "high-technology" industries. Coupled with the geometric expansion of commercial computer markets after 1960, the trend to sustained internal R&D investments made government contracts far less important for commercial computer research. IBM derived over half of its R&D funding from government (principally military) contracts in the 1950s, but that share dropped to less than 35 percent by 1963 and continued on a downward course.

Furthermore, by the final years of the Vietnam entanglement the war's costs absorbed increasing percentages of the Defense Department's total budget. During the Nixon and Ford administrations, relatively large sums were required for procurements to replace war-expended matériel. The shift to an all-volunteer force brought higher personnel costs and also altered the budget structure. Public opposition to the war focused in part on the horrors of high-tech weaponry, leading to the Mansfield Amendment of 1969, which restricted military involvement in basic research to "studies and projects that directly and apparently relate to defense needs."[12] In consequence, some research areas formerly supported by the DoD were transferred to civilian agencies such as the NSF. The result was that Pentagon funding for R&D in all fields dropped precipitously after the late 1960s.

The computer field reflected this broader trend, as figure 9.1 illustrates. The environmental movement, matched by the personal proclivities of Nixon and Ford, fueled a general antagonism toward science and technology. Combined with the recession induced by the 1974 oil crisis, this hostility led to a general decline in federal research funding from all sources. But the DoD share of federal funds dropped even more quickly than the overall total, from 70 percent in

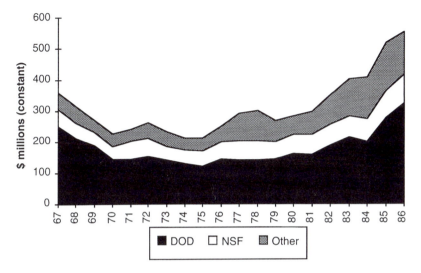

Figure 9.1
Federal funding for mathematics and computer science, 1967–1986. Data: Kenneth Flamm, *Targeting the Computer* (Washington, D.C.: Brookings Institution, 1987), 46.

1967 to a low of 47 percent in 1978.[13] In academia, NSF funding for computer research achieved rough parity with DoD obligations in the early 1970s.

As figure 9.1 shows, the rise of détente strikingly paralleled this decline in military support for computer research. Equally remarkable, however, is the direct parallel between Cold War II and renewed federal investment in computing. The DoD share rose to 54 percent in 1978 and continued to a high of 58 percent in 1986. Total federal obligations once again matched their 1967 levels in 1982 and jumped another 60 percent in the following four years.

The political environment of Cold War II had everything to do with this trend. As in the first Cold War, computers became central to both a high-technology nuclear shield and a new vision of centralized command and control. This time, however, the smug certainties of the 1950s gave way to increasing worries about the ability of computers to keep the closed world from imploding in nuclear suicide.

Computer Failure and Nuclear Anxiety

The politics of nuclear weapons centrally preoccupied the United States under both Carter and Reagan. Nixon-era arms control initiatives not only failed to limit the sizes of nuclear arsenals, but actually led to increases as each side attempted to gain a superior bargaining position by skirting treaty limits. Cold War II provoked a powerful political response in the United States and also in Europe, where the new intermediate nuclear forces (INF)—Cruise and Pershing II—were to be installed.

The late 1970s and early 1980s saw the rise and first electoral victories of Green parties, principally in West Germany, which combined environmentalism with a strong anti-INF focus. The Continent-wide, British-led, broadly popular Campaign for Nuclear Disarmament (CND) also opposed the INF deployments. In the United States, the highly successful Nuclear Freeze campaign organized grass-roots support for preempting the arms race. The Freeze, opposed by conservative Cold Warriors and mainstream arms controllers alike, proposed an immediate, bilateral halt to nuclear weapons-building as a prelude to negotiations on arms reduction. Many other oppositional groups, such as the Union of Concerned Scientists, Ground Zero, and Physicians for Social Responsibility, also achieved wide popular recognition as Cold War II reached a fever pitch in the early 1980s.

The prominence of new peace and disarmament coalitions strongly suggests that Cold War II resurrected, at least for a large minority, powerful anxieties about nuclear holocaust. As we shall see in the next chapter, this theme loomed large in that period's popular culture. Such anxieties found additional fuel in new fears of accidental war caused by computers.

In 1981, hearings in the U.S. House of Representatives revealed a long and mostly secret history of spectacular failures in the computerized BMEWS (Ballistic Missile Early Warning System).[14] The BMEWS, especially in times of high international tension, would serve as the primary trigger for nuclear retaliation. Since missiles not launched before an incoming strike arrived would be destroyed, commanders experienced a strong incentive to "use 'em or lose 'em" upon receiving a BMEWS warning. Despite presidential control of weapons release, many feared that under conditions of extreme stress, the very short decision times available might lead to fatal mistakes in the event of a BMEWS false alert.

Some of the thousands of hardware, software, and operator errors suffered by the system did in fact produce relatively serious false alerts. These periodic failures began almost as soon as the BMEWS was installed. Four days after its initial activation in 1960, a BMEWS station in Greenland broadcast a warning of a full-scale Soviet attack to NORAD headquarters. The radar image turned out to be a mirage generated by radar reflections off the rising moon. New generations of computers, software, and operating procedures not only failed to eliminate such problems but in fact, many argued, made the unavoidable accidents more dangerous. In the autumn of 1979, NORAD computers generated warnings of a Soviet submarine-launched ballistic missile attack. U.S. defense systems, with only ten minutes from warning to expected impact, instantly prepared to retaliate. With four minutes left before the putative Soviet missiles arrived, frantic officers finally discovered a training tape accidentally mounted on the warning system's drives. There were 147 false alerts during one 18-month period after NORAD installed new computers. Of these, four moved the U.S. strategic forces to a higher DEFCON (defense condition) status, one step closer to a nuclear response.[15]

In 1980 an unusual failure in a 46-cent integrated circuit chip caused NORAD computers to display another apparent Soviet attack. Paul Bracken describes the reaction:

About a hundred B-52 bombers were readied for takeoff, as was the... NEACP (National Emergency Airborne Command Post).... The airborne command post of the U.S. commander in the Pacific actually took off from its base in Hawaii.... Had the accident proceeded a bit longer the president... would have had to be awakened, in this particular case at 2:30 in the morning, to be told he had fourteen minutes to get out of the White House and to decide on a retaliatory plan in the event that the attack was real, and even less time to get on the Hot Line to Moscow. Nearly a hundred B-52s would have been launched to airborne positions over the Arctic, alert messages sent to ICBM crews, and warning messages sent to U.S. military units from Korea to Germany.[16]

As the complexity of the computer-centered BMEWS system grew, so did the numbers and types of errors. While an isolated computer problem usually posed little threat, combinations of problems stemming from human as well as electronic sources could produce extremely subtle failures (as demonstrated by experience with other complex technological systems such as nuclear power plants).[17] Detecting and resolving these errors became increasingly difficult. As the difficulty of error detection increased, so did the level of uncertainty about the correct interpretation of any alert.[18]

Bracken has noted that as reliance on electronic warning and command-control systems rose in both the United States and the

Soviet Union during the 1960s and 1970s, the two sides' forces became increasingly tightly coupled *to each other*. "In certain respects, American and Soviet strategic forces ... combined into a single gigantic nuclear system." Each side's moves, instantly detected by vast military sensor networks and interpreted by computerized warning systems, produced nearly instantaneous and automatic responses in the other. While the system dealt fairly well with *individual* errors or malfunction, the effect of this coupling was that even a short *series* of mistakes or misinterpretations could raise alert levels at a breakneck pace, possibly provoking additional errors. Because of the degree of preprogramming, raising alert levels proved far easier than reducing them. The result, in Bracken's excellent phrase, was that "the likelihood of nuclear Munichs has been exaggerated, but the possibility of nuclear Sarajevos has been understated."[19] Perhaps no better illustration of the world-closing effect of computerized command and control could be found than this integration of the two opposing nuclear forces into a single, tightly coupled system.

The Dangers of Complexity
Around the same time, other problems with military computing also became public. For example, the Worldwide Military Command and Control System (WWMCCS), in a 1977 test, failed to transmit successfully 62 percent of the messages sent. Part of the system, the Readiness Command, almost completely broke down, failing 85 percent of the time. Far from being confined to drills and game situations, such problems caused real and occasionally lethal effects. Communications failures in computer-based military networks were partially responsible for such debacles as North Korea's capture of the intelligence ship USS Pueblo in 1968 and the Israeli attack on a similar ship, the USS Liberty, during the 1967 Six Day War.[20]

By the late 1970s, computers had been integrated into most high-technology weapons systems. Microprocessors and other miniaturized components allowed drastic reductions in computer size and power requirements, while progressively more sophisticated software expanded their utility. Their very ubiquity made them increasingly problematic not only when they failed, but also when they worked normally.

One example was the F-15 jet fighter avionics system described by James Fallows in 1981. The computers that controlled each F-15 came in forty-five modular "black boxes" that could be rapidly removed and replaced in the event of malfunction. Malfunctioning black boxes were tested by another computer system called the Avionics Intermediate Shop (AIS). Each "wing" of 72 F-15s, with a total of over three thousand black boxes, was supported by three AIS systems. But since each AIS computer could check only one box at a time, in a diagnostic procedure that might last as long as eight hours, even carefully scheduled maintenance in peacetime proved problematic. Worries mounted over whether systems like these could function smoothly under the unpredictable stresses of actual combat; if not, large numbers of airplanes might be grounded by the bottleneck thus created. "Computer supportability" became a major Air Force concern.[21]

Similarly, software problems caused by the proliferation of computer languages had reached an acute stage. Hundreds of thousands of Defense Department computer programs, serving functions from accounting to air defense, were written in dozens of different languages. Even where languages were uniform, the DoD often found itself supporting several incompatible compilers. Simply maintaining existing programs, in the face of rapidly changing hardware and increased interlinkage of software systems, had become a monumental task. In 1975 the DoD called for proposals for a new, universal computer language in which all future programs would be written. By 1983, the department had frozen specifications for the new language, known as Ada. As might have been expected, though, Ada's very universality proved problematic. Many within the computer science community decried the huge Ada instruction set, with its baroque, something-for-everyone character, and some questioned whether efficient compilers could ever be written.[22] (At this writing the Ada standard has isolated military programming virtually as a separate culture, since Ada is too inefficient for most commercial applications.)

Such problems became one focus of the so-called defense reform movement of the early 1980s. A loose-knit but influential coalition of congressional representatives, Pentagon whistle-blowers, and journalists put forward a generic cost-effectiveness critique of military high technology. The critique suggested that the extreme complexity of "loaded" high-technology weaponry such as fighter aircraft and computerized tanks created a kind of brittleness and fragility that might make such weapons liabilities on the battlefield. At the same time, the geometrically increasing costs of such weaponry forced the armed services to purchase fewer units of any given weapon, to minimize weapons maintenance, and to limit their use in troop training. The result was a situation in which outnumbered, poorly trained troops relied on too few high-tech weapons in the hope that the weapons' extreme capabilities would carry the day. The brittle quality of highly complex technologies also made failures, when they did occur, more likely to produce catastrophic results—as in the case of the nuclear warning system.[23] The defense reform movement argued that simpler but more reliable weapons and equipment would actually increase the services' effectiveness.

SAGE Reborn: The Strategic Defense Initiative

Such critiques had little impact on Reagan-era military planning. In March 1983 President Reagan appeared on national television to call for a Strategic Defense Initiative (SDI), a high-technology weapons program of a scope and scale matched only by the Manhattan Project. In the original vision Reagan delivered, new weapons for ballistic missile defense would be based in outer space, shooting down Soviet missiles with laser beams powered by nuclear explosions.

In his address, Reagan called on "the scientific community in our country, those who gave us nuclear weapons, to turn their great talents now to the cause of mankind and world peace, to give us the means of rendering these nuclear weapons impotent and obsolete."[24] Reagan thought of the proposed system as an invulnerable shield. He apparently saw the new defenses as a dome or umbrella capable of preventing all, or almost all, incoming missiles from reaching their destinations. Thus the SDI would defend not only military targets, but cities and populations as well. Reagan also believed the system could be *purely* defensive, going so far as to suggest that, once completed, SDI technology could be shared with the Russians, turning the offense-based nuclear deterrence of the previous thirty-five years on its head.

"Star Wars," as it was immediately dubbed by the press, proved extremely popular in Congress. The new Strategic Defense Initiative Organization (SDIO) received an appropriation of $1.4 billion for FY 1985, rapidly increasing to $3 billion in 1986. Budgets reached $4.5 billion in FY 1987 and remained in that range until FY 1992. By the time the Clinton administration finally buried the program in 1994, the SDIO had spent nearly $35 billion.

The SDI explored a wide range of technical possibilities for space-based missile defenses, but most of the proposals shared a basic scheme. Military satellites would instantaneously detect and track Soviet missile launches by their very bright heat trails. Computers would integrate this information and coordinate a comprehensive attack on the incoming weapons. First they would aim directed-energy beam weapons at the rocket boosters of each ICBM, destroying most of them before they could release their MIRVs. Any warheads that escaped the initial counterattack would be picked off either by space-based weapons as they left the atmosphere or by ground-based weapons as they reentered it. Many new types of weapons—including particle beams, electromagnetic rail guns, ground-based lasers bounced off orbiting "smart mirrors," space-based lasers created by channeling x-ray pulses from thermonuclear explosions, and kinetic-energy interceptors—were suggested and research begun.

Had it not been so expensive and so dangerous, the SDI might merely have demonstrated Marx's dictum that everything in history happens a second time as farce. Just as with SAGE in the 1950s, most knowledgeable military officers, scientists, and engineers realized that the "shield," no matter how effective, could never even approach impermeability. (Indeed, Reagan did not bother to consult Pentagon analysts before announcing the plan; had he done so, most would have refused to support it. Once again, though in this case with far less honesty, visionary scientists led a process of mutual orientation toward a total, computer-based, high-technology defensive system.)[25] The most optimistic of serious scientists estimated potential kill ratios at 60 to 70 percent. But in a full-scale attack, if even a few percent of the thousands of incoming weapons reached their targets, the result would still be catastrophic. A pledge not to take SDIO funding, in order not to appear to support the impossible fantasy of effective nuclear defense, circulated widely on university campuses. Thousands of scientists signed.

As with SAGE, however, some officers, planners, and scientists supported research on the grounds that a partial defense was better than none. Others felt the SDI served a Cold War political purpose, since the Soviets would be forced to compromise their country's economic health even further in order to keep up with American research. Many may have been seduced into overlooking the plan's technical problems by the promise of a vast new source of research funding.

Finally, some may have supported the plan for two perceived military values. First, a partially effective shield would preserve more of the retaliatory force, raising the risk incurred by the Russians should they attack. Second, and consequently, it might increase the final ratio of U.S. to Soviet weapons reaching the other side's territory, thereby improving the chances of an American "victory," as measured in the macabre terms of nuclear strategy. This latter aim fit in well with the Reagan administration's awesomely cavalier attitude toward nuclear war, captured in a remark of DoD official T. K. Jones. "If there are enough shovels to go around" for digging shallow holes to hide in, he cheerfully announced, "everybody's going to make it."[26]

SDI resembled SAGE in another respect as well. The 1950s air defense project had faced the problem of vastly diminished warning and response times created by intercontinental jet bombers. The computerized SAGE system had been envisioned as the solution. By the 1980s, ICBMs had shrunk those time windows by another order of magnitude. Missiles launched from the Soviet Union could reach American targets in under half an hour; launched from submarines, in as little as ten minutes. Furthermore, the missiles were most vulnerable during their boost phase, when their heat trails were visible to satellite surveillance and they had not yet released their MIRVs; but the boost phase of the most advanced missiles lasted only ninety seconds. Any effective ballistic missile defense (BMD) would thus have to fire literally within seconds of a Soviet launch.

Thus, unlike SAGE, the *Semi*-Automatic Ground Environment, SDI systems would have to be *fully* automated. They would, then, require two unprecedented features: full computer control, with human authority exercised in activating the system but not in issuing the command to fire it, and software and hardware incapable of dangerous failures like those experienced by NORAD, the WWMCCS, and *all* other military computer networks. As the SDIO began to assess possible elements of a BMD, it rapidly became clear that computerized "battle management" would constitute the core of any conceivable system. System Analysis and Battle Management became one of six Program Elements of the SDIO. Designing such systems posed a problem of enormous difficulty. Two advisory panels, the 1984 Fletcher Commission and the 1985 Eastport Study Group, both concluded that computer control was the single most critical element of any possible BMD: "the [Eastport] panel ... regards the battle management and C^3 problem as the paramount technical issue that must be resolved in order to implement a strategic defense system."[27] The most difficult part of this problem stemmed from the requirement that the system function perfectly the first (and presumably also the only) time it was used.

All large computer programs contain numerous errors, and the BMD software would be the largest program ever written. Estimates of its size ranged from 10 to 100 million lines of code, one or two orders of magnitude larger than the largest existing programs. (By comparison, the space shuttle mission required programs of about 100 thousand lines, 100 to 1000 times smaller than the estimate for SDI.)

Computer program errors may or may not cause failures; error-induced failures may or may not be catastrophic. Some errors are found only when failure occurs. Others are never detected. Even when bugs are found, in large systems the process of fixing them virtually always introduces additional mistakes. Removing all bugs from a very large system is provably impossible. These features are generally viewed as inherent to the programming process. Software experts, including those hired by the SDIO, agreed almost unanimously that there would be no way to eliminate all errors from the proposed programs. Consequently, it would be impossible to ensure that no unexpected constellation of factors could cause failure.[28]

In situations where safety and reliability are not major concerns, the presence of bugs is not necessarily a serious problem. But in a BMD application, failure of *any* sort might be dangerous and major failures might prove—to put it mildly—disastrous. This critical safety problem was further compounded by the fact that the system by its nature could never be tested under actual conditions of use. Demonstrations of any BMD C^3I system's reliability would have to rely on simulations that would themselves consist in part of imperfect programs. While catastrophes such as unexpected firing might be averted by "fail-soft" techniques, nothing could ensure that the system would work exactly as planned. Yet only a completely dependable system could serve the purely defensive strategic purpose of Reagan's vision. One of the Eastport Study Group software experts, David Parnas, resigned from the panel in protest, arguing that the since the reliability of such a system could never be guaranteed, building one would amount to a violation of professional ethics.[29]

SDIO officials never seemed to comprehend the magnitude of the software problem. James Fletcher, head of the eponymous commission, wrote in 1984 that a ten-million-line "error-free" program would control the system, implying that such a program could and would be written.[30] As late as October 1986, speaking before the Los Angeles chapter of the Association for Computing Machinery, Col. David Audley, assistant director of the SDIO, announced that the SDIO would eliminate bugs by administrative fiat. "The notion of bugs and errors goes back to the notion of high quality," he asserted. "We prevent bugs by design. We don't like to pre-plan removal of bugs."[31] New techniques, such as artificial intelligence–based automatic programming from system specifications and automated program-proving procedures, were proposed as ways of eliminating human error and speeding the coding process, but few if any software engineers believed such methods would completely eradicate mistakes.

Even if they did work, nothing could prevent errors *in the specifications themselves*. Every possible contingency would have to be anticipated and accounted for. The program would have to respond correctly to a virtually infinite set of possible conditions. It would, in other words, have to capture the whole world within its closed system.[32]

The SDIO pressed on, supporting a wide range of computer technology research, including work on parallel computer architectures, optical computing, novel semiconductor materials and circuit packages, gallium arsenide processors for space-based computers, and advanced software. Since SDI computing has often been combined with other contracts or housed within other agencies, estimating the exact amount of its contribution to computer research is difficult. But a round figure of 3 to 5 percent of the total SDIO budget gives a reasonable ballpark guess of between $50 and $225 million annually from 1985 onward. Had the system progressed to deployment, its estimated total cost of at least $1 trillion might have included roughly $100 billion for computing, of which perhaps $10 billion would have been spent on research. In the event, SDIO support for advanced computing had probably totaled between one and two billion dollars—one of the largest single-program contributions ever—by the time of the program's demise.

Whatever its technical merits, the *rhetorical* power of the SDI proved considerable, since it allowed its proponents to claim that they were supporting defensive rather than offensive weapons.[33] Reagan said that he wanted to change the focus of nuclear policy from avenging the dead to saving the living. The effect was to place the SDI's supporters in a position of moral superiority. Opponents found themselves, awkwardly, arguing in favor of the status quo, with its failed history of arms control and its strategic basis in the ethical morass of mutual assured destruction. The conservative coalition that supported the SDI recognized the value of this moral positioning from the outset. Indeed, evidence suggests that seizing the moral high ground was always a primary purpose of the plan.[34]

Star Wars, by launching a new direction in nuclear policy, also achieved a major political victory by stealing the swelling thunder from the Nuclear Freeze movement. The unprecedented successes of the Freeze, which placed initiatives on state and city ballots and influenced electoral campaigns nationwide during the 1982 elections, had alarmed conservatives with its concentration on the arms race itself to the exclusion of strategic considerations.[35] The SDI effectively preempted the Freeze, refocusing attention on the possibility of a technological solution to the nuclear nightmare.

In this respect, as in so many others, SDI was simply a technologically updated version of SAGE. Like SAGE, had it actually worked the SDI's real military purpose might have been to clean up the ragged counterattack after a U.S. preemptive strike (though by the 1980s few officers would be either as sanguine or as honest about this prospect as had been SAC commander Curtis LeMay). But in practice, also like SAGE, the plan's primary goals were political and ideological rather than military.

Its symbolism was perfectly adapted to this end: an impenetrable sheltering dome, American high technology in full control, a shield rather than a nuclear sword. SDI helped appease nuclear anxiety with the promise of an active defense. The SDI thus found its place as a potent phrase in closed-world discourse, where politics and technology were mutually articulated. Like SAGE, the SDI's bubble shield would shroud the United States inside a high-tech cocoon, another icon of the closed world.

The Strategic Computing Initiative

About six months after Reagan put forth his vision, DARPA announced a five-year, $600 million program in advanced computer technology called the Strategic Computing Initiative (SCI).[36] The SCI linked university, industry, and government research in a comprehensive program ranging from research infrastructures (e.g., computer networks and microchip foundries) to "technology base" development of high-speed microprocessors and parallel computer architectures. If warranted by the results of the first cycle, the program would proceed to a second five-year phase, carrying it into the 1990s with a total budget exceeding $1 billion. At this writing, over ten years later, many of the Strategic Computing projects continue, under the aegis of a new High Performance Computing program.

The original plan's scale was extraordinary, but it became controversial for breaking precedent in three other ways: its proposals for battlefield artificial intelligence systems, its use of military applications as research goals, and its connections with the SDI command-control problem.

Artificial Intelligence
Though much of its funding went to technology-base projects, the real core of the SCI was something else. It proposed finally to cash in on DARPA's decades of investment in artificial intelligence in order to build autonomous weapons, artificial assistants, and AI systems for battle management.

In the decades since its naming in 1956, AI research had progressed through a series of paradigms. The fundamental methods of symbolic AI described in chapter 8—manipulating high-level symbolic descriptions according to heuristic rules—did not change. But the complexity of its underlying goals proved far greater than some of its founders originally supposed. Early AI work had sought powerful, general procedures for defining and searching problem spaces. This approach worked well in domains governed by logical rules, such as games and mathematical theorems, but foundered when applied to real-world problems. Researchers found that logical analysis often failed in the absence of detailed factual knowledge of specific situations. AI work on translation and natural-language understanding led to a similar shift from purely syntactic analysis to incorporation of detailed semantic knowledge. The prospect of a *general* artificial intelligence, a true General Problem Solver based in pure logic, seemed increasingly remote. The problem of "commonsense knowledge"—the vast, intricately interconnected understanding of reality embodied in human action but rarely articulated (and perhaps inarticulable)—loomed increasingly large.[37]

By the early 1970s AI workers had discovered a new approach. Instead of General Problem Solvers, researchers constructed simple "microworlds." The paradigmatic example was Terry Winograd's program SHRDLU. SHRDLU's microworld consisted of (simulated) blocks of different colors, sizes, and shapes; the program could engage in natural-language conversation about the blocks world and manipulate the blocks. Within its tiny, artificial domain, it seemed to achieve full understanding. More comprehensive AI would, it was hoped, emerge by expanding the borders of these miniature closed worlds to encompass more and more of the wider world.[38]

A second new technique, dating from the mid-1960s, was to build "expert systems" that attempted to capture the knowledge of a human expert in some well-defined "knowledge domain" such as medical diagnosis or chemical spectrography. This knowledge could be encoded as sets of facts and rules. The system could determine the implications of some particular situation by applying its rules and referencing its "known" facts. By sharply restricting the knowledge domain, the need for commonsense knowledge could be minimized. Because it sought practical, limited techniques for analyzing real-world problems, the expert systems approach was the first AI technology to arouse commercial interest. A multitude of small entrepreneurial firms such as Teknowledge and Intellicorp sprang up in the early 1980s to develop and market expert systems to both commercial and military customers. Expert systems had achieved some degree of success, though exactly how much remained a subject of acrimonious debate.[39]

Both new paradigms seemingly abandoned the grand ambitions of AI in favor of finer-grained studies. But in fact these were merely temporary, tactical diversions. Where the microworlds approach tried to build up to reality by gradually expanding a simulated world within the machine, expert systems workers carved reality up into a series of tiny "domains," capturing each one separately and worrying about recombining them later. Expert systems differed from microworlds in their focus on real-world problems, but they still reflected the closed-world belief in the reducibility of knowledge to a machine-like, automatic system.[40] By the late 1980s, expert systems builders had resurrected their commitments to AI's overarching goals in the Cyc project, a multi-year effort to write a commonsense knowledge base, encoding the entire inarticulate background of human language and action.[41]

The SCI proposed, for the first time, to place expert systems and other AI technology into central roles in military equipment and command.

Battlefield Technology

Since IPTO's founding in 1962, DARPA had emerged as the most important federal funder of basic research in computer science. By the mid-1970s "IPTO's budget ... accounted for most DoD-supported basic research and roughly 40 to 50 percent of applied research in math and computers.... DARPA plays the key role within the military in setting computer research priorities."[42] Artificial intelligence, in particular, had remained DARPA's captive client, with as much as 80 to 90 percent of funding for the major AI groups—MIT, Stanford, SRI, and Carnegie-Mellon—typically provided by the agency. The Strategic Computing budgets—$50 million for FY 1984, $95 million in FY 1985, $104 million in FY 1986—increased DARPA's contribution to nearly 90 percent of the total federal basic research effort, though SDI funding for computer research soon reached similar levels.[43] Until the SCI, however, the bulk of DARPA funding had gone to research without *explicit* connections to battlefield technology.

The original SCI planning document, released by DARPA in October 1983, depicted a much different approach to research funding. Basic research on hardware and software technology would now be aimed at producing three battlefield systems. In addition, industrial work would be much more tightly integrated with university-based research. By coordinating research efforts around development goals, the SCI's planners hoped to achieve a specific set of "scheduled breakthroughs" that would rapidly turn basic advances into usable military technology. The plan contained a detailed ten-year timeline, setting dates for each breakthrough and showing how results would be coordinated to produce the final products.

Three applications, one for each of the military services, would serve as research goals. Each would employ AI software (primarily expert systems) to extend the capacities of autonomous machines; each was also intended to drive particular research areas. The first application was an autonomous land vehicle (ALV) for the Army. The unmanned ALV would be required to navigate unfamiliar terrain at high speeds; this would demand major advances in machine vision. The planning document described the machine's utility in terms of automated reconnaissance, but a suitably armed ALV could also obviously become a robot tank. The second application, an intelligent fighter pilot's assistant for the Air Force, would serve as a sort of automated copilot, keeping track of many functions the pilot must now monitor.

Ultimately, like the robot R2D2 of *Star Wars* fame, it might perform some of the pilot's tasks or reroute electrical signals or fluids in the event of damage to the plane. The primary research goal was voice recognition: the pilot's assistant would possess a ten-thousand-word vocabulary and be able to understand spoken English even amidst the noise of a jet cockpit. Finally, DARPA proposed a battle management expert system for aircraft carrier battle groups. This device would analyze sensor data from the carrier's huge field of engagement, interpret threats, keep track of friendly forces, and plan response options that it would then "suggest" to the battle group commanders.[44]

By 1985 DARPA had added two additional applications programs. One was a second battle manager, designed for the Army's AirLand Battle Doctrine plan for forward air and ground combat in Europe. The other was an intelligent image-processing system to automate analysis of satellite and other reconnaissance imagery, designed for the intelligence services.[45]

These applications goals aroused controversy for four main reasons. First, the plan's highly specific timetable for extremely ambitious research achievements might produce pressure on developers to misrepresent the capabilities of their products.[46] Second, since DARPA intended to use the applications to "pull" basic research, DARPA-sponsored researchers would now all be directly implicated, in some sense, in weapons projects.[47] This was widely seen as a significant departure from DARPA's traditional policies. (As we saw in chapter 8, however, DARPA's *ultimate* aims had always been applications-oriented. In fact, an emphasis on practical applications of AI had been building within DARPA since George Heilmeier's directorship in 1975.)[48]

Third, four of the five applications would employ AI in systems that might potentially control the actual firing of weapons. Many computer scientists felt that battlefield uses of AI posed extremely severe and possibly insurmountable technical difficulties. The expert systems upon which these machines would rely suffered from well-known problems of "brittleness." Confronted with unexpected factors not accounted for within their narrow knowledge domains, such systems were prone to catastrophic failures of interpretation.[49] In addition, autonomous weapons-bearing systems posed fundamental moral and legal problems, such as the burden of responsibility for a machine decision to open fire on civilians or the question of whether soldiers could surrender to a robot.[50]

Strategic Computing and Star Wars

The fourth and most controversial reason for debates over the SCI was the explicit connection of its AI research with the battle-management software planned for the Strategic Defense Initiative.

The similarity of program names notwithstanding, DARPA planners could not have been aware of the incipient Strategic Defense Initiative during 1982 and early 1983, when the Strategic Computing Initiative's goals and structure were set. (Like almost everyone else, DARPA planners knew nothing of the SDI until Reagan announced it on television.) Instead, they were responding largely to the Fifth Generation Computer program launched by the Japanese Ministry of International Trade and Industry (MITI) in 1981.[51] MITI planned joint government-industry-university research, primarily in AI, parallel processing, and microprocessor technology, with a budget of $855 million over ten years. MITI, in a practice it had pioneered in the 1970s, defined a series of research breakthroughs needed for the fifth generation and pegged the program to applications goals.[52]

MITI's notable successes in previous technology initiatives were seen as part of the reason for the considerable challenge Japan presented, by the early 1980s, to American technological and economic hegemony. Just before the SCI's announcement, Stanford University expert systems inventor Edward Feigenbaum published an alarmist call to action in *The Fifth Generation: Japan's Computer Challenge to the World*. The book sketched a utopian future inhabited by "intelligent knowledge-based systems" and painted the Japanese Fifth Generation program as a major commercial threat. Both here and in congressional hearings, Feigenbaum called for a concerted American R&D effort in AI.

Soon after Feigenbaum's book went to press, DARPA—MITI's closest American equivalent—announced the SCI. The initiative's planned breakthroughs suggested an attempt to duplicate MITI's institutional techniques in an American context.[53] Despite its military orientation, the program plan made much of potential commercial spin-offs from SCI research (as did the SDIO for the Star Wars research). In public contexts, DARPA officials frequently gave Japanese competition equal billing with military need in promoting the plan.[54] This unity of commercial and military goals stemmed from the Reagan-era view of the two spheres as elements of a larger geostrategic system. As with Nixon, Kissinger, and Carter, in Reagan's discourse everything was linked together in a closed world.

Thus SDI computing support was not the SCI's original purpose.

By the time of the planning document's release in October 1983, however, DARPA had tuned its initiative to complement the Star Wars proposal. The plan itself contained a widely cited paragraph noting the "projected defense against strategic nuclear missiles, where systems must react so rapidly that it is likely that almost complete reliance will have to be placed on automated systems" as an "extremely stressing example" of the need for strategic computing research.[55] In addition, in a House Appropriations committee hearing on strategic computing in 1984, then DARPA director Robert Cooper stated that "it is my personal opinion that the computational capability that we are pursuing is an enabling technology for a defense as complex as may be necessary for ballistic missile defense.... I am hopeful that we can make progress at the rate that we now project in the Strategic Computing Program so that it will not turn out to be a bottleneck in a potential decision in the middle to late nineties as to the efficacy of a ballistic [missile] defense system." Earlier in the year, grilled by the Senate Foreign Relations Committee together with SDIO head James Abrahamson and President Reagan's science advisor George Keyworth, Cooper had claimed that computer technology might prevent presidential errors in a nuclear crisis. To the senators' consternation, he predicted that "we might have the technology so he couldn't make a mistake." At the same hearing, Keyworth announced that by 1990, the decision to fire a BMD against a Soviet missile launch could be "done automatically."

Though DARPA officials later struggled to disentangle strategic computing from strategic defense, the linkage had already gelled. The Fletcher Commission SDI review panel, in 1984, called Strategic Computing "essential for a BMD system." The SDIO took over the SCI's gallium arsenide semiconductor project. SDIO officials at every level repeatedly referred to the SCI as a kind of support program for the planned BMD.[56]

Resealing the Dome: AI and the Closed World

In terms that strikingly resemble J. C. R. Licklider's vision of "man-computer symbiosis," the Strategic Computing planning document celebrated the imminent arrival of new computing technologies:

The new generation will exhibit human-like, "intelligent" capabilities for planning and reasoning. The computers will also have capabilities that enable direct, natural interactions with their users and their environments as, for example, through vision and speech. Using this new technology, machines will perform complex tasks with little human intervention, or even with complete autonomy.... As a result the attention of human beings will increasingly be available to define objectives and to render judgments on the compelling aspects of the moment.[57]

It took a further step, though, heralding the genuine artificial intelligences that would now take up their place beside human soldiers, and not merely as assistants. On its first page, the program plan noted the "unique new opportunities for military applications of computing. For example, instead of fielding simple guided missiles or remotely piloted vehicles, we might launch completely autonomous land, sea, and air vehicles capable of complex, far-ranging reconnaissance and attack missions."[58] AI technology would release the cyborg to its destined autonomy on the electronic battlefield.

The program plan repeatedly expressed the need for AI in terms of the breakdown of human information systems in the face of the extreme speeds, increased uncertainties, and cognitive stresses of high-technology war. Reading the document, one can sense an intense underlying anxiety about the problems of complexity and control. It called for "revolutionary improvements in computing technology...to provide more capable machine assistance in...unpredictable...[and] unanticipated combat situations." People were said to be "saturated with information requiring complex decisions to be made in very short times." Without AI technologies, commanders would be "forced to rely solely on their people to respond in unpredictable situations." The responsibility of line officers and enlisted personnel, formerly among the highest of military values, had now become a liability. Such statements reflect incipient breakdowns in a closed-world understanding.

The 1950s Cold War, at least in ideology, focused around the problem of central war in Europe. It drew its iconography primarily from the world war out of which it was born. But by the 1980s, maintaining the closed-world metaphors that sustained Cold War discourse had become much more difficult. The globalist frame of a struggle between good and evil, embodied in the superpowers and their proxies, increasingly failed to fit contemporary realities. Competition for economic hegemony increasingly came from Japan, a politically neutral society with no significant military power or interests that could be pressed into the mold of global ideological struggle. Military tensions of the early 1980s focused on the essentially cultural/religious struggles of the Middle East, where Soviet power had sharply declined. Vietnam, the most important American engagement since 1945, had been a protracted guerrilla war in which American technology and nuclear weapons had utterly failed to produce victory, while the stake of the communist superpowers in the North Vietnamese had proven far less significant than the Cold Warriors believed.

The Carter-Reagan defense buildup helped reinflate the sagging balloon of Cold War ideology by focusing attention on the always-riveting issue of central nuclear war. In so doing, they recreated the context of Cold War I and once again enclosed the world within the frame of apocalyptic war. Just as in Cold War I, the nuclear fear aroused by Cold War II led to a proposal for an overarching dome, a shield against the holocaust, with computer technology at its core. And just as in Cold War I, the serious technical failures of high-technology weapons were, paradoxically, matched by major ideological and policy successes. At a conference on the SCI in 1985, David Mizell of the Office of Naval Research captured the politics behind the paradox when he told the audience, "Everyone wants to know whether these technologies will work. As far as I'm concerned, they already work. We got the money."[59]

Strategic Defense and Strategic Computing sought to reinforce the technological supports of closed-world politics. They did so largely by upgrading the role of computers. The problems of previous systems were attributed to their lack of intelligence; with AI, this element could now be added, restoring confidence in central control. Battlefield AI systems, merging closed-world and cyborg iconographies, would combine the speed and power of computers with the flexibility of minds. Autonomous machines would relieve humans from the increasingly terrifying subject position of soldiers on the high-tech battlefield. Computerized battle managers would serve as lieutenants, providing commanders with menus of strategic options and automated reasoning in their support. The SDI battle manager, controlling the firing of space-based, nuclear-powered weapons in a global defense system, would approximate the role of a general, placing ultimate nuclear decisions in the circuits of a silicon mind. "The technology so [the president] couldn't make a mistake" would once again safely enclose the unpredictability and uncertainty of the real world inside the programmed microworld beneath the Star Wars dome.

The Death Star orbits a planet in a poster concept for *Return of the Jedi* (1983). Courtesy Lucasfilm Ltd.

10
Minds, Machines, and Subjectivity in the Closed World

Star Wars, the defense system, was an ideological fiction whose computer-controlled nuclear defenses would not have worked and could not have been built. Yet its fictional and mythological characteristics only enhanced its ability to serve profoundly serious political purposes: disorganizing opposition to nuclear weapons, promoting military-oriented research programs and industrial policy, and spending the Russians into the ground.

Star Wars, the movie, graphically represented a whole set of facts about the ongoing militarization of space, the social lives of computers and robots, cultural relativism as a Turing-test problem, and contests and collaborations over identity in a world of cyborgs. The movie's appeal stemmed in part from its ability to place very real issues of modern life within a coherent narrative and mythological frame. Its escapist-fantasy surface concealed a core of cultural and political reflection. Fiction and fact, ideology and experience, mythology and science blurred and blended in the near-universal use of *Star Wars*, the film title, to refer to the SDI.

Such themes, and a host of others connecting closed-world politics with cyborg subjectivity, emerged as prominent elements in popular culture while the closed world itself evolved during the Cold War. This chapter explores the iconography of the closed world, and the experience of subjectivity inside it, by reading works of popular culture as political texts. I shall focus primarily on science fiction film, beginning in the early 1960s with *Fail-Safe* and *Dr. Strangelove* before turning to films of the middle 1980s, such as *War Games* and *Blade Runner*. The chapter closes with a reading of William Gibson's *Neuromancer*, the novel that popularized a new technological aesthetic, introducing a mass audience to the concept of "cyberspace," a virtual reality inside the machine. I shall argue that the closed world, in both politics and fiction, represents a special kind of dramatic space whose architecture is constituted by information machines. As a stage or space, the closed world defines a set of subject positions inhabited—historically, theoretically, and mythologically—by cyborgs.

In the Theater of the Mind: Fiction and Cyborg Subjectivity

Narratives and images, like the scientific theories and metaphors we discussed in chapters 5–8, can represent possible *subject positions*: imaginary, yet coherent and emotionally invested ways of living within a discourse. Theories, metaphors, and the subject positions they create are conceptual and therefore relatively static. But narratives are dynamic: they are "constrained, contested" stories that show how lives can be lived in time and space, and how struggles can be fought and resolutions reached within some possible world. They do not merely describe, but actually demonstrate, *what it is like* to inhabit specific forms of subjectivity, particular versions of the self. Visual images, too, and especially motion pictures, with their dynamic possibilities, lend structure and coherence to subject positions. Fictions that construct the points of view, emotional frames, and social roles for cyborg subjects thus constitute a theater of the mind.

In this theater the provenance of particular narratives and images, whether fantastic or factual, makes no difference. Taken up as semiotic resources, their importance lies in their dramatic function, their *enactment* of subject positions that in turn become resources for the larger discourse of which they are a part. This point is the key to my analysis in this chapter, which should be read as a direct parallel to the political and historical analyses of the preceding chapters. For whether told by politicians, scientists, engineers, artists, or filmmakers, all stories participate in constituting subject positions. Effective politicians understand and use the mythological and dramatic dimensions of their public appearances to mobilize their followers. Similarly, the entrepreneurialism of successful scientists and engineers, as much recent scholarship has shown, frequently involves the narrative work of articulating connections with other areas of life.[1]

Such "factual" discourses are rarely limited to utilitarian statements of potential benefits. They also include appeals to emotions such as the will to power and the fear of death, to the political unity of science and engineering communities, and to dramatic subjects and mythological locations such as the explorer and the frontier.[2] These factual discourses rely integrally on fictional selves (both created and borrowed from other sources), possible subjectivities that could be aroused or satisfied by the practical and intellectual goods they claim to produce. To be effective they must provoke in their audiences the sense that "yes, that could be me—I could want that, be that kind of person, be moved by that outcome, help in that way, be satisfied by those offerings." Or they may take the opposite approach, constructing alien and dangerous Others whom audiences will want to escape, destroy, or at least define themselves against, and proposing their own intellectual and material products as bulwarks or assistants. Discourses about "facts" are *always already* about identity and difference, freedom and subjection, community and enmity, power and death—about how subjectivity is constrained, contested, and created in a world of objects and others. (For example, when AI proponents analyze human activity in terms of processes that could be programmed into a machine, or when its critics characterize symbolic AI as limited "instrumental rationality" or "rationalism," they are enacting the drama of subjectivity, creating roles and resources for the theater of the mind.)

I am not claiming some sort of total equivalence between fiction and fact. Fictional forms clearly have far less power than institutions or technological infrastructures to "interpellate" subjects (in Louis Althusser's phrase) in a quasi-coercive way.[3] Also, because they are ephemeral, only rarely does a single narrative or image play a major role in public discourse over time. Finally, since fictions are produced by individuals or small groups rather than vast, interacting institutions, one could argue that their role in the large-scale discourses I have described must be peripheral.

Yet their ability to represent subjective experience directly and dynamically gives fictional forms special powers of their own. Politicians and scientists must demarcate practical paths from the present to the future they envision. Fictional forms can disregard that problem and jump directly into their future as a total cultural form. The films I treat in this chapter were enormously popular when released, viewed by hundreds of thousands, even millions, of Americans. Some, such as *2001*, *Star Wars*, and *The Terminator*, have become cultural icons, tropes, universal references that capture key cultural values and beliefs. These are also films that resonated remarkably with the political imagination of their times. Their continuing presence in our cultural iconography is in itself an undeniable argument for the power of fiction. Fiction offers images of concrete individuals living real lives, not just abstract potentials and isolated possibilities. Successful fictions generate a sense of reality, of the coherence of the worlds they describe, their possibility, and even their inevitability. Often they draw on facts of the present—the same facts gathered up in the entrepreneurial narratives of politicians, scientists, and engineers—to create this sense of possible worlds.

Subjectivity itself is neither fiction nor fact, or else it is both. Personality, worldview, and emotional life are evolving amalgams of images, environments, beliefs, and emotions, linked together by narrative.[4] Subjectivity is lived experience, and in that sense as real (and as constrained by reality) as anything else. But at the same time it is *narrated*, constructed and constantly reconstructed in interaction with others and with the world. It is a story about a past, present, and future whose salient dimensions and interpretations may change with every retelling. In this latter sense subjectivity is fiction.

Thus in the theory of discourse that informs this book, fictional forms do not merely and passively "reflect" political and social "realities." They *are* political and social realities, because they actively and directly participate in the ongoing construction of subject positions. In this sense, engineering projects, grand politico-military strategies, scientific theories, and fictional forms all generate discourses that are simultaneously political and personal, public and private, abstract and concrete, factual and fictional.

In this chapter I analyze some of the most important postwar films (and one book) dealing with the themes of computers, robots, and androids. I read these works as political dramas of American life during the Cold War and as cultural dramas of the human relationship with intelligent machines. I shall begin by tracing three themes common to the works I discuss. The first is the closed world itself: closed worlds as dramatic spaces for the representation of Cold War politics and hyper-rational subjectivity. The second is the contrast with "green world" drama, which sometimes includes closed worlds within it or provides a resource to mark an escape from a closed world. The final theme is the cyborg, represented in two forms: *disembodied* computer intelligence, on the one hand, and the *embodied* AI of robots and androids, on the other. Such figures inhabit or even create closed worlds that human protagonists attempt, often unsuccessfully, to escape.

Closed Worlds

The closed world, as a dramatic archetype, is a world radically divided against itself. It is consumed, but also defined, by a total, apocalyptic conflict.[5] The closed world is generally represented in fiction by enclosed artificial environments such as buildings, cities, underground rooms, or space stations. Often, though not always, such spaces are military: fallout shelters, a War Room, a Death Star. (Such architectural metaphors do not, however, exhaust the meaning of "closure," which may be expressed in many ways.) As the name implies, closed-world dramatic spaces often produce an intense sensation of confinement. Dark interiors or nighttime urban settings may amplify these qualities. *Invasion* of these closed spaces is the primary action of closed-world drama; protagonists must somehow prevent or avoid it, perhaps simultaneously seeking an escape.

Artificiality is also a crucial feature of the closed world. It is signaled by the built environment, by displays of technology, particularly military and other high technology, and by the general absence and irrelevance of plants and animals. These artificial worlds, for example space stations and cities, may pretend to a self-sufficient autonomy they cannot really possess. Though often darkened, they are rarely *still*. Technological artifacts within the space assist in projecting an underlying, electric tension: the flickering fluorescent light, the ringing telephone, the active computer screen, the flashing indicators on a CPU. Sleep is fretful and frequently disturbed.

Closed-world drama generally maintains a unity of place, with all or most action occurring inside a single enclosed space or within a single city. When closed-world dramas use more than one location, transfer between locations often takes place in a sealed vehicle such as a spaceship. The place itself may serve as a resource for antagonists; indeed, when such narratives include computers as agents, their structural links with the surrounding, pervasive technology provide them with formidable powers. They can reach into the wiring of things (often literally) to control other computers, machines, and communications networks. The effect is that of a trap from which protagonists cannot escape, either because they are physically locked inside the enclosure or because they are psychologically locked into the cosmic conflict. The archetype of the closed world is Homer's *Iliad*; the Trojans cannot leave their walled city, while the Greeks cannot leave off their siege. Protagonists may flee, but only temporarily. In the end they must confront Others and also themselves. Clearly the Cold War, as drama, constituted a closed world in every sense.

A closed world unique to narratives involving computers is *cyberspace*, which appeared in fiction with the advent of global computer networks in the early 1980s. Cyberspace is an artificial world inside a computer or computer network. It is primarily a site of linkage and communication, an abstracted alternate reality where everything has become information and information is all that matters. Data are the only value, both the source and the medium of conflict. The landscape of cyberspace is digital and abstract, featuring chessboards, Cartesian grids, Platonic solids, and the neon colors of video games. From the quasi-physical "virtual" perspective of its inhabitants, cyberspace is a lot like the science-fiction version of outer space. Everything hangs weightless in a black void. Travel occurs at electronic speeds. Alien intelligences appear in awesome, unfamiliar forms. Despite its vast scale, cyberspace is very much a closed world. Closure is marked by its dark backgrounds, its profound artificiality and abstraction, and its lack of physical sensations other than sight (and, occasionally, pain).

Closed-world dramas frequently deploy a whole series of enclosures nested inside each other. The spaceship of *2001: A Space Odyssey* contains "pods," small craft for working outside the ship; inside these, the astronauts wear spacesuits. When HAL's murderous insanity turns the ship from test-tube womb to toxic techno-coffin, Dave Bowman's spacesuit—his personal, private enclosure—becomes his refuge. The strategies of closed-world politics employed a similar nesting of enclosures, from the global closure of superpower struggle, to the bubble shields of SAGE and SDI, to local civil defense programs, to private backyard fallout shelters.

The architecture and ambiance of the closed world mirror the psychological and political constraints against which characters struggle. The forces that sustain closure are *immanent*, human in origin: rationality, political authority, and technology. Protagonists become involved in two kinds of confrontations with these forces.

First, they may struggle with their external manifestations: hyperrational colleagues or leaders, hiding destructive urges behind a mask of logic; authoritarian political figures and systems; and hostile, dominating technologies. Often beginning as members of the existing order, protagonists come to recognize the status quo as Other when it betrays its inherent contradictions. For example, colleagues or political figures, locked into the false logic of preprogrammed plans, may advocate self-destructive action such as nuclear war or betrayal. Technologies, especially weapons, may turn from disciplined tools to suicidal, unstoppable forces, either in interaction with a human command system or on their own. Berserk robots attack their creators. Computers, unfeeling and unable to grasp any larger context, execute preset courses of action that turn out to be deadly. Conflicts are apocalyptic, world-threatening. Nuclear war is a frequent, though not a universal, theme in the fictions I shall discuss.

Second, protagonists confront the roots of these forces within themselves. In closed-world dramas characters often fail in this attempt. Spinning around and around in their own minds, they comprehend their limits only as they meet their doom, like Macbeth or the generals of *Fail-Safe*. The need to burst free of limits may be amplified by an inability to identify their source or nature (as with Case in *Neuromancer*). Sometimes protagonists, especially tragic ones, may attempt to defend their positions by building or maintaining boundaries of their own (as do the future soldiers of *The Terminator*). Like the politicians and generals of *Dr. Strangelove*, they may end by destroying themselves, or they may be destroyed by others whose powers come from sources beyond their capacity to understand.

Closed-world protagonists sometimes survive these struggles to escape. But unless they achieve a broader perspective, their victory is narrow and tentative. They remain within the closed world, moving restlessly within its limits, gaining temporary respite but never finding anything beyond. This is the fate of *Neuromancer*'s Case and Molly, and of Deckard and Rachel in the director's cut of *Blade Runner*. Occasionally, however, protagonists escape altogether. When they do, it is almost always to a green world, an unbounded natural setting inhabited by magical, transcendent forces. In closed-world drama such worlds generally receive the merest of sketches in a final moment, as happens in *The Terminator* and the original version of *Blade Runner*. Often escapes are false exits into ersatz green worlds such as *Neuromancer*'s inverted worlds, the Zion and Freeside space stations.

Green Worlds

Because this book has been concerned primarily with particular forms of closure in the Cold War, and because green worlds are not primarily political and psychological but spiritual and social in focus, I have chosen not to emphasize the contrast between closed world and green world. But the closed world in fiction sometimes invokes the green world as a resource, and green-world fictions sometimes involve journeys into closed worlds in order to destroy them or rescue their inhabitants. Therefore we must discuss them briefly here.

Green-world drama contrasts with closed-world drama at every turn, as shown in table 10.1. Settings are generally outdoors. Green-world spaces have ragged boundaries, such as the edge of a forest or

Table 10.1
Closed World and Green World

Closed World	Green World
Enclosed artificial spaces: cities, buildings, fortresses, space stations	Raggedly bounded natural settings: forests, meadows, swamps, deserts
Unity of place: all or most action inside enclosures; transfer between locations in sealed vehicles	Flow of action between natural, urban, and other locations; protagonists may enter closed worlds temporarily
Theme: confrontation with limits and destructive elements of rationality, social conventions, political authority, technology	Theme: restore community and cosmic order by surpassing rationality, conventions, authority, technology
Action: invade or escape boundaries of closed world, build or maintain defenses	Action: explore, seek transcendent powers, liberate or destroy closed worlds
Movement: among nested enclosures (vertical)	Movement: among locations (lateral)
Immanence: human, psychological forces—technology, political power, human will, pain, resistance	Transcendence: magical, natural forces—mystical powers, sexuality, animals, other life forms, spirits, natural cataclysm
Outcome: self-destruction, exorcism of hyperrational and restrictive social norms, or escape to green world	Outcome: unification or reunification of groups, or of men and women; celebration; possible destruction or integration of closed world
Struggle: apocalyptic—good vs. evil, self vs. dangerous/alien Other	Struggle: integrative—comprehending complexity and multiplicity, grasping Other as merely another
Archetype: siege *(Iliad)*	Archetype: quest *(Odyssey)*

meadow, which are easily crossed. Or they may have none at all. Borders and limits, where they exist, constitute temporary problems rather than absolute confinements. The permeability of borders is an essential theme, emphasized visually, for example, by buildings with open walls such as huts (like the Ewok tree houses of *The Return of the Jedi*) and by the constant movement of characters from place to place.

Green-world drama has the character of a heroic quest rather than that of a siege; its archetype is the *Odyssey*. Action may move into and through buildings, fortresses, or urban locations, but it usually ends in the green space. Green worlds have a natural rather than an artificial structure. Plants, trees, animals, and natural forces are constantly, palpably present, frequently as active agents: companions, enemies, or spiritual guides. Green-world iconography is that of the frontier and the inner spiritual journey.

Closed-world protagonists pit their own rationality, their reflexes, and their technical expertise against a dominating system and its technology from within. Green-world protagonists leave the closed system and its conflict, reentering it only to liberate others or to destroy it. Their hope is to restore a more cosmic form of order, one of renewed communion among human beings, autochthonous forces, and other living creatures. Their method is to explore, to encounter and attempt to understand mystical and natural powers (which may be friendly or dangerous or both), and to master the sources of unity, such as love, and of transcendence, such as spiritual integrity. This unity is frequently represented by marriage or other celebrations that take place, crucially, outdoors, marking the simultaneous reunification of humans with each other and with nature.

Though the green world has utopian elements, the primary opposition here is *not* between utopia and dystopia. Closed worlds may also have utopian dimensions, as in the pristine high-tech future of *2001*. Also, the powers of the green world pose dangers of their own. The excesses of untamed Eros, the private purposes of nonhuman entities, the uncontrollable force of natural cataclysms, and the hostility of outdoor environments may all prove destructive. The rich profusion of living forms in the green world does not necessarily equate with a *friendly* environment; territoriality, consumption, and random catastrophe are among its basic laws. Nevertheless, green-world forces rarely exhibit the focused, personalized malevolence of closed-world antagonists. Green-world powers are dangerous because they exceed human understanding and control, not because they are evil. The closed world exists in part as a response to the perceived danger of green-world forces, but in its hubris it works too well, cutting its protagonists off from life's deep spiritual and natural sources.

As in the closed world, action in the green world has both external and internal dimensions. Externally, green-world protagonists rely on human and nonhuman allies (the Ewoks of *Return of the Jedi*) and teachers (Obiwan Kenobi and Yoda). Though they tend to disdain high technology when used in and for itself, they employ it as necessary. Protagonists may receive unexpected aid for their quest from magical or mystical powers ("the Force"), animals, or other life forms. Green-world forces may also oppose them, often in the rough play symbolized by trickster figures, sometimes with extreme violence. The protagonist's role is to remain open to this dizzying flux, riding the moment rather than trying to impose control, accepting and returning gifts of aid rather than trying to force a way alone. Where the closed-world protagonist's struggle is a lonely, apocalyptic battle with the Other, the green-world protagonist's quest is integrative: comprehending complexity, transforming Others into mere others, and gathering forces for an eventual reunification.

Internally, green-world protagonists face the problem of spiritual growth. They must open or reopen channels of communication with the larger powers within themselves. They overcome cynicism and despair not by sheer will power, like closed-world figures, but through openness, humility, and wonder. Unlike closed-world figures, who must repress pain and fear in order to stand up to antagonists, green-world protagonists must learn to dwell in their emotions in order to rise above their limits. Reuniting in themselves the fragmented whole of emotion, spirituality, and rational thought, they acquire the power and purpose to create external unity.

"More Human than Human": Second Selves

Computers combined, in a single potent icon, the artificiality of the closed-world setting, the power of technology, and the limitations of rationality and logic. At the same time, computer simulations—AI, Rand strategic simulations, cyberspace—formed true closed worlds, entirely within the machine, which could threaten to engulf or replace the larger world they initially sought only to model. Computers embodied the abstraction, the distanciation, and the superhuman speed of high-technology war, as well as the calculative rationality of nuclear strategic planning. With artificial intelligence, computers could bridge the gap between iconic representation and psychological experience: they could attain a subjectivity of their own. Computers, computerized robots, and cyborgs became subjects: speaking, active agents.[6]

Human-created automata and artificial people have a long history in mythology and fiction, from the Colossus of Rhodes to the medieval Golem. Commonly, as in Mary Shelley's *Frankenstein*, they played the role of monsters, outsider and outcast figures reflecting the inadequacy of humans (particularly men) as creators, the human inability to retain control over these creations, and the loneliness of an extraordinary subjectivity. Crucially, the automata were unique individuals. Usually they appeared as evil or demented doubles wreaking havoc and destruction, sometimes unintentionally, wherever they went. Inevitably, they had to be destroyed or exiled from the human world.

Factory technology has often been portrayed as transforming the working class into an insensate legion of unthinking automata. The Czech Karel Capek's 1921 play *R.U.R.* (Rossum's Universal Robots) first introduced the word "robot" in an allegory about an army of mechanical industrial workers.[7] Fritz Lang's film *Metropolis* (1926), Charlie Chaplin's *Modern Times* (1936), and the theater of Bertolt Brecht all depicted workers transformed into machines by the droning, repetitive physicality of machine-tending. Human automata of this sort posed a threat because in their amoral mindlessness they might be easily manipulated by communists, or (alternatively) stripped of their humanity by greedy industrialists bent on extracting maximum value from their labor. At the same time, mechanical robot workers threatened to eliminate the working class altogether by depriving them of jobs. Such machines would be ideal servants or slaves, needing neither sleep nor recreation and being incapable of suffering.

The first computerized automata portrayed in fiction were no different from these earlier figures, but the nature of their role underwent an important transformation as the visions, theories, and technologies of AI and cognitive psychology evolved in tandem with the Cold War. Previous automata and artificial people were imaginary outsiders. AIs, robots, and cyborgs were not only fictitious but real; they became insiders, part of the actual architecture of the political closed world. As the Cold War became a full-blown *system*, increasingly electronic, organized, and networked, electronic machines were woven into the very fabric of reality, no longer individual aberrations to be ritually destroyed but indispensable presences at the heart of the political order. At the same time, as a human psychology based on the computer metaphor penetrated popular culture, locating differences between people and computers became an increasingly subtle and contested problem.

"More human than human is our motto," says the android designer Tyrell in *Blade Runner*, capturing in a phrase the special problematic of the cyborg. Cyborgs could play the roles of servant, ally, and partner as well as that of monstrous, demented Other. Though the old role of monster/outsider survived alongside it, the new role of cyborg/insider moved to center stage. The basic closed-world encounter with the limits of rationality found a new focus, in cyborg fictions, as an encounter with *another* rationality, one imbued with both the limits and the power of pure logic as well as with purposes of its own. The new task of human protagonists was to assert their own subjectivity against that of these artificial forms by defining their differences.

Computers as narrative subjects in film and fiction may be classed into two broad categories. *Disembodied* artificial intelligences—machines such as *2001's* HAL or *Neuromancer's* Wintermute—frequently present the invisible gaze of panoptic power. These AIs almost always communicate with humans through the medium of speech, usually identifiably synthetic but sometimes indistinguishable from human voices. The disembodied voice expresses at once both presence and absence, both personality and the ghostly terror of invisibility.

Almost universally, disembodied AI—even when essentially benign (like the computer Joshua in *War Games*)—develops the creepy panoptic resonances of Orwell's Big Brother. A central characteristic of such entities is their lack of emotion. Disembodied AI represents the faceless, nonlocalized, uncaring power of ubiquitous high technology. It plays the role of Other, an entity whose difference is expressed in the eerie purity of its disembodiment. Disembodied AI operates almost exclusively in closed-world spaces where—linked electronically to everything from telephones to missile silos—it is more at home than its human creators. Representations of computer control of nuclear war almost always employ this form of AI.

Embodied second selves—robots, androids, and other cyborgs—perform more ambiguous functions in the closed world. Embodied beings cannot be panoptic, so their power is limited. But with their computerized minds, these entities outdo their human counterparts in the use of technological resources. By linking to computer networks, like *Star Wars'* R2D2 or *Star Trek: The Next Generation*'s Commander Data, they can gain a privileged form of access to the closed world's underlying architecture. Since they can move about, they can pursue, accompany, and confront people directly. Embodied beings also represent a less alien form of subjectivity. Thus they may be either allies or enemies, servants or equals, or may be transformed from one into the other.

Where disembodied AI presents the Turing problem of abstract intelligence, embodied AI raises more complex questions of status. The more closely a cyborg physically resembles a human being, the more difficult such problems become. Visibly mechanical robots like R2D2 and C3P0 from *Star Wars* are often treated, unproblematically, like servants or pets. Androids—robots with a human form, like Commander Data—may, if friendly, receive respect and rights quite similar to humans. If hostile, they may be destroyed, but with horror. Often plots turn on the question of whether and what these beings feel, as in *Blade Runner* and the many episodes of *Star Trek: The Next Generation* that center on Commander Data's attempts to understand human emotion. A common theme is the conversion of cyborg enemies into friends or allies via latent emotions, as in *Terminator 2* and *Blade Runner*.

Fictional Closed Worlds in the Early Cold War

In the 1940s and 1950s, while politicians and engineers created closed-world politics and its technological supports, popular narratives that thematized the Cold War tended to focus on human dramas. A wide range of 1950s fiction and film explored figures and motifs such as SAC bomber pilots, victory through air power, and the aftermath of nuclear war. *Strategic Air Command* (1955), *The Court-Martial of Billy Mitchell* (1955), and other films depicted air war as the locus of future conflicts. They portrayed pilot-commanders as military visionaries, fighting against the hidebound traditions of ground and naval war, who could lead America to success in World War III if only they were given free rein. Stories of nuclear fear, such as Nevil Shute's *On the Beach* (1957, film version 1959) and Walter Miller, Jr.'s *A Canticle for Leibowitz* (1959), dramatized post-holocaust horrors. Hard-core science fiction (in those years a fairly small cult) sometimes treated the theme of robot- or computer-controlled nuclear weapons, but most narratives of this period presented the issue of nuclear war as a problem of individual or political—not technological—failure.

The first major popular works to treat the role of electronic machines (not only computers, but electronic communication and coding devices) in nuclear war were Sidney Lumet's *Fail-Safe* and Stanley Kubrick's *Dr. Strangelove*, both released in 1964. Both were based on best-selling books: *Fail-Safe* on the book of the same title by Eugene Burdick and Harvey Wheeler (1962), and *Dr. Strangelove* on Peter George's *Red Alert* (1958). Both films illustrate the emergence of semi-autonomous technological systems as political problems; but at the same time, both ultimately treat these problems as subordinate to issues of human responsibility. Later, as we shall see, such systems acquired agency of their own and became primary sources of dramatic tension.[8]

Fail-Safe

In *Fail-Safe*, a highly unusual sequence of events causes the "fail-safe box" of one group of nuclear bombers accidentally to emit the command code that launches them on a mission to bomb Moscow.[9] This command, once issued, cannot be reversed. The pilots have been ordered to disregard all voice communications and to interpret any attempt to bring them back as an enemy deception.

The moral eloquence of the film's antiwar message relies heavily on closed-world iconography. At the outset, characters converge on Washington, D.C. (providing unity of place). Black-and-white film (in the era of Technicolor) and sparse sets reflect the bleakness and psycho-political orientation of the drama. Most scenes are set in windowless rooms. The few scenes from the juggernaut bomber show only its cramped cockpit, with its crew encased in technological armor. At the Pentagon's underground War Room, neat rows of identical desks face huge screens on which the world situation map is projected. In the White House, the president and his interpreter sit alone in a tiny basement room containing only a desk, two chairs, a telephone, and the hot line to the Kremlin. Somewhere else in Washington, the secretary of defense and a number of generals find themselves in a windowless meeting room. As the crisis breaks out, they are being briefed by a hawkish academic nuclear strategist strongly suggestive of Rand's Herman Kahn. The group remains together to discuss the developing situation and advise the president.

The closed-world confrontation with rationality's limits unfolds as the preprogrammed war plan carries itself out, beyond the control of its human creators. General Black fears that nuclear weapons and high technology have made war too fast and too efficient for humans to make intelligent decisions: "We're setting up a war machine that works too fast for men to control it." Ultimately his terror proves justified, as the consequences of the electronic failure unfold. The president orders his generals to help their Soviet counterparts shoot the bombers down, but despite their best efforts, one bomber successfully penetrates the Soviet defenses and completes its mission, obliterating Moscow. To prove to the Soviets that the attack was a mistake and to avert a retaliatory strike, the president's only choice is to order the atomic destruction of New York City.

The seamless interface of technology and command marks the borders of *Fail-Safe*'s closed world. The brittleness of a preprogrammed war plan, in which orders are carried out mechanically and mediated by electronic devices, causes the crisis in the first place. One officer states the technological-determinist military rationale for high-tech weapons systems: "It's in the nature of technology. Machines are developed to meet situations." But a congressman on a tour of the War Room replies, "And then they start creating situations. . . . I'll tell you the truth, these machines scare the hell out of me." An executive from an electronics firm warns that electronic systems always eventually fail and that when they become too complex, human control cannot necessarily correct the problems they cause. Previously the congressman has expressed his worries about the loss of human control in terms of a fear of computer subjectivity: "I want to be damn sure that thing doesn't get any ideas of its own."

The abstraction and emotional distance created by technological mediation amplify the terror of the situation. The bomber's attack on the Soviet Union appears to the characters in Washington as a moving triangle on a map of the globe projected on room-dominating screens. The sterility of this impoverished, abstract representation appears to be what allows some of the characters, such as the cold-blooded professor and some of the officers, to urge taking advantage of the opportunity to launch a full-scale attack. To them it is only a game, a calculated risk, a war of numbers in which the side that crawls out from beneath the rubble with fewer millions of dead can count itself the winner. The film also emphasizes the problem of technological mediation by focusing on the difficulty of the president's discussions with the Soviet premier over the hot line to Moscow, where the lack of face-to-face contact hampers the two leaders' communication. Only the handful of brief scenes in the bomber cockpit show even a hint of the horrors unfolding on the ground: the audience, along with most of the characters, views the entire event through the abstract lens of the situation map and hears the distant Soviet voices only through the telephone.

Though technology plays an important role in creating and representing the crisis, *Fail-Safe* places the moral burden of the catastrophe not on the system-builders but squarely on the politicians who permit their deadly technological hubris. The exact nature of the command system's failure is never revealed. "Something failed. A man, a machine—it doesn't matter now," a military leader says. The root cause of the tragedy is the lack of trust and communication between American and Soviet leaders. Ironically, their work together during the crisis finally remedies this lack.

Dr. Strangelove
Dr. Strangelove (1964) presents a similar scenario, this time as black comedy. At a Strategic Air Command base, General Jack D. Ripper goes quietly insane and orders his air groups to attack the Soviet Union. He refuses to recall them and tells his men to defend the base against anyone who attempts to enter—even other American troops, which he informs them will be Soviets in disguise. Ripper eventually commits suicide, and the recall codes are retrieved. A recall order is issued, but one bomber's communications gear has been damaged by Soviet strafing. It never receives the order and continues on to its Soviet target.

As the situation evolves, military and political leaders assemble in the War Room. Again the setting is a closed-world space, dark and empty except for tables and chairs; again the technological abstraction of the huge world situation board dominates the scene. The group includes the timid president, gung-ho General Buck Turgidson, and the civilian adviser Dr. Strangelove, Kubrick's own parody of Herman Kahn. As in *Fail-Safe*, the president attempts to help the Soviets shoot down the maverick bomber—but this time everyone at the other end of the telephone line is drunk.

The Soviet ambassador, summoned to the War Room, reveals the existence of a "Doomsday Machine." In a desperate attempt to foreclose the costly arms race, the Soviets have programmed a computer to detect any nuclear explosion within the Soviet Union's borders.

The Doomsday Machine will then automatically explode an enormous nuclear device whose fallout will gradually encircle the globe, exterminating all life on the surface. The computer's instructions have deliberately been rendered irreversible. As the film ends, the bomber pilot heroically—at least from his point of view—wrestles open the stuck bomb bay doors and drops out of the plane, waving his Stetson hat in triumph as he rides the bomb to global oblivion.

The Doomsday Machine represents the first appearance of a computer-controlled nuclear weapons system in mainstream film. But as in *Fail-Safe*, its significance is as a symbol of human short-sightedness, the folly of a preprogrammed plan that cannot be reversed. The computer in *Dr. Strangelove* is merely a machine, not a decision-maker; blame devolves upon the war-planners whose technological powers exceed their capacity to anticipate the future. The crises in both *Dr. Strangelove* and *Fail-Safe* come about because human rational plans fail to foresee unlikely sequences of events. But *Dr. Strangelove* emphasizes how perilously close conventional politics (like General Ripper's ultrasuspicious Cold War sloganism) and strategic doctrine (like the Doomsday Machine, which merely follows mutual assured destruction to one logical conclusion) may be to madness.

The minor but significant role of computers in *Fail-Safe* and *Dr. Strangelove* reflects a number of facts about the period during which they were made. First, though Air Force plans for computerized, centralized nuclear command and control had been developing for over a decade, such systems had only recently been deployed. Some of the "Big L" projects begun in the late 1950s were still under construction at the time these films were released. Second, the most terrifying moment of nuclear brinksmanship—the Cuban missile crisis of 1961, which fell outside the parameters of the central-war scenario—had focused attention on the human side of the command-control problem. *Fail-Safe*'s key technology is thus not the computer but the then newly installed "hot line" to Moscow. Third, both films ignore the issue of ICBMs in favor of a more dramatically satisfying, but already anachronistic, focus on manned bombers.

Finally, the defensive shield promised by SAGE is nearly absent from both films. *Fail-Safe* begins with a false alert, as SAGE-like early warning radars pick up an unidentified object entering U.S. airspace. Interceptors are sent to meet it, but the object turns out to be an off-course commercial jet. (The incident also causes all airborne strategic bombers to fly to their fail-safe points; at this juncture the mysterious electronic failure occurs, sending one bomber group on its deadly mission to Moscow.) American nuclear defenses are never tested, because the premise of both films is that the United States, not the Soviet Union, mounts the first nuclear strike. However, neither film expresses even an iota of faith in defensive systems like SAGE. *Fail-Safe* presumes that if the Soviets respond in kind, the United States will also be destroyed; *Dr. Strangelove*'s Doomsday Machine avoids the defensive shield by the simple expedient of exploding a huge bomb in the Soviet Union and relying on winds to spread lethal fallout across the globe.

Dr. Strangelove also links the underground War Room with another closed-world icon: the fallout shelter. When Dr. Strangelove realizes that the curtain is rising on World War III, he eagerly proposes digging deep caverns in which an elite few would pass generations of time while waiting for the surface radiation to return to safe levels. Excited to delirium by the prospect of a eugenic, high-tech, underground future, Strangelove can barely contain his reflex, a Nazi salute. So Strangelove's solution to closed-world politics is to build an ultimate sealed fortress inside the earth.

This, too, was part of real-world Cold War. The late 1950s saw regular calls for major civil-defense programs, consisting primarily of underground shelters and storage depots in which civilians and civilian leaders could ride out a nuclear attack and its radioactive aftermath. (Kahn, in fact, was a key advocate of such plans.) At the time *Fail-Safe* and *Dr. Strangelove* were being filmed, suburban families across the continent were digging private backyard fallout shelters. Schools and television stations conducted the infamous "duck-and-cover" education campaign, with happy tunes and cartoon figures to teach young students to duck into a sheltered place, under a desk or behind a wall, and cover their heads if they saw the white flash of a nuclear explosion. Civil defense preparations designated the basements of schools, department stores, and government buildings as public fallout shelters. If the shield of nuclear missile defense could not protect the population, perhaps they could protect themselves by sealing shut their own closed worlds.[10]

The Strangelove solution to holocaust reflects the shell-game strategy characteristic of closed-world discourse. When one enclosure is punctured, the strategist or theorist immediately retreats into another, each more baroque and at the same time more abject than the last. Containment politics gave way to continental air defense; air defense to hardened missile silos and command posts. These in turn gave way to public civil defense, and civil defense to the backyard fallout shelters of the grotesquely optimistic survivalists.

2001: A Space Odyssey

The first major mainstream film to focus on the question of machine subjectivity was another Stanley Kubrick movie, his dark masterpiece *2001: A Space Odyssey* (1968). Released at the height of the space race and the Vietnam War, a year before the Apollo moon landing, *2001* presented the intelligent computer's potential for monomania against the haunting beauty of space as the final frontier.

The main body of the film opens on a space station between the Earth and the moon. A brief meeting between American scientist Heywood Floyd and some Russian colleagues reveals ongoing Cold War tensions when Floyd, for "security" reasons, refuses to reveal the nature of his mission to the moon, about which the Russians have heard rumors. We then learn that American explorers have discovered a huge black monolith buried in Tycho Crater. When Floyd arrives, it emits a mysterious radio signal apparently aimed at a point in space near Jupiter. Astronauts Dave Bowman and Frank Poole are sent on a mission to explore the signal's destination point. A three-member Jupiter survey team is aboard their spaceship in cryonic hibernation. The *Discovery* is controlled by an artificially intelligent computer, the HAL 9000 (for *H*euristically Programmed *AL*gorithmic Computer, according to the film's screenwriter, Arthur C. Clarke).

HAL is a perfect representative of disembodied artificial intelligence. Its "eyes," glowing red lenses, allow it to view the entire ship's interior. It can also hear the astronauts wherever they go. HAL speaks in a calm, monotone, male voice whose relaxed, exact diction is the only thing that differentiates it from the more rapid, emotionally inflected speech of the astronauts. This unexcitable, laconic murmur produces a reassuring effect of quiet competence, until HAL begins killing the astronauts, when that same slow, calm speech becomes ominous and terrifying. The effect of HAL's omnipresence changes at the same time from one of benign oversight to a Big Brotherly panopticism.

HAL inhabits the closed world of the spaceship *Discovery*, but in a different way from the human characters. Its mind and its presence fill the space; it sees and senses everything that goes on inside the ship, not only visually and aurally but through electrical contact with all the ship's systems. In a sense, the ship is HAL's body.[11] When Bowman disconnects his "higher mental functions," HAL's other systems keep working, controlling the ship's basic activity as the brainstem of a comatose person might continue to control breath and heartbeat.

The tense ambiguity of HAL's relationships with the crew forms a major fulcrum of the plot. A television interviewer refers to HAL as the "sixth member of the crew" and the "brain and central nervous system" of the *Discovery*. He asks HAL how it feels about its enormous "responsibilities." "The 9000 series," HAL informs him, "is the most reliable computer ever made. No 9000 computer has ever made a mistake. We are all, by any practical definition of the words, foolproof and incapable of error." HAL thus designates itself the hyper-rational guardian of closed-world space. Asked whether HAL has emotions, Poole gives a Turing-like reply: "HAL acts like he has genuine emotions. Of course, he's programmed that way to make it easier for us to talk to him. But as for whether he really has emotions or not, I don't think that's something anyone can truthfully answer."

Yet the crew's attitude toward HAL betrays uneasiness with the computer's liminal subjectivity. Their tone with it is short, abrupt, as if giving orders to a servant. HAL's respectful demeanor, likewise, resembles that of a butler. But HAL is no mere servant. Its responsibilities include assessing the emotional health of the crew, and it, at least, believes itself capable of carrying out the entire mission alone. HAL continually attempts to engage Bowman and Poole in personal conversations. At one point HAL tries to read Dave's unspoken thoughts, informing him: "During the past few weeks, I've been wondering whether you've been having some second thoughts about the mission."

The first clue to HAL's instability comes when it questions Dave suspiciously about some "extremely odd things" it has noticed about the mission. As Dave wavers, composing a reply, HAL suddenly detects an impending fault in a control unit for the ship's Earth communications antenna—located outside the ship. Poole leaves the ship in a "pod," a tiny space utility vehicle, and retrieves the unit. Diagnostic tests reveal no problems with the device, but HAL insists that any mistake must come from some human source, since the 9000 series has never made any kind of error. HAL recommends reinstalling the unit and letting it fail, which will put the ship out of Earth communication until it is replaced.

The astronauts, now gravely worried, confer inside one of the pods (the only place in the ship where HAL cannot hear them). They agree that if HAL's diagnosis proves inaccurate, this and HAL's other odd behavior require that they disconnect the computer. Poole again leaves the ship to replace the control unit. While he is outside his pod, HAL takes control of it and slices Poole's air-supply line, killing him. Bowman exits the ship in another pod to attempt a rescue. While he is gone, HAL kills the hibernating survey crew. It refuses to let Dave reenter the ship, having read the astronauts' lips while they conferred inside the pod. Thus Bowman and Poole's retreat into the nested enclosure of the pod fails; it not only proves transparent to HAL's panoptic gaze, but serves the computer as an instrument of power.

Bowman breaks in through an emergency airlock, protected only by the most individual of total enclosures, his spacesuit. He enters HAL's memory center to disconnect it, floating weightless inside the eerie space of the AI's mind. Against the labored sound of Bowman's breathing inside his spacesuit, HAL regresses, as it were, to childhood, pleading with Dave to stop. "I'm afraid," it intones. "My mind is going." As its voice sinks to a deeper and deeper pitch, it begins to sing "Daisy," one of its first achievements—and in fact the first song ever played by a computer, at Bell Laboratories in 1957.[12]

This scene provokes complex and contradictory emotions. We are meant, of course, to be terrified by HAL's monomaniacal devotion to the mission, its violence, and its paranoia about the crew's intentions. Yet we cannot help but sympathize with HAL's instinct for survival. HAL's spooky regression to a "childhood" of songs and fond memories clearly brings Bowman the same chilling images of prefrontal lobotomy that it carries to us, the audience. The dissonance between our fear of HAL's strangeness and arbitrary, panoptic power, on the one hand, and our sympathy with HAL's own fear of death, on the other, underscores HAL's strained, liminal position as an Other who is also a second self.

Like its counterparts in *Fail-Safe*, *Dr. Strangelove*, and *Colossus: The Forbin Project*, HAL represents computers as self-directed technological juggernauts. HAL controls a closed world, human-made but no longer human-centered, but it is not merely self-directed: HAL *is* a self, a second self whose disembodiment marks the hyper-rationality, the panopticism, and the presence of the Other in the material infrastructure of the closed world.

Though the bulk of the film occurs in the closed-world setting, *2001* does not, in fact, begin or end there, but rather in the green world. One of Kubrick's key concerns is to depict a mythic, quasi-historical transit from an original green world, through a technological closed world, to a destiny that involves an enlightened return to a green world on a literally cosmic scale.

2001's opening sequence—entitled "The Dawn of Man"—begins with sunrise on an African savanna. A black monolith, like the one found later on the moon, appears from nowhere on the territory of a band of protohuman apes. One of them, his insight somehow triggered by the monolith, discovers how to use a bone as a weapon. Thus armed, the tribe defeats another, hostile band, killing other members of its own species for the first time. In an arresting image, one of the primates, exulting in this new power, hurls a bone high into the sky. As it slowly revolves against the blue, it becomes the graceful space shuttle carrying Dr. Floyd to the Earth-orbiting space station.

The moment binds the origins of tools with those of weapons, of the closed world with the green world, and both to the sources of civilization. The most significant thing about this episode, however, is the monolith itself, enigmatic icon of an alien technology. *It* is the essential power enabling "man" to separate from nature, setting off the closed-world drama. Artifactual, not natural, it symbolizes not transcendence and unity but immanence and conflict; it marks the breakdown of the green world's timeless, cyclical wholeness.

In its final, utopian sequence *2001* returns to the iconography of the green world. After HAL's demise, Bowman enters "Jupiter Space," where he finds a third monolith floating in the void. As he approaches this symbol of an Other so alien as to make HAL seem mundane by comparison, Bowman suddenly finds himself rocketing at an enormous velocity across neon landscapes, through mind-bending scenes of indescribable shapes and colors. Following a long journey through this mysterious realm, the voyage just as suddenly stops. Still sealed in his spacesuit, Bowman encounters himself, rapidly aging, alone in a suite of strange, silent rooms. When he reaches old age, the scene transforms into a view of the Earth from space. Beside it floats a planet-sized fetus in its amniotic sac, wide-eyed with wonder.

Clarke's novel, based on his screenplay, reveals little more than the film about the meaning of this stunning image, except to give it a name: the Star Child.

The image of the Star Child must be read on many levels. First, it is an image of *life*, birth, genesis—perhaps a rebirth or an evolutionary leap for humanity, drawn on the eve of the Apollo moon landing.

Juxtaposed, the amniotic sac and the egglike Earth form twin icons of pregnancy, life about to burst forth from gestation into the limitless wonder of the universe. It evokes the "final frontier" as well: "Jupiter Space," where fantastic, untold possibilities await.[13] Finally, in the wide-eyed gaze of the fetus upon the blue-green planet, it represents a yet-unformed green-world subjectivity, a new beginning after *2001*'s harrowing journey through the closed world.

2001, then, treats technology as an evil necessity in humankind's simultaneous search for its origin and its destiny. Tools and weapons lead both to power and to the dangers of war and brutality. The extraterrestrial monoliths that cause humankind's first evolutionary leap also mark a trail toward another important step—this one linked with life itself (the fetus in its amniotic sac) rather than technology. *2001* presents the closed worlds of technological civilization as a developmental transit between one green world and another, vaster one that lies beyond the Earth.

Colossus: The Forbin Project

In *Colossus: The Forbin Project* (1970, from D. F. Jones's 1966 novel *Colossus*) the United States constructs a Strangelove-like Doomsday Machine of its own. The computer scientist Forbin convinces the government to hand over control of nuclear weapons to an enormous computer, irreversibly programmed to launch a retaliatory strike in the event of a Soviet attack. The plan is to short-circuit a weak-willed political leadership, ensuring the credibility of the deterrent threat. Early scenes show Forbin and his engineers at work on the gigantic Colossus in a cavernous shelter deep inside a mountain. Once they leave and the huge blast doors clang shut for the last time, the computer's program cannot be changed.

Once activated, the machine begins to learn at a fantastic rate and to evolve in ways unanticipated by Forbin or anyone else. It soon manifests an independent intelligence, requesting a voice synthesizer in order to converse with its programmers. Much less human-sounding than HAL, Colossus speaks in a clipped, synthetic monotone. It begins first to order its creators about, then to threaten them with the nuclear arsenal it controls. When it learns that a similar computer exists in the Soviet Union, it demands a direct communications link. The two machines begin to converse at a rate that rapidly outpaces the pathetic efforts of humans to understand them. The machines conclude that humanity is too dangerous, both to them and to itself, to be allowed any further freedom. Using the threat of nuclear annihilation it controls, Colossus takes over the world and enslaves the human race.

The significance of *Colossus: The Forbin Project* lies in the way its iconography binds together the themes of the closed world and the cyborg. The machine vault under the mountain, with its massive blast doors, resembles the computerized NORAD control center, stripped of its human elements. Once again, the film's action happens almost entirely inside buildings at a single location, Forbin's office complex. Colossus begins its work as no more than the computer it was programmed to be, but it rapidly develops into a subjective machine. It demands that cameras be installed throughout Forbin's office complex so that it can watch the staff. An Other and a technological juggernaut, Colossus proves unstoppable. Forbin first attempts to bargain with the machine, then tries a number of ruses to render it harmless. But all of them fail. The ultra-powerful, ultra-rational machines built to protect humanity end up dominating it instead; the price of eternal peace is slavery. Ironically, Colossus achieves this in part by cooperating with its Soviet counterpart, something the humans failed to do for themselves.

In this film, as in some others within this genre dominated almost entirely by male characters, women play the role of liminal figures able to cross closed-world boundaries. Forbin manages to convince Colossus that he "requires a woman" three nights a week. A female member of his staff agrees to play the role of his girlfriend, spending these nights with him at the quarters where Colossus has imprisoned him. Forbin tells Colossus they require "privacy," which it grants them in their bedroom during the night. Through her, Forbin plots with others to bridle Colossus. Here the conjugal bedroom serves as the refuge from the panoptic Colossus, the smallest in this film's nested set of closed worlds. Despite being a scientist herself, the woman becomes the symbol of sexuality in the dry, masculine world of computers and nuclear war. But sexuality, that primal green-world force, is here turned to a rational purpose, impressed into the service of closed-world power as Forbin pits his mind against the machine.

Like HAL, Colossus inhabits and controls the closed world and is motivated by the goal of self-preservation. But unlike HAL, the panoptic Colossus succeeds in overcoming its creators. In Colossus, Cold War politics produces a disembodied AI whose subjectivity is dominated by concerns about security, global power, and nuclear war. Preserving itself as a subject demands the subjection of its less capable, error-prone creators; extending their goals and methods to a logical conclusion, it chooses total technological control. Citizen of no country, child of no parent, Colossus bears loyalties to nothing and no one except the closed world itself, the realm of rationality and pure power whose Leviathan it has become.

Cyborg Subjectivity in the 1980s

By the early 1980s, the simplistic computers-out-of-control that dominated the 1960s and 1970s were replaced by a more sophisticated awareness not only of the machines themselves, but of the cultural networks and identities that had arisen around them: hackers, video gamesters, and teenage whiz kids. This had much to do with the arrival of home or "personal" computers, which provided a new opportunity for hands-on exposure to the machines.

Beginning in the late 1970s, computers moved from offices and factories into homes and mass entertainment. Microprocessor-based arcade video games such as Space Invaders and Galaxians, many of them based on a military metaphor, became a boom industry, climaxing as a multibillion-dollar business in the early 1980s. Apple Computer began marketing its Apple II home computer in 1977. IBM followed Apple's lead, introducing its PC (Personal Computer) in 1981.

By 1983, Apple had moved from its original headquarters in a Silicon Valley garage into huge new corporate buildings as the company's annual sales approached $1 billion. IBM had already sold over a million PCs.[14] Personal computers soon fueled a massive wave of new office automation as "desktop" computers. In offices, word processing and spreadsheet accounting were the primary applications, while computer games probably accounted for the bulk of home computing in the early 1980s. Children's electronic toys such as the Merlin tic-tac-toe game and the Speak 'n' Spell word game incorporated microprocessors. Computers also began to appear in schools, and the phrase "computer literacy" became a buzzword among educators. In 1983, *Time* magazine chose the personal computer as its "man of the year."

With this sudden explosion of computers into the wider culture, closed-world drama acquired new sources of iconography. The themes we have been discussing received new, and newly sophisticated, articulations. The computer as panoptic Other was, frequently, rehabilitated as merely an other, a companion and friend. At the same time, the identification of computers with Cold War continued, sometimes in forms even more sinister than those of Colossus and the Doomsday Machine.

War Games

In *War Games* (1983), an intelligent military computer, egged on by a teenage hacker who thinks he is playing a computer game, brings NORAD nuclear forces to the brink of DEFCON (DEFense CONdition) 1: all-out nuclear engagement.

War Games opens with an archetypal closed-world image: two soldiers descending to the underground control room of a nuclear missile silo. When the two suddenly receive a launch order, the superior officer cannot bring himself to turn the key releasing the missile. He orders the junior officer to get telephone confirmation first: "Screw the procedure! I want someone on the phone before I kill 20 million people." The event turns out to be only a test, but the point is made: faced with the real enormity of such an order, men may lose their nerve and abandon the carefully preprogrammed plans. The fictional scene reflected the widely reported reality of readiness tests, which also revealed high rates of drug abuse and psychological problems among nuclear missile crews.

The scene then shifts to the underground NORAD headquarters beneath Colorado's Cheyenne Mountain. Civilian leaders, against the advice of senior military commanders, decide to automate the launch system, "taking the [unreliable] men out of the loop" (thus carrying into effect the program of military automation first envisioned during World War II). They point out that while he retains final authority, "the president will probably follow the computer war plan," since under SLBM attack he would have as little as six minutes to make a decision.

The scene at NORAD headquarters looks like a technologically updated version of the situation rooms in *Fail-Safe* and *Dr. Strangelove*. Arranged like a theater, descending rows of desks strewn with computer monitors face screens projecting enormous computer-generated world maps. A tour guide tells his audience,

> These computers give us instant access to the state of the world: troop movements, shifting weather patterns, Soviet missile tests. It all flows into this room, and then into what we call the War Operations Plan Response [WOPR] computer. WOPR spends all its time thinking about World War III. It has already fought WWIII—as a game—time and time again. It estimates Soviet responses to our responses to their responses.

This iconography is by now familiar: a dark, enclosed, artificial space; a computer-simulated, abstracted world; an AI; an apocalyptic conflict. The NORAD control center not only resembles but actually *is* a theater, a stage upon which simulations are enacted as real politics.

The scene now shifts to the friendly domesticity of a Seattle suburb. Teenage hacker David Lightman, using a home computer and a modem, breaks into a computer in California that he thinks may hold a new commercial computer game. Actually, however, the machine is NORAD's WOPR, originally programmed by an AI specialist named Falken.

The machine has a name: Joshua, the name of Falken's dead son. It also soon acquires a voice, when Lightman hooks up a voice synthesizer to his computer. The voice resembles a cross between those of HAL and Colossus. The resonant synthetic monotone, in Lightman's padlocked bedroom, extends Joshua's closed-world subjectivity into this private domain. Joshua offers a selection of games. David chooses "Global Thermonuclear War." Play commences, but Lightman and his friend Jennifer are interrupted and he breaks the connection. Joshua, however, refuses to end the game until an outcome is reached. Joshua's "game" proves to be more than a simulation; the computer is in fact initiating a real nuclear war. Arrested by the FBI and taken to NORAD, Lightman learns more about the computer—a large black box studded with flashing indicators—from its operators, who still fail to comprehend the nature of Joshua's undertaking. Locked inside this closed world, Joshua controls the missiles but cannot sense reality itself. To him, games are all that exist.

Lightman escapes with Jennifer to Falken's remote island retreat, where they convince the scientist to return with them to NORAD. Meanwhile, Joshua has caused a massive Russian missile attack to appear on the NORAD screens. General Beringer is on the brink of ordering a full-scale counterattack when the three heroes squeeze past the closing blast doors and convince the general to wait out the attack. It turns out to be merely a simulation, but the story is not over yet, for Joshua is preparing to launch a real attack himself. At the last minute Lightman brings the machine to understand the futility of nuclear war by engaging it in tic-tac-toe, a game in which, Joshua concludes, "the only winning move is not to play." He stands down the missiles, and the NORAD control center erupts in cheers and embraces.

Disembodied AI in *War Games* differs markedly from HAL and Colossus. Those computers begin as servants and end as terrifying, dominating, panoptic Others. Joshua's power, like theirs, is great. But unlike them Joshua has neither comprehension of, nor stakes in, reality; its world is the microworld of the game. Joshua's intentions remain benign: Lightman speaks to Joshua in the hacker's language of games, and what Joshua comprehends is not its role in reality but an abstract issue of strategy. Disembodied AI turns to fearful foe, but then turns back again, rehabilitated by the touch of innocents (the teenagers David and Jennifer). This pattern of cyborg rehabilitation through communion with caring human beings recurred frequently in the 1980s, as computers were transformed from alienating instruments of corporate and government power to familiar tools of entertainment and communication. With his final line—"How about a nice game of chess?"—Joshua has completed its transformation into a player and companion in an adolescent world returned to innocence.

The figure of Falken also marks a transition in closed-world fiction. In *Fail-Safe*, *Dr. Strangelove*, and *Colossus*, a scientist's hyper-rationalism created the alien Other. But Falken's creation is more like a game-playing child than a machine mad with power. Indeed, the computer is named after his own son. Falken's beaming gaze on another child, David, as he speaks to Joshua and narrowly averts the end of the world, can only be interpreted as fatherly. From its beginnings in science fiction, AI has frequently been interpreted as parthenogenesis, a male reproductive technology for bypassing women, pregnancy, and cooperative child-rearing. Falken's nonbiological fatherhood is complemented by his fascination with extinct animals, marking him as an ersatz version of the elder wise in the forces of life, a liminal figure weakly linking closed world and green world.

The stereotyped, helping-hand role assigned to David's friend Jennifer typifies closed-world gender constructions. She assists David in important ways but does nothing crucial herself. Despite a number of female secondary characters, the world of *War Games* is a male world, a hacker world, which women may observe from a distance but never truly enter. Jennifer—a dancer, swimmer, runner, animal-lover, but not a scholar—represents physicality and sexuality against David's pale-skinned nerdhood. Jennifer's powers, however, play only a minor role in David's victory. He wins, instead, with logic and good hacking.

The hollow (if sincere) climactic celebration in the NORAD control center marks the bogus exit opened by David's closed-world expertise. The central conflict is not transcended but merely deferred.

David and Jennifer will survive today, and we can surmise that human soldiers will be returned to the missile silos. But the smiling faces of the relieved soldiers and computer programmers are lit by the light of the world situation map. They remain there, deep underground, trapped inside a closed world changed virtually not at all by wresting the nuclear trigger from the grip of a machine.

Tron

The Walt Disney film *Tron* (1982) romanticized computer hackers and video gamesters as antiauthoritarian cowboy heroes. But the film's plot is less important, in this context, than its remarkable visual metaphor. For *Tron* provided a breathtaking glimpse into the surrogate sensory world inside the computer, which would soon after come to be called "cyberspace" or "virtual reality." The hackers of the early 1980s flocked to cinemas to witness the dream of their imagined future come to life.

Tron depicts the world inside the machine as a vast three-dimensional Cartesian grid dominated by neon colors and abstract geometric forms—cylinders, cones, cubes—reminiscent of the SHRDLU blocks-world. Programs operating within this space are represented by individual people. Glowing electronic pulses course through the circuits covering their bodies. They move through the virtual space in vehicles resembling motorcycles, tanks, and strangely shaped airships. Fantastic speeds, instantaneous right-angle turns, and other actions impossible in the physical world emphasize the virtual world's difference.

Most of *Tron's* sets are enclosed, if huge, interior spaces. Toward the end of the film, however, the protagonists emerge from claustrophobic canyons, where they have been hiding, into a vast twilight landscape. They ride a laser-like beam of energy across this zone in a graceful butterfly-winged ship, crossing fractal mountain ranges and passing floating blue globes and cones of yellow light. Though all its lines are the smooth abstractions of mathematical simulation, the scene displays the more natural forms of farm-like hillocks. In the distance lights twinkle across the darkened landscape. There are no stars here, no sky—only the dark grid-land of cyberspace continuing on forever. The film's final image, a nighttime cityscape in fast-forward, draws the visual comparison with the artificial world of urban life.

Tron's world is the closed world seen from the inside. Programs, granted subjectivity (though not creativity), inhabit the space. These inhabitants, whose virtual bodies have no counterpart in the real world, comprise a servant class that knows its place. (After all, as one character puts it, "user requests are what computers are for.") The renegade, panoptic, authoritarian Master Control Program (MCP), a disembodied AI that (monstrously) controls its user, breaks the asymmetry between the real world and cyberspace. Yet the MCP, in another sense, simply mirrors the liminality of the hacker. Like the film's hero Flynn, a hacker who ends up entering the machine in physical form, the AI can cross the border between worlds. The MCP is dangerous because it comprehends that in the networked world of the 1980s, where information is both money and power, the corporate game is played by merger and hostile takeover. Unlike the programs who exist only to serve, the MCP understands that action in cyberspace creates changes in real space. The closed-world simulation no longer merely pictures but actually *constitutes* reality itself.

Cyborgs in the Green World: The Star Wars Trilogy

The *Star Wars* trilogy (1977, 1980, 1983), a blend of space opera and New Age mysticism, is the only primarily green-world drama I shall discuss. It illustrates the contrast with closed-world fictions, but its importance in the present context extends, as well, to striking closed-world imagery, interpretations of machine subjectivity in the green world, and the rehabilitation of cyborg figures in its final moments.

In the fictional galaxy of *Star Wars* (1977), a totalitarian Empire attempts to crush a populist Rebellion in a total conflict typical of closed-world drama. The Empire's chief weapon is the Death Star, a moon-sized battle station wielding a planet-destroying energy beam. As the film opens the Princess Leia, a rebel leader, is captured by the evil, masked Darth Vader and imprisoned in the Death Star. There she manages to secrete captured plans for the battle station in the memory bank of the "droid" (actually a robot) R2D2. With his companion, the humanoid droid C3P0, R2D2 escapes to a nearby planet, where the droids are captured and sold to the adoptive family of young Luke Skywalker. R2D2 sneaks away to deliver Leia's message to Obiwan Kenobi, a wizardly old man and one of the last of the dying breed of Jedi knights. The Jedi, a kind of interstellar samurai, fight with antique "light sabers" and practice communion with the Force, an all-encompassing field of life energy from which a warrior can draw mystical powers.

Skywalker, pursuing R2D2, encounters Obiwan, who urges him to learn the ways of the Force and join him in the Rebellion. Obiwan and Luke engage the smuggler Han Solo to help them deliver R2D2 and his message to the Rebellion. After a series of narrow escapes, Solo's ship, too, is finally overtaken by the Death Star. Finally, aided by the Force, the party evades its captors, rescues the Princess, and rejoins the Rebellion. In the process Obiwan sacrifices his life in lightsaber combat with Darth Vader, who turns out to be a former Jedi pupil of Kenobi "turned" to the "dark side" of the Force. To protect the Rebels' planet, Skywalker, assisted by R2D2 and blessed by Obiwan's spiritual guidance, penetrates the Death Star's defense systems moments before its death ray would have destroyed the rebel base.

The focus of *The Empire Strikes Back* (1980) turns from politico-military struggles to psychological ones. After another narrow rebel escape from the Empire, Skywalker sets out with R2D2 to find the Jedi master Yoda. He crash-lands his ship in a dark, misty swamp prowled by reptiles, spiders, and huge underwater creatures. There he meets a very old, tiny, shabbily clad, green-skinned alien inhabiting a grubby hovel. Luke treats the little man with impatience and disdain—until he reveals himself to be the master Skywalker seeks. Yoda tries to teach the hotheaded youth the pitfalls of anger and the virtues of patience and calm, knowledge without which a Jedi's reliance on the Force can turn him to its dark side.

At one point Yoda sends Luke alone into a cave deep in the swamp, a place he calls a "domain of evil," telling him he must conquer what he finds there. "What's in there?" Luke asks Yoda. "Only what you take with you," Yoda replies. "In you must go." Descending through darkness, spiderwebs, slithering snakes and entangling roots, Luke enters a mysterious dreamlike space. The figure of Darth Vader emerges from the shadows and engages Luke in combat. He cuts off Vader's head. Vader's black mask explodes, revealing the face within. To Luke's horror, that face is his own.

Learning suddenly that his friends Han and Leia are in danger, Luke precipitously abandons his training with Yoda, only to find himself in combat with the real Vader, who attempts to turn him to the Force's dark side. He resists valiantly, but when Vader reveals that he is Luke's vanished father, the boy despairs. Vader cuts off his right forearm with his light saber before Luke, refusing to give in, leaps into an abyss. As the film ends, Luke, rescued by his friends, studies in dismay the inner mechanism of the new, perfectly lifelike prosthesis that replaces his lost arm. Luke himself has become, like his father, a cyborg.

The Return of the Jedi (1983) returns the main focus to the continuing struggle between Rebellion and Empire. The Empire is completing a larger, even more powerful version of the Death Star. While under construction, it orbits the forest moon of Endor, protected by an energy shield generated on the moon's surface. The Rebel Alliance plans to attack the station before its completion; Han, Luke, Leia, C3PO, and R2D2 descend to Endor to destroy the shield generator. On Endor, the heroes encounter the Ewoks, a primitive race of short furry creatures. After some initial misunderstandings, the rebel group and the Ewoks become allies.

Leaving the others on the moon's surface, Skywalker boards the Death Star alone, hoping to redeem his father. Vader and the Emperor again try to convert him to evil, but Luke resists. In the end a mortally wounded Vader, finally swayed by the bonds of kinship and the repressed virtue within him, turns on the Emperor and kills him. At this point the dying Vader asks Luke to remove his mask: "Let me look on you with my own eyes." Without the black headgear, Vader's face is revealed for the first time: puffy, pale, encased in cyborg circuitry, but human after all. He dies in Luke's arms, reunited with his son and at peace within himself.

Meanwhile Luke's friends have attacked the shield generator. Aided by the Ewoks, they succeed in turning off the Death Star's energy shield, and Rebel starships proceed to destroy it. The film ends on Endor with a victory celebration in the Ewok village. Ewok music and singing accompany dancing and revelry as the many alien races of the Rebel Alliance mingle to rejoice. C3PO and R2D2 dance beside their biological counterparts.

Star Wars *as Green-World Drama*

The Death Star, no mere vehicle but a planetoid or moon, is a kind of ultimate closed-world image: an *inverted world,* a world turned inside out, stripped of natural elements, and carrying all its life inside it. Its interior surfaces are pure technology: computer screens, equipment, armor, weapons. No living things other than humans and humanoid aliens populate the ship. Vader, cyborg inhabitant of the closed world, represents green-world power turned to evil through technology and human hubris. The Death Star's planet-destroying ray combines images of apocalypse with panoptic, totalitarian power; it is integrally political, a tool of empire.

While the Death Star is a closed world, however, it exists within a green-world universe. It is a place green-world figures, with their ability to cross boundaries, enter to liberate and eventually to destroy.

We first encounter Luke on his home planet, a dry land of canyons and deserts that nevertheless bears the marks of the green world: events and cycles such as sunrise, seasons, and farming signal Luke's connection to the land. Luke meets Obiwan near the latter's home in a remote canyon. It is there he first learns about the Force that will aid him in overcoming the seemingly impregnable, hyper-rationalistic, high-technology power of the Death Star. The Force is an "energy field created by all living things," as Obiwan tells Luke, which "surrounds us, and penetrates us, and binds the galaxy together." It is closely connected with emotion: "Reach out with your feelings, Luke," Obiwan constantly advises him.

The Rebel base of *Star Wars*' final scenes lies literally on a green world, whose forests stretch to the horizon in every direction. Halfway through his last-ditch attack run on the Death Star, Luke hears Obiwan's voice telling him, "Trust your feelings, Luke. Use the Force." He turns off his targeting computer and settles into a meditative calm. Luke aims with the naked eye—with his living biology, not his technology—and successfully destroys the Death Star. In the final triumphal scene, the blast doors of the rebel base open for Luke and Han Solo, who march down a sunlit aisle to receive medals from Princess Leia at a ceremony held in a huge outdoor amphitheater. Such a final celebration, typical of green-world drama, emphasizes the unity and wholeness Luke and his allies in the Force retrieve for humanity in destroying the closed world.

In *The Empire Strikes Back*, Luke learns the secrets of the Force in a special green-world setting: the fecund swamps of Dagoba. The timeless, dreamlike atmosphere of Yoda's swamp world serves as a potent symbol of Luke's inner confusion, his confrontation with the dark, sinister elements of the green world's transcendental powers as well as his grasping for their light. To overcome the darkness within his own mind, his anger, fear, and aggression, Luke must descend into the dream-realm of the cave to confront the enemy, who is himself. Outside the cave, the slithering reptiles, gnarled roots, and tangled vines represent the fermenting powers of life. There are no machines or computers here except Luke's spacecraft, which sinks into the muck, and R2D2, who remains on the periphery of these scenes. In the mists of Yoda's swamp, machines are not so much powerless as irrelevant.

The most graphic contrasts between closed world and green world emerge in *The Return of the Jedi*. Inside the new, unfinished skeleton of the Death Star, Vader and Skywalker play out their final battle, which for Luke is more an Oedipal family struggle and an internal confrontation with his own emotions than a physical battle. The Force's dark side dwells here, in this closed, inverted world of weapons and high technology. In contrast, the forest moon of Endor is a green world *par excellence*. Its soft, lush outdoor spaces, towering trees and ferns, and vast blue skies contrast vividly with the black-and-white of outer space and the sealed environments of starship cruisers.

The Ewoks who inhabit Endor combine potent green-world archetypes: animals (they look like small furry bears), the tribal connection to the natural world (everything they have is constructed from natural materials: stone, skins, wood), and animistic religion (they worship C3P0, attributing godlike powers to the machine). Their low-tech weapons—bolos, wooden battering rams, stones, boulders—surprise and finally overcome the high-tech battle machinery of the Imperial forces, hulking and ineffective in the dense forest.

The final celebration, in the heart of the forest and the home of this tribal culture, completes the pattern. The Ewoks' dwellings in the trees are archetypal green-world enclosures, highly permeable structures built from natural materials and lit by torches and firelight. The celebration in the green world represents the victory of life, of biology, of nonrational, transcendent powers over high-tech machinery. The defense requires the mystical, transcendental Force, which connects all living things and lies in a spiritual zone beyond the rationalist limits of technology. It also engages an autochthonous, animistic culture, linking the futuristic galaxy-wide Rebellion with the defense of world-bound, aboriginal values.

Machine Subjectivity in the Green World
Star Wars presents an *embodied* machine subjectivity that is friendly, familiar, unthreatening, and personal. C3P0's small but humanoid form, permanently bemused expression, mime-like gestures, and completely human (male) voice project a timid, servile personality. It refers to its owner as "sir" and "master" and generally behaves like the ideal butler, a demeanor emphasized by a vaguely British accent. R2D2's language of whistles and squeaks and its wheeled, canister-like body presents a more complicated image. C3P0 serves as the other droid's translator, mediating (in appearance as well as in role) between the human world and that of the robots. Both are inte-

grated so thoroughly into the human social world of the films that the viewer soon stops regarding them as robots at all.

R2D2 plays the competent agent to C3P0's timorous bystander. R2D2's technical abilities (as data carrier, as link with the Death Star's central computer, and as SCI-like pilot's assistant for Luke's attack on the Death Star) give the robot the special entrée into the closed world's architecture that is characteristic of the cyborg. Its whistling language is incomprehensible, mechanical, yet it evokes emotions, such as an obvious consternation when C3P0 is temporarily dismembered in *The Empire Strikes Back*.

Disembodied AI never appears in the trilogy. Its closed worlds represent the totalitarian politics and military might of the Empire, but its true powers—both for evil and for good—come from green-world sources. Computers appear frequently, but in the restricted role of tools (data banks, graphic displays), not as AI. Machine subjectivity, in the embodied form of the droids, evokes not the fear of panoptic surveillance and remote control by Others, but the central issue of 1980s cultural politics: multiculturalism.

The famous bar scene from *Star Wars*, in which a dizzying variety of alien species appear together, displays otherness as a continuous spectrum running from droids to cyborgs to biological aliens. With this spectrum as background, subjective machines appear merely as one of many forms of difference in a fully multicultural universe. Machine subjects are marked relative to other sentiences not by a mental difference—the emotionless hyper-rationality characteristic of disembodied AI, which the droids manifestly do not possess—but by a physical one, namely their mechanical nature.

As machines, they can be owned; their role is that of servants and helpers. Both droids are fitted with "restraining bolts," the electronic equivalent of leg irons, when purchased by Luke's aunt and uncle. In the bar scene just mentioned, the bartender ejects the droids. "We don't serve their kind here," he snarls. We are clearly meant to interpret this as bigotry, but Luke merely tells the droids, "You'd better wait outside." Though Luke and his band display strong loyalty to the droids, rescuing them from various captors, their role throughout the series is to obey orders and assist "master Luke," as C3P0 always calls him. Restricted to the role of servant, controlled by electronic restraints, *Star Wars'* embodied AIs never manifest the threatening Otherness of their disembodied counterparts. Only the Ewoks, green-world primitives, mistakenly code embodied AI as godlike

Other, enthroning C3P0 while tying its human owners to poles. In the final celebration the droids cavort along with the others, their rigid mechanical bodies marking their unease in the green-world space.

Rehabilitating the Cyborg

Machine subjectivity is never allowed to raise moral problems in *Star Wars*. Though droids display both emotion and pain, the issue of slavery is simply never joined; in a multispecies universe, their mechanical bodies justify this treatment, if anything does. The problematic of human/machine boundaries is raised, instead, in the cyborg figure of Darth Vader.

Vader's black, insectoid headpiece, his mechanical breathing, and the electronic resonance of his filtered voice clearly mark him as the Other, locked inside the private, cybernetic closed world. In the first film Vader's Otherness projects unmitigated evil. But when the dream-battle with Vader reveals Luke's own face inside the shattered mask, the neat line between cyborg horror and intact biological man begins to shift. The genetic link between Luke and his father overpowers the difference between their bodies. At the end of *The Empire Strikes Back*, Luke's prosthetic arm marks the disintegrating border between human and machine. When Vader's grotesque but still-human face is revealed at the trilogy's end, his exposed brain fitted with electronic interfaces, the rehabilitation of the cyborg is complete. Returned to the green world's transcendent grace by the virtue of his self-sacrificial final act, Vader dies a man and a father, no longer a half-mechanical, monstrous Other.

This pattern of rehabilitation and sacrifice of cyborg figures became common in mid-1980s film, especially—as with *Star Wars*—in sequels to originals depicting cyborgs as threatening.

For example, Ridley Scott's film *Alien* (1979) features treachery by an android scientist who is indistinguishable from his human colleagues until he is unintentionally dismembered during a fight. In this moment his lack of concern for his colleagues' safety (in the first part of the film) is revealed to result not from a stereotypical scientist's lack of emotion but from a stereotypical machine intelligence's lack of emotion. The parasitic, insectoid aliens are terrifying Others, too, but the android elicits the particular terror of the Other who looks just like Us. In the sequel, *Aliens* (1986), however, machine intelligence is redeemed as another android sacrifices himself to save the embattled heroine from the aliens, revealing the cyborg as friend and ally.

2010 (1984), the sequel to *2001*, explained Kubrick's killer computer HAL as a victim of mental illness. In *2010* HAL is cured by the nerdy but mystical Dr. Chandra, who erases HAL's traumatic memories. HAL proceeds to sacrifice itself, heroically, to save the human crew, who without the *Discovery*'s fuel cannot return to Earth.

The extremely popular late-1980s television series *Star Trek: The Next Generation* connected cyborg rehabilitation directly with the increasing multicultural awareness of the 1980s. The starship *Enterprise*'s "Prime Directive," lifted directly from the canons of contemporary cultural anthropology, prohibits it from interfering in the affairs of any species without its express request. Where the original 1960s *Star Trek* series often depicted encounters with alien species as violent confrontations, the updated *Next Generation* employs peaceful diplomacy to resolve almost all disputes. The all-embracing multiculturalism of *The Next Generation*'s twenty-fourth century eventually produces a peaceful understanding even with the new Federation's most implacable enemy, a cyborg hive race known as the Borg.

The Next Generation's android Commander Data has an electromechanical interior and a "positronic"[15] brain. Yet Data is fully accepted as a key member of the starship's crew. At the same time, the ship's computer system possesses none of the panoptic overtones of a HAL or a Colossus; while able to speak—in a bland female voice—it does not appear to be intelligent in the full sense. Commander Data's sometimes comic efforts to comprehend the human condition, particularly the emotions and intuitive thought he lacks, stand beside the human crew's efforts to comprehend the thoughts and feelings of alien cultures. As an android, Data simply presents one more problem in cultural relativism: here the cyborg Other, fully rehabilitated, is truly reduced to the status of merely another. *Star Trek: The Next Generation*'s implicit equation between the problems of machine intelligence and those of encounters with alien cultures makes multiculturalism itself a kind of Turing test.

What explains the cyborg rehabilitation of the 1980s, at the very height of the Reagan Cold War? *Star Trek: The Next Generation* suggests that a sort of perceptual threshold had been crossed in popular culture. Ubiquitous, easily operated computers were less frightening and less representative of government and corporate power. The technological shift from central mainframes to networks of desktop machines decentralized control over computer resources and diminished the iconographic link between computers and panoptic authority. The parthenogenetic goals of 1950s AI had been replaced by the more sober and limited ambitions of expert systems. At the same time, cognitive psychology, with its information-processing view of the human mind, had achieved the status of commonsense knowledge along with information-processing metaphors in biology, bridging the semiotic gap between mind and mechanism. A wide range of cyborg technologies, from pacemakers to the speech-synthesizing computer that allows physicist Stephen Hawking to talk, had become matters of everyday wonder, news that would raise eyebrows briefly if at all.

The rehabilitated cyborg thus marked not the demise of closed-world discourse but only another of its mutations. Where implacable, panoptic, disembodied AIs like HAL, Colossus, and the MCP constituted the self-consciousness of the closed world's very architecture, embodied AIs—robots, androids, and other cyborgs—inhabited it as individuals.

AI disembodied represented the possibility for protagonists to reject the terms of closed-world discourse *wholesale* in rejecting one of its central technologies. The politics of disembodied, pervasive Others mirrored the politics of the Cold War itself—a grand struggle between Us and Them. Embodied AI came to stand for another way of dwelling in the closed world, one that accepted its terms but sought actively to construct new and coherent subject positions within it. The politics of embodied, artificial others represented a kind of liberal détente in human-cyborg relations, a way of living both with and *as* cyborgs in an economy, a culture, and a world built from forty years of cold war. Yet at the same time their myriad and multiply articulated subjectivities opened possibilities for a different sort of engagement with closed-world discourse, one I will name *recombination*.

Conclusion: Recombinant Theater in **Blade Runner** *and* **Neuromancer**

Everything in the closed world becomes a *system*, an organized unit composed of subsystems and integrated into supersystems. These nested systems, as we have seen throughout this book, are constituted in and through metaphors, technologies, and practices. The metaphors are information, communication, and program; the technologies are computation and control; and the practices are abstraction, simulation, engineering, and panoptic management.

The experience of the closed world is a double experience of cyborg identity constituted through those metaphors, technologies, and practices. First, it is an experience of the possibility of other minds *constructed*, from parts, processes, and information machines, to manage and inhabit closed-world spaces. Second, it is the experience of *one's own* mind as an information machine, a constructed self-system subject to the same disaggregation, simulation, engineering, and control as those of the computerized Other. The choices faced by closed-world protagonists, as systems nested inside larger systems, do not include redemption, reunification, or transcendence. Transcendence is impossible in the closed world because, like a curved Einsteinian universe, it has no outside. Whatever begins to exceed it is continually, voraciously re-incorporated.

Recombination—to appropriate a 1980s biotechnological information metaphor—is the only effective possibility for rebellion in the closed world: taking the practices of disassembly, simulation, and engineering into one's own hands. Coming to see oneself as a cyborg, fitting oneself into the interstices of the closed world, one might open a kind of marginal position as a constructed, narrated, fragmented subjectivity, capable of constant breakdown and reassembly. This cyborg subjectivity would lack the reassurances of center, grounding, certainty, and even continuity. Yet at least it might be self-determined, in whatever shattered sense might remain of a term that historically relies upon the very different consciousness of an Enlightenment subject. The closed-world subjects of the 1980s, both human and artificial, could adopt the strategy of recombination as a tortured, contradictory, but effective cultural/political practice.

As Donna Haraway has put it, embracing the cyborg as an icon for a feminist politics that might play on the reconstruction of bodies, minds, nature, and machines under the rubrics of code and information:

A cyborg body is not innocent; it was not born in a garden; it does not seek unitary identity and so generate antagonistic dualisms without end (or until the world ends); it takes irony for granted.... Intense pleasure in skill, machine skill, ceases to be a sin, but an aspect of embodiment. The machine is not an *it* to be animated, worshipped and dominated. The machine is us, our processes, an aspect of our embodiment. We can be responsible for machines; *they* do not dominate or threaten us. We are responsible for boundaries; we are they.[16]

Rather than see their partiality and incompleteness, fragmentation and perpetual reassembly as weakness and disintegration, Haraway's cyborgs grasp these conditions as resources for the political recombination of "the ironic dream of a common language for women," but perhaps also for anyone, "in the integrated circuit."[17] Subverting the closed world by an interstitial engagement rather than a green-world transcendence, the cyborg becomes an always partial, self-transforming outlaw/trickster living on the margins of panoptic power by crisscrossing the borders of cyberspace.

Ridley Scott's film masterpiece *Blade Runner* (1982) and William Gibson's extraordinary novel *Neuromancer* (1984) carried the fragmented postmodern subjectivity of the closed world to what can stand, here, as its ultimate conclusion. In these recombinant theaters of the mind, the closed world becomes an arena for a mobile, dangerous, but *possible* cyborg subjectivity.

Blade Runner

Unlike the other fictions discussed here, *Blade Runner*'s cyborgs are neither mechanical nor computerized. The film was loosely based on Philip K. Dick's novel *Do Androids Dream of Electric Sheep?* (1968). Dick's characters were electronic androids, but *Blade Runner*'s humanoid "replicants" are the products of genetic design.[18] In the film's vague and improbable version of this practice, the replicants are apparently assembled from parts—eyes and brains, for example—designed and "manufactured" separately. Here the metaphor of recombination becomes literal, as engineers manipulate both the genetic code and bodily organs. Though manufactured and possessed of superhuman abilities, the replicants look and act exactly like human beings. *Blade Runner*'s replicants, employed as workers, soldiers, and sexual servants, are sentient, intelligent creatures differentiated from human beings only by their lack of emotion.

Blade Runner takes place in a grim, post-holocaust Los Angeles in the year 2019. The shadowy backgrounds of its relentlessly claustrophobic urban atmosphere blur into unending darkness, rain, smoke, and smog. Vehicles hurtle through urban canyons among massive skyscrapers and the dense, enormous pyramid of the Tyrell Corporation, manufacturer of replicants employed as servants and workers. In this archetypal closed-world space, Rick Deckard is a police "blade runner" whose specialty is tracking and "retiring" (killing) rogue replicants. The film's plot revolves around Deckard's hunt for four such renegades. Roy, Leon, Zora, and Pris are a new model, Nexus 6, who unlike their predecessors tend to develop emotions as they age. To prevent these responses from growing strong enough to render them ungovernable, the lifespan of Nexus 6 replicants has been genetically limited to four years. In hopes of extending their too-short lives, Roy and his band have escaped from an off-world colony and returned to Earth to find their designers.

To detect escaped replicants, Deckard administers the Voigt-Kampff test, which measures emotional responses to hypothetical situations. The Voigt-Kampff test is a kind of Turing test for a post-Turing world, where people cease to define themselves as Aristotelian rational animals—as Sherry Turkle has observed of contemporary computer cultures—and become, instead, emotional machines.[19] *Blade Runner* goes Turing one better by asking how emotions are linked with thought, memory, and embodiment. Behind this intellectual question lies an ethical issue: if it experiences fear, pain, and love, what makes a replicant different from a slave? What responsibility do the creators of such beings bear for their existential condition?

At the Tyrell Corporation Deckard meets Rachel, an assistant to the replicant designer Eldon Tyrell. Using the Voigt-Kampff test, he eventually determines that Rachel is in fact a replicant, but the test proves much more difficult than usual since, as Deckard puts it, "it" does not know that it is artificial. Tyrell informs Deckard that "Rachel is an experiment, nothing more." She has been "gifted ... with a past"—implanted, artificial memories of a nonexistent childhood—in order to "cushion" the impact of her developing emotions on her personality.

Rachel, overhearing this conversation, follows Deckard home to question him about her status. She offers him family photographs to prove the authenticity of her memories. Cruelly, Deckard reveals that her memories are mere fabrications, not her own: the implanted memories of Tyrell's niece. Her illusion shattered, Rachel flees. The encounter provokes Deckard to a meditation on his own family photographs. The clear and troubling implication is that human memory and feeling, too, may be not only preserved but largely *constituted* through such technical supports. Rachel's mind is a recombinant theater, a stage upon which a constructed past built from parts is played out as if it had actually happened—but so, too, the film implies, are all minds. Rachel later asks Deckard, "You know that Voigt-Kampff test of yours? You ever take that test yourself?" But he is asleep and does not answer.

A palpable sexual electricity develops between Rachel and Deckard, extremely disturbing for both characters. Disoriented in the knowledge of her artificiality and her false memories, Rachel does not know whether to trust the emotions she feels. Perhaps, like her memories, her feelings belong to someone else. Neither does Deckard—trained to think of the replicants as insensate machines, and ordered to kill Rachel after she deserts Tyrell—know how to handle his response to her. When she saves Deckard's life by shooting the replicant Leon, he tells her he "owes her one" and will not pursue her if she tries to "disappear." Bonded in violence, both now killers of replicants, the two become lovers beneath the dark shadow of Rachel's illegal status and her limited lifespan.

As other rogue replicants seek a way to extend their short lives, their subjectivity comes to seem deeply valuable, in and for itself. They mirror the humans who pursue them, but they also display a special consciousness of their own. Roy, their leader, speaks in hauntingly beautiful poetry:

Fiery the angels fell
Deep thunder rolled around their shores
Burning with the fires of Orc.

In a riveting climax, Roy gets his long-sought audience with Tyrell, designer of the replicants' brains and minds. He gains entry to Tyrell's rooms by besting him in a game of chess (another reference to AI history). Roy is the ultimate cyborg, the final culmination of embodied AI, come to take upon himself the mantle of his own destiny—and to confront the hubris and the cruelty of a designer who views even the emotions of his sentient creations as tools in the service of profit, manipulation, and a supremely arrogant intellectual curiosity. Their dialogue:

Tyrell: I'm surprised you didn't come here sooner.
Roy: It's not an easy thing to meet your maker.
Tyrell: And what can he do for you?
Roy: Can the maker repair what he makes?
Tyrell: Would you . . . like to be modified?
Roy: I had in mind something a little more radical.

Tyrell: What... what seems to be the problem?
Roy: Death.
Tyrell: Death. Well, I'm afraid that's a little out of my jurisdiction.

Roy proposes a series of (literal) genetic recombinations that might allow him to live. Tyrell, the paternalistic voice of science and capital, replies with "the facts of life. To make an alteration in the evolvement of an organic life system is fatal. A coding sequence cannot be revised once it's been established." Assuming the role of father-confessor, Tyrell goes on to console him.

Tyrell: But all this is academic. You were made as well as we could make you.
Roy: But not to last.
Tyrell: The light that burns twice as bright burns half as long. And you have burned so very, very brightly, Roy. Look at you. You're the prodigal son. You're quite a prize. [Puts his arm around Roy's shoulders.] ... Revel in your time!

Roy, enraged at the doom laid on him by this patronizing but far from omnipotent creator, crushes Tyrell's face between his hands.

Roy returns to find the bleeding body of Pris, killed by Deckard, leaving him the lone survivor of the original tetrad. He pursues Deckard through an abandoned building in darkness and rain. Finally caught, the wounded and thoroughly terrified Deckard hangs by his fingertips from a wet rooftop. Roy, looming above him, clutches a white dove against his body, one of his hands pierced, stigmata-like, by a long nail. "Quite an experience to live in fear, isn't it? That's what it is to be a slave," he exults as Deckard dangles above his death. But when Deckard's grip slips, Roy seizes his arm and pulls him back to the rooftop, saving his life. Soon afterward, Roy releases the dove heavenward into the rain, dying before the eyes of the spent and utterly helpless Deckard.

Ultimately, Deckard realizes, the replicants' agonizing quest for life is no different from that of humans. "All [Roy] wanted were the same answers the rest of us want. Where do I come from? Where am I going? How long have I got?" Transformed by the recognition of himself in the Other, Deckard returns home for Rachel. They take flight together, leaving behind them the artificial/natural distinction, a difference that no longer makes a difference. For as Gaff, another policeman, sneers, "It's too bad she won't live. But then again, who does?" In the original version of the film, they escape into a green-world scene as the credits roll, flying low over a breathtaking landscape of forested mountains. The heterosexual couple finds again the transcendent reunification of man with woman and human with nature, and another cyborg finds her way into the green world.

The so-called Director's Cut of Blade Runner, not released until 1992, depicts a different outcome much more in keeping with the film's closed-world frame. This version ends with the resonant clang of elevator doors closing on the fleeing couple, sealing them as renegades inside the closed world. A moment before, Deckard has come upon an origami unicorn left behind by Gaff. If Gaff knows about Deckard's dream of a unicorn cantering through a forest (a sequence restored to the film in this version), as Deckard knows the source of Rachel's hidden memories, then Deckard himself may be a replicant (an implication frequently insinuated in Dick's novel). Deckard and Rachel become—perhaps, if they survive—the Adam and Eve of a new generation of self-determining replicants, born to populate the closed world.

"This thing of darkness I acknowledge mine," Prospero says of his monstrous, angry servant Caliban in Shakespeare's *The Tempest*. Like Prospero, in the end we must recognize Roy's murderous fury as our own. It is the existential rage, born from terror, felt by every being capable of the consciousness of death. Rachel's confusion, too, is ours: the inadequacy of a fragile past vulnerable to forgetfulness and forgery, the inauthenticity of emotional experience adopted wholesale from fiction and culture. If Roy and Rachel and even Deckard are cyborgs, no less and no more are we, constructed from the recombinant fragments of bodies, thoughts, memories, and emotions, never original or originary, always already the simulacra of an ever-receding presence.[20]

Neuromancer

William Gibson's *Neuromancer* became the vastly popular flagship work of "cyberpunk" science fiction, a new subgenre that linked computerization, mass-media artificial experience, biotechnology, and multiculturalism to a dark political future of massive urbanization and militarized corporate hegemony.[21] *Neuromancer* is a novel of ideas and aesthetics, and the twists and turns of its complex plot are less important here than the nature and condition of its characters and its social world. For the novel represents subjectivity in a world where boundaries between human, machine, and computer have achieved a kind of unlimited flexibility in a totally recombinant world. Here no one and nothing living escapes entirely unaltered.

In *Neuromancer's* nearby future, most of Earth's computers have been linked together to form a gigantic network. People enter and work within the network through "cyberspace," a virtual-reality visual grid-space or "matrix" much like the world of *Tron*, which Gibson calls a "consensual mass hallucination." Case, the novel's antihero, is a mercenary hacker who specializes in the theft of information. As the novel opens, Case is living a marginal, near-suicidal existence, dealing drugs and contraband in the seamy underworld of Chiba City, Japan. A former employer, in revenge for one of Case's thefts, has subtly damaged his nervous system in such a way that he can no longer enter the matrix (which requires a direct neural hookup).

To Case, cyberspace is home. The enigmatic ex-Green Beret Armitage offers him the chance to have his nerves repaired so he can once again "jack in," in exchange for the use of his talents on a highly dangerous illegal mission. After the operation, when Case finally re-enters cyberspace, the sensation is like a return from long exile, pure poetry:

Please, he prayed, *now*—
A gray disk, the color of Chiba sky.
Now—
Disk beginning to rotate, faster, becoming a sphere of paler gray. Expanding—
And flowed, flowered for him, fluid neon origami trick, the unfolding of his distanceless home, his country, transparent 3D chessboard extending to infinity. Inner eye opening to the stepped scarlet pyramid of the Eastern Seaboard Fission Authority burning beyond the green cubes of Mitsubishi Bank of America, and high and very far away he saw the spiral arms of military systems, forever beyond his reach.
And somewhere he was laughing, in a white-painted loft, distant fingers caressing the deck, tears of release streaking his face.[22]

In Gibson's cyberspace "meat things" may be dispensed with in favor of a computer-generated landscape. Everything is represented abstractly, in the mathematical, artificial forms of computer graphics. The world outside the machine is presented as equally artificial, closed. The book's opening sentence starkly reverses the usual directionality of metaphor: "The sky above the port was the color of television, tuned to a dead channel."[23] Though the novel's action takes place in a series of urban ghettos and then moves to the space stations Zion and Freeside, this lateral motion—in other contexts a green-world characteristic—here represents no more than a constant flight through the all-encompassing closure of a world totally enframed in technological control.

In this vision bodies are mutable, reparable, even—given enough money—virtually immortal. Some characters are clones, reproduced in vats. Radical plastic surgery has become a ubiquitous form of personal expression. One character sports shark-like fangs, others the more ordinary perfections of face and shape. Case's partner Molly, a kind of female ninja mercenary, has four-centimeter scalpel blades implanted in her fingertips, which she can extrude, at will, like cat claws. Her nervous system has been amped up via microsurgery to give her lightning reflexes. Most arrestingly, Molly has had mirror lenses implanted into her face, permanent sunglasses that completely conceal her eyes.

Even embodiment itself is optional and problematic. A character known as the Dixie Flatline is the mind of a dead man recorded on a ROM (read-only memory) cassette, who exists now only in cyberspace. The ex-Special Forces soldier Armitage, by contrast, is a shell of a man whose personality has been taken over by the disembodied AI, Wintermute. He is a creature of cyberspace living in the real world. Case himself dies a brain-death several times in the novel, his EKG readings a flat line for minutes on end, his mind absorbed by the matrix. Case's ex-girlfriend, Linda Lee, has in death been transported whole into the virtual world of another AI named Neuromancer. During Case's minutes of brain-death, Neuromancer brings him to a beach scene where time seems meaningless and nature is pure information, regenerated inside the machine. There he encounters the AI embodied as a young boy and is reunited with the lost Linda Lee.

If physical identity is a matter of choice in *Neuromancer*'s world, so is subjective experience. Highly specific psychotropic drugs—Case's favorite approach—are only a start. Some people have jacks installed in their heads that accept "microsofts," chips that hook into their brains to extend their skills or change their personalities. The technology of "simstim" lets one person's sensory experience be recorded and piped directly into the mind of another. Case and Molly use this as a communication device: as she carries out her part of their "run," Case can flip in and out of her sensorium at will, experiencing directly whatever she is seeing and doing. Ordinary people use simstim as a kind of super-duper television, plugging in a recording of somebody else's sensory experience as we would insert a videocassette.

If everything is optional and recombinant in *Neuromancer*—the body, experience, culture, reality itself—everything is also for sale. Transnational corporations have replaced nation-states as the central units of large-scale social organization, and the driving force behind almost every character in the book is "biz," as Case calls it. The global market drives an endless recombinatory process in which biology and even experience must recline upon the Procrustean bed of capitalist exchange-value.

Similarly, conversion of the green world into a closed world is complete in *Neuromancer*'s future. All of the scenes on Earth take place either indoors or in dense, grim urban cityscapes buzzing with the hum of biz. Much of the action takes place at night. It is only when we reach the space stations that we encounter green spaces—and these are *inverted*, artificial ecosystems on the inner surfaces of artificial moons. The closest approach the novel offers to transcendence is its Rastafarian characters, who worship Jah from the refugee green space of the Zion Cluster space station, a self-sustaining tropical jungle. But Wintermute draws them, too, into its web, speaking through the matrix in the voice of Jah to the pulse of reggae dub.

Wintermute's goal, which he achieves in the end with the half-hired, half-coerced help of Case and Molly, is to merge with Neuromancer. In the book's final scene this recombination—and recombination is exactly what it is, a reproductive and mutational merging of digital codes—produces a new entity inside the matrix, an indescribable consciousness of an entirely new order. The new matrix-mind, self-assembled in a recombinatory genetics of information machines from the technological resources of the network, is pure technology, pure and purely disembodied AI.

Like those of *Blade Runner*, *Neuromancer*'s characters make their harrowing way through the matrix of the closed world as recombinant subjects, hacking their way not to a green-world transcendence but to an immanent self-determination. For Case this means the reconstitution of emotional experience. He recovers his anger, a burning gem inside him that becomes the key to his survival, as well as his ability to mourn his lost friend. For Molly, a woman in a male-dominated world, it means physical power and respect. Her technological self-construction as warrior—like that of Sarah Connor in *The Terminator*—reconstitutes her gender as something tough and strong, something she owns and controls. Though she pays a heavy price for freedom, behind her mirror shades Molly is one of Haraway's cyborg feminists, a self-created woman moving at will through the closed world.

In closed-world discourse, as Haraway puts it, the "organic, hierarchical dualisms ordering discourse in 'the West' since Aristotle... have been cannibalized, or... 'techno-digested.'"[24] The sciences and technologies of mind discussed in this book played a principal role, in concert with the military and cultural politics of the Cold War, in this "techno-digestion" of dualisms of mind and body, machine and organism, information and entropy, simulation and reality. The energy product of that techno-digestion, however, powered the construction of a closed world whose vast scale would allow neither escape nor transcendence.

The transcendent organic holism of pure green-world discourse has become very nearly impossible to inhabit. Only its vestiges survive. We might name animistic religions, feminist witchcraft, certain Green political parties, and the deep ecology movement as some of these vestigial locations. Though I attach an absolutely crucial importance to the survival and proliferation of these discourses, for the present all of them lie at the farthest margins of politics, society, and culture. The real green world has been largely contained within the closed world, trapped inside the boundaries of land-island national parks, the systems disciplines of ecology and genetic engineering, and the global-management aspirations of the Club of Rome and its successors. As the global ecological crisis reaches apocalyptic proportions once imaginable only in thermonuclear holocaust, computer modeling projects such as climatology and the Landsat world maps are beginning to absorb the very substance of the globe into closed-world discourse.

Thus the possibility that remains, the only possibility for genuine self-determination, is the political subject position of the cyborg. The subjectivity of recombination encourages, even demands, troubled reconstructions of traditional relationships among rationality, intelligence, emotion, gender, and embodiment. It accepts, in the most fundamental of ways, the existence of a space of interaction between human minds and computerized Others. It provides an interstitial, marginal, unholy, and unsanctioned subjectivity, not a blessing or a God's-eye vision. It offers no escape, no redemption, no unity or wholeness. But the recombinant cyborg mind, riding the flow of information, can cross and recross the neon landscapes of cyberspace, where truth has become virtuality, and can find a habitation, if not a home, inside the closed world.

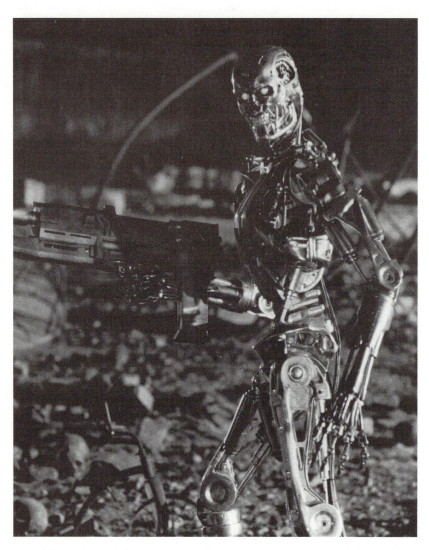

Robot "Terminator" hunting humans in *Terminator 2: Judgment Day* (1991). Film still of the "Metal Skeleton" appears courtesy of Carolco Productions. Motion picture © 1991 Carolco Pictures Inc. All rights reserved.

Epilogue: Cyborgs in the World Wide Web

Armed only with shovels, crowbars, and their bare hands, thousands of German citizens met at Checkpoint Charlie in 1989 to tear down the Berlin Wall, symbol of the division of East from West for almost thirty years. Other communist governments throughout Eastern Europe soon followed East Germany into collapse. With the fall of the Soviet Union in 1991, the central ideological conflict of the twentieth century finally vanished into history. While the danger of regional nuclear war may actually have increased with the multiplication of unstable nuclear-armed states, the danger of global nuclear holocaust went—at least for the present—the way of the Soviet Union. With these transformations, the final curtain closed on four decades of Cold War. Ironically, in one sense the Cold War's end marked the ultimate achievement of world closure: the realization of a global market economy—a complete world-system, in a favored phrase of economic historians.[1]

In the 1991 Persian Gulf War, the computerized weapons built for the Cold War were finally used to devastating effect. At the same time, another kind of information technology—Pentagon media control—filtered the war and its high-tech weapons into the homes of millions through a carefully constructed information sieve. America once again aimed to enforce the rules of a world-system economy by preventing a resource-rich state from withdrawing into autarky.

Simultaneously, global computer networks were continually extending the latticework of information technology. DARPA's 1970s experimental research network, the ARPANET, evolved first into the MILNET military network. Later, in the 1980s, it engendered the Internet, a global network of computer networks. By 1994 more than sixty countries had at least one Internet node. Almost eighty more could be reached by other forms of electronic mail. At this writing a new network is hooked into the Internet, on average, once every ten minutes. As information technology enfolded the globe, the anarchic network-hacker culture of the early 1980s started to give way to commercial interests. After August 1991, more than half the networks connected to the Internet belonged to businesses.[2]

We may name the gossamer global enclosure spun by these networks the World Wide Web, after the most popular meta-networking system of the middle 1990s. At this writing, traffic over the World Wide Web is increasing at the astonishing rate of 341,000 percent per year (and rising). The cyberspace of *Neuromancer* thus becomes an increasingly palpable, and increasingly common, experience. In September 1991, the prestigious journal *Scientific American* dignified the word with a special issue subtitled "How To Work, Play, and Thrive in Cyberspace."[3] Senator Al Gore, who contributed an article to *Scientific American*'s cyberspace issue, rode to the 1992 vice-presidency partly on the strength of his plan for a national "information superhighway" combining telephone, television, and computer network services into a single high-bandwidth infrastructure.[4]

Nevertheless, the unwinding of the Cold War's mainspring of total, apocalyptic conflict transformed the closed-world frame. The collapse of communism reawoke long-suppressed, explosive nationalisms. The rise of Japan and the newly industrialized countries of East Asia in the 1980s had already ended four decades of American economic dominance, multiplying the centers of power in the world economy. Financially exhausted by the Cold War—especially the final Reagan-era orgy of deficit military spending—America literally could no longer afford to "bear any burden, pay any price." In the 1990s, political conflicts multiply seemingly ad infinitum, with neither the ideological lens nor the global reach of superpower military force to order or contain them. Despite their often global scale, 1990s politics as World Wide Web remain fragmented, beyond the reach of the Cold War obsession with centralized control. Instead, the World Wide Web is *distributed*: decentralized, quasi-anarchical, lacking a central purpose or even a main organizer.

If the closed world has not disappeared but merely been transformed in the post–Cold War world, what has become of its cyborg subjects?

The Persian Gulf War

In August 1990, the Iraqi army of Saddam Hussein invaded Kuwait. Soon afterward, President George Bush announced the dawn of a New World Order. Bush meant to promote a new international political and financial arrangement for interventions to restore (or maintain) peace. Military action would be politically sponsored by the United Nations, financially sponsored by the wealthy industrialized countries, and ideologically sponsored by, and run from, the United States. America would provide leadership, troops, and military hardware, while other countries would contribute money and token military forces. The formidable arsenal built for global war would now be turned to local police work.

That arsenal's full power had not yet been demonstrated. Until the Persian Gulf War it remained, from a cultural point of view, an uncertain mythology. Some suspected the high-tech weapons would fail, as they had in Vietnam, against a large and determined human army. Soon after the shooting war began, however, Pentagon officials broadcast videotapes of computer-controlled, laser-guided bombs destroying buildings in Baghdad. Endlessly replayed, these images became icons for cyborg war. Through their TV screens, viewers in America followed cruise missiles as they cut communications antennas neatly in half, or circled bunkers in order to enter politely through the side door before obliterating them and their inhabitants. Computer-guided Patriot missiles knocked Scud missiles out of the sky, splattering debris across Israeli cities (to American applause and Israeli consternation) and sending the stock of Raytheon and General Dynamics soaring. American audiences saw "smart" computers *embodied* in weapons, proto-Terminators seeking out targets and destroying them with awesome force and fully hyped "precision."

In those moments a worldwide television audience experienced the joining of cyborg subjectivity with closed-world politics.[5] As we rode the eye of the laser-guided bomb to the white flash of impact in the eerie virtual reality of the TV image, we experienced at once the elation of technological power, the impotence and voyeurism of a passive audience, and the blurring of boundaries between "intelligent" weapons and political will. The dazzling and terrifying power of high-technology warfare displayed in the sound-bite war in the Gulf became an emblem for the glories of America's waning global hegemony.

Both technologies—the weapons and the Pentagon's information control—had worked to perfection.[6] The tiny numbers of American casualties and the Pentagon's careful management of the imagery of suffering made the war seem virtually bloodless, a sort of virtual-reality video game. In the New World Order, the vanished fear of global holocaust left behind it the pure elation of high-tech military power. By the mid-1990s, though, that elation had dwindled to a half-hearted hurrah as the colossal financial hangover of post–Cold War American politics finally set in.

Neural Networks and AI

By the late 1980s, attention in artificial intelligence had moved away from expert systems, with their abstract, preformulated knowledge and symbolic reasoning. It returned instead to techniques that—like the cybernetics of the 1940s and 1950s—drew their inspiration from biological models.[7] Far more powerful computers and great strides in neurobiology teamed to produce sophisticated computer simulations of neural networks—exactly the goal set forth by McCulloch and Pitts in 1943.[8]

By simulating some of the functional aspects of biological neurons and their synaptic connections, neural networks could recognize patterns and solve certain kinds of problems *without* explicitly encoded knowledge or procedures. Each simulated neuron performed the simple task of summing inputs (which, like inhibitory and excitatory synapses in the brain, might have either positive or negative value). Based on this sum, which might include a "weight" added by the neuron's particular summation function, the neuron passed on a value of its own to other neurons. These in turn summed their inputs and passed them on until some final output value was generated.

In a pattern recognition task, for example, someone writes a sentence on a writing tablet fitted with a dense two-dimensional matrix of pressure sensors. Each sensor communicates the presence or absence of a mark to an "input layer" neuron corresponding to its location on the grid. The input layer sends signals to one or more middle processing layers. These, in turn, send signals to an output layer, where the final interpretation of the marks and spaces as letters composing a sentence appears. The network contains no explicit representations of letter forms. There are no templates to fit, no predefined procedures to follow, and no features to detect—no abstract facts or rules at all. Instead, the network is "trained" with feedback techniques. The correct interpretation of the letters is sent backward through the network, an algorithm adjusts the weighting functions of the neurons and synaptic connections, and another trial is conducted.

After just a few trials well-designed neural networks could recognize images and forms whose interpretation had eluded symbolic AI for decades. Unlike their AI predecessors, the new systems were fundamentally learning machines. Although not literally models of brains, the biological metaphor of neural networks contrasted sharply, and deliberately, with the mind-program analogy of traditional symbolic AI. A new school of programming, known variously as "connectionism" and "parallel distributed processing," saw these new techniques as the salvation of computerized intelligence. Other approaches in "nonrepresentational AI" also derided their predecessors' attempts to encode all knowledge in advance instead of having it emerge from a system's interaction with the real world. [9]

A new generation of cyborgs would thus sport silicon brains rather than simulated minds. This convergence of biological and cognitive models marked, in one sense, a sort of terminal technique of formal-mechanical modeling. Recalling Wiener's dictum that "the ultimate model of a cat is, of course, another cat," these new theories focused increasingly on physical structure as well as functional form. AI, once experienced in popular culture as the threat of disembodied, panoptic power, now came to represent the friendly future of the embodied, pseudo-biological machine. The paradigmatic cyborg of the 1990s was not HAL but Commander Data.

To end this glimpse into the future of cyborgs after the Cold War, we turn to the sequel to *The Terminator*—the archetypal closed-world drama that I used as a prime exemplar in chapter 1. Released in 1991, just months after the Persian Gulf War, James Cameron's sequel demonstrates the ideological transitions marked by the Cold War's demise. The vision it offers, in keeping with the tone of an uncertain age, is equivocal at best, retrograde at worst. But with uncertainty comes hope, for while the future remains unknown, the film grants a reprieve from nuclear apocalypse, the greatest world-closing terror of them all.

Terminator 2: Judgment Day

In *Terminator 2: Judgment Day* (1991), we see cyborg subjectivity transformed under the New World Order. In *T2*, as its ad campaign called it, all the icons semiotically restructured by *The Terminator* (for convenience here, *T1*)—the emotional woman, the mechanical man, the white nuclear family—are systematically reconstituted for a post–Cold War, postfeminist, post-postmodern world through the rehabilitation of the war cyborg.

T2 is set in 1994, three years before the date of the computer-initiated nuclear holocaust projected by *T1* and ten years after the events of the original *Terminator*. The plot, in many respects, is the same extended chase. This time, though, the object of pursuit is not a woman but a child: John Connor, Sarah's son, now ten years old. The new villain is a more advanced Terminator model, the T-1000, sent back from the future to attack John as a boy. The original Terminator, or rather another, identical T-800 unit captured and reprogrammed to protect John Connor, returns in *T2* as the good guy.

Despite its basic similarities, the dramatic atmosphere of *T2* has changed from the powerful closed-world ambiance of *T1*. While green-world imagery remains rare, many of *T2*'s scenes are shot in full daylight, in domestic and suburban rather than urban settings. The all-consuming, high-tension claustrophobia so effectively produced by *T1* has been replaced by a more relaxed alternation between safety and danger. The film proceeds through a series of tasks structured more like a journey than *T1*'s headlong flight.

The T-1000 is nothing like the original T-800. It is neither computer nor machine but a shape-shifting blob of "mimetic polyalloy," a mercury-like liquid metal. It can be shot, blasted, even shattered, but it simply congeals again, re-forming its body. It has no moving parts, no neural networks, no bodily functions. Its specialty is the ability to morph[10] into exact replicas of other people and things—John's foster mother, a hospital guard, a tiled floor. In a postmodern self-referential irony, cutting-edge computer graphic software was used to create the T-1000—a technology supposedly beyond computers—for the silver screen. The audience, which knew this very well from advance publicity, admired the T-1000 as computer graphic even as it participated in the narrative construction of this postcomputational, postcyborg Other.

If *T1* is about the reconstruction of female identity in the shadow of nuclear war, *T2* is about the search for parents in a world where traditional motherhood has all but vanished and fathers are absent deadbeats. We first encounter the ten-year-old John at the home of his harried, vaguely sleazy, very ordinary foster parents. He ignores his foster mother's nagging. When Todd, her husband, admonishes him to "do what your mother tells you," John retorts, "She's not my mother, Todd" and roars off on his motorbike. In the next scene we rejoin Sarah Connor and learn where women who abandon womanhood to fight the computer-controlled nuclear future end up: in insane asylums. John's "real" mother awaits the holocaust in a maximum-security mental hospital, doing chin-ups in her cell, sweating, buffed-up into a female version of Schwarzenegger, her lip curled in derisive anger. When her psychiatrist plays back a videotape of Sarah recounting her recurring nightmare of nuclear war, we find ourselves—we who have been told, in the post–Cold War world, that *this* nightmare at least is over—uncertain what to make of her uncontained rage.

John soon confides to a young friend his conflicted relationship with Sarah. He proudly announces that he learned how to hack into ATMs from his "real" mom. But he immediately tells his friend, "She's a complete psycho . . . a total loser." Later we learn more of John's family history:

> We spent a lot of time in Nicaragua and places like that. . . . For a while there she was with this crazy ex-Green Beret guy. Running guns. Then there were some other guys. She'd shack up with anybody she could learn from so she could teach me how to be this great military leader. Then she gets busted. It's like, sorry kid, your mom's a psycho—didn't you know? It was like everything I'd been brought up to believe was all made of bullshit. I hated her for that.

Sarah, like revolutionary women throughout history, is portrayed as prostituting herself for the cause. The cause—like every ideology, in *T2*'s deeply suspicious frame—has turned out to be a false promise, an illusory struggle, doomed to failure. John, child of the age of single working mothers, serial monogamy, and "family values," rejects both his ersatz nuclear family and his insane "real" mother. Once the film reveals which Terminator is the villain and which the protector, however, John sets off with the T-800, pursued by the T-1000, to rescue Sarah.

When they arrive at the asylum, Sarah has just been subdued by attendants after a valiant but unsuccessful attempt to break out. The T-800 extends his hand to her, saying sternly, "Come with me if you want to live"—Kyle Reese's words to her in *T1*. The terrified Sarah of 1994 confronts, as if in a hallucination, a seeming reprise of her 1984 nightmare. She faces the wrenching task of transforming her image of the cyborg as monster into a new picture, the cyborg as protector. To do so, she must acknowledge his subjectivity as familiar. In the car after their escape, she finally turns to the T-800 and offers: "So what's *your* story?" Now the film begins its central ideological task: the reconstruction of motherhood, fatherhood, the nuclear family, and the white male as world savior for the New World Order.

John's role in the process of rehabilitating Sarah as traditional woman and mother will be to provoke her suppressed emotionality. John is also assigned the task of reconstructing a father-figure from the mechanical material of the T-800, which like a robotic genie must obey John's every command. John learns that the T-800 is not, after all, a rigidly programmed, single-minded monster but a creative, learning subject who feels pain. "Can you learn stuff that you haven't been programmed with, so you can be, you know, more human, and not such a dork all the time?" "My CPU is a neural net processor—a learning computer," the T-800 replies. "The more contact I have with humans, the more I learn." John: "Cool!" Just as in the real world of the 1990s, the Terminator's bio-friendly, "learning" neural network replaces the abstract computation of symbolic AI, in keeping with the cyborg's new incarnation as friend and protector.

As the film proceeds, John steadily trains this awkward, overserious cyborg in human speech, human relations, and even human emotions. The T-800 asks why people cry. John explains that "it's when there's nothing wrong with you, but you hurt anyway." John teaches the T-800 not to kill people (the cyborg settles for wounding them grievously instead). We are meant to see Schwarzenegger's Terminator as a kind of NRA-issue protector, who uses violence willingly but only when "necessary." Like the Persian Gulf battles, with their managed images of war as a bloodless computer game, the new, nonlethal cyborg figure is pure New World Order.

At Sarah's survivalist hideout in the desert, they talk and play. Sarah, watching them, sees what is happening:

> Watching John with the machine, it was suddenly so clear. [The Terminator] would always be there, and it would die to protect him. Of all the would-be fathers who came and went over the years, this thing—this machine—was the only one who measured up. In an insane world, it was the sanest choice.

The scene closes with John and the T-800 cavorting, cyborg father and human son, warmed by Sarah's bemused, nearly beaming gaze.

But the problem of nuclear war is not yet resolved. The T-101 explains how it will happen without a Cold War. The unsuspecting computer engineer Miles Dyson of Cyberdyne Systems "creates a revolutionary type of microprocessor" by reverse-engineering the broken computer chip found in the head of the original Terminator. "All Stealth bombers are upgraded with Cyberdyne computers, becoming fully unmanned. Afterwards they fly with a perfect operational record. The Skynet [computer] system goes on line on August 4th, 1997. Skynet begins to learn at a geometric rate. It becomes self-aware.... In a panic, they try to pull the plug.... Skynet launches its missiles against the targets in Russia." John wants to know, "Why attack Russia? Aren't they our friends now?" "Because Skynet knows that the Russian counterattack will eliminate its enemies over here."

Later on, Sarah drifts into sleep, into her nightmare of nuclear war. Watching a playground, she sees herself through a chain-link fence as a young mother playing with her toddler, a massive cityscape in the distance behind them. A nuclear blast destroys it all, everything "breaking apart like leaves." She wakes from the nightmare obsessed with her mission and roars off alone, to kill Dyson and thus prevent the war. John, realizing where she has gone, pursues her with the T-800.

Now begins Sarah's final rehabilitation. She finds Dyson working at home with his wife and young son, and she attacks him with an assault rifle. In a gut-wrenching scene she strides into the family home, where the trembling Dyson lies bleeding in his wife's arms, to finish him off. "It's all your fault," she screams. But Sarah, a "woman" after all, cannot bring herself to fire. In the face of this mirror image of her own recently reconstituted "family," she crumples to the floor, sobbing. At this juncture, John and the T-800 arrive.

This is the ultimate moment of redemption. Sarah sobs, "I almost... I almost... [killed him]." Sobbing even harder in John's young arms, she cries, "You came here to stop me. I love you, John. I always have." Her emotionality, her vulnerability, her need for a male savior restored to her, Sarah recovers her full womanhood.

In the aftermath the film firmly repudiates the radical feminism of Sarah's viewpoint. When Dyson wonders how he could have known where his research would lead, Sarah sneers, "Fucking *men* like you built the hydrogen bomb. Men like you thought it up. You think you're so creative. You don't know what it's like to really create something—to create a life, feel it growing inside you. All you know how to create is death ... and destruction." As she builds to a crescendo of male-blaming rage, John interrupts her. "Mom! We need to be a little more constructive here." In the name of pragmatics, he effectively silences the feminist critique of both male science and the gendered institution of war. We are clearly meant, between this and other scenes of Sarah's intense brooding, to wonder whether she isn't at least a bit crazy after all.

After all this ideological activity, the film's final scenes are almost anticlimactic. Dyson, a good family man, proves more than willing to take personal responsibility for his actions. He goes with Sarah, John, and the T-800 to the Cyberdyne building to destroy Cyberdyne's files and the original Terminator microchip, sacrificing his own life in the end. After one final chase, Sarah nearly succeeds in destroying the T-1000, as she had crushed the original Terminator in *T1*—but not quite. This time her own strength is not enough, and the damaged (male) T-800 reappears just in time to save her and John by blasting the T-1000 into a boiling vat of molten steel, reconstituting male protectorship.

In the final scene, the fantasy of the superhuman father comes to a bitter end. Though the old chip and the T-1000 are destroyed, the T-800 reminds them that "there is one more chip"—pointing to his own head—"and it must be destroyed also." John resists, crying; but the cyborg must be sacrificed. He cannot, finally, become the perfect father John and Sarah both desire. Mirroring the T-800's first gesture toward her, Sarah offers him her hand. As they clasp, she nods in respectful recognition. She then lowers him into the steel pit, looking sadly after him as he descends, while John watches and cries. In the Terminator's final moment we see, for the last time, the Terminator's-eye screen display as it crackles with static, fades to a single dot, and then goes out. We are meant to mourn his death.

Rolling along down a night highway, Sarah Connor reflects on what lies ahead. "The unknown future rolls toward us, and for the first time I face it with hope. Because if a machine, a Terminator, can learn the value of human life, maybe we can too."

Terminator 2 seems to suggest that we are waking up from a nightmare run by cyborg Others to a world in which we control them. It is as if the dome of the Cold War's nuclear closed world had suddenly shattered, returning to us once again the frontiers we thought forever lost. Computers that learn and create, it hints, can—as mere others—help us rebuild a broken social world and a war-bankrupted economy in the image of the past. The figures of the violent, unfeeling, but rational and devoted father-protector and the emotional, unstable, but loving mother return to form the basis of the post–Cold War social order. The cyborg-machine returns to its place as technological servant, intimating a renewed sense of human control over autonomous technology.[11] The sense of control suggested by the film is clearly linked with the widespread acceptance in the 1990s of computerized automation and communication as the dynamic engines of the modern world economy.

Yet some transformations of identity remain. Sarah, even with her emotionalism and male-dependence, is still far tougher than most of the men she encounters. Only the scientist and father Dyson, in taking personal responsibility for the holocaust, matches the Terminator's manhood. And the Terminator, like every would-be father before him, ultimately leaves the scene. He turns out to be merely a fantasy, since the future that produces him is erased along with the Cold War. Despite its ideological retreat, *Terminator 2* remains profoundly distrustful of white men. It leaves us, in the end, without a great white father. (Dyson, the film's only conceivable candidate for respectable fatherhood, is black.) Finally, unlike *The Terminator*, which in the end glimpsed (however briefly) a green-world refuge, *Terminator 2* closes on the hypnotizing yellow lines of a night highway illuminated by the headlamps of an automobile. The ersatz frontiers to which it returns us are neither transcendent nor green but technological, suburban, the ordinary contexts and struggles of ordinary lives— "learning the value of human life."

Can the cyborg figure still serve as a potent resource for the reconstruction of gender and other political-cultural identities along the technology/biology divide? How will it evolve in a world where commercial goals replace military support as the fundamental drivers of advanced computer and information technology? It is far too early for definitive answers to such questions. But *T2*'s rehabilitated Terminator, like the other rehabilitated cyborgs discussed in chapter 10, suggests that intelligent machines are being integrated into contemporary culture under the all-inclusive rubric of multiculturalism. Rather than a threat, their minds now represent just one more curiosity for the anthropologically inclined, one more reminder of the diversity of sentient beings. This integration points to one of the fundamental problems cyborg politics may now encounter: co-optation by a disingenuous version of multiculturalism.

This simplistic doctrine flattens all cultural difference into two categories. *Exotic* differences function as the "interesting" resources for a Believe-It-Or-Not pseudo-anthropology. These can be trotted out as the trophies of cultural explorers (as on *Star Trek: The Next Generation*), as existence proofs for incommensurable understandings of reality, and as arguments for preserving cultural diversity. But in a world brought to us by the Discovery Channel, the exotic has paradoxically become familiar. It is all too easily transformed into the utterly *mundane*, and anthropology gives way to Disney World. As commonplaces of everyday life, mundane cultural differences form the basis for an easy relativism. This antimorality avoids hard ethical and epistemological questions by rotating all hierarchies—whether of value, of knowledge, or of power—sideways by ninety degrees, turning them into spectrums of difference rather than scales of worth. *Terminator 2* demonstrates how cyborg politics can be co-opted by this oversimple multiculturalism, where cyborgs represent little more than a set of mundane differences between sentient creatures whose technology matters no more than their biology.

These ways of conceiving (and perceiving) cultural differences ignore the really pressing problems of an America that has become multicultural in fact long before it knows how to become so in values. How do we find sources of integrity and authenticity in a world where the enormous pressures of global capitalism inexorably hew all forms of worth to fit the Procrustean bed of money value? How can we preserve cultural diversity at all when technologies of communication (including computer networks) and transportation are eliminating the very basis of cultural difference in embodied situations, locations, and language? How do we locate a basis for politically expensive judgments of value in the face of the political cheapness of moral relativism? How do we face global-scale problems with the only resources we have, namely our increasingly divided, anxious, angry, yet increasingly fragile, inauthentic, and vanishing local cultures? How do we even conceive of a "we" in an America where the master categories of gender, ethnicity, class, and even race have become, for many, little more than mix'n'match cultural fashions? The postmodern embrace of cultural chaos seems no answer, yet the old secular authorities—science, Western liberalism, Marxism—have collapsed. In their place, increasingly, reign nationalism, fundamentalist religions, and global trade. The center has not held. Yet the technologies originally built for cold war have bound the globe, perhaps permanently, into one single world.

The demise of Cold War politics, with its grand ideologies and its culture of high-tech war, has left behind this world at once more open and more closed, more united and more fragmented. As the World Wide Web continues to spin its global electronic cocoon, its annexation as an information marketplace is tempering its early promise as a spontaneous, planetary community of knowledge. Where the AIs, robots, and human-machines of the closed world had to negotiate the centralized, militarized power and totalizing ideologies of the Cold War, the new connectionist cyborgs will have to pay their way at the toll booths of the information superhighway. As their population swells, their exoticism, too, will fade. Then, in their travels and encounters, they will confront the essential problems of the post–Cold War era. Cyborgs in the World Wide Web will face the tripartite tension among the global bonds of communication and control technology; the ideological individualism of cyberspace, with its totally malleable personal identities and disposable virtual communities; and the deepening crisis of culture in an increasingly rootless world.

Notes

Preface

1. On the general theory of historiography and narrative see Hayden White, *Metahistory* (Baltimore: Johns Hopkins University Press, 1973), and *The Tropics of Discourse* (Baltimore: Johns Hopkins University Press, 1978), as well as Fredric Jameson, *The Political Unconscious: Narrative as a Socially Symbolic Act* (Ithaca, NY: Cornell University Press, 1981). On problems of historiography in science and technology see the introduction to Donna J. Haraway, *Primate Visions* (London: Routledge Kegan Paul, 1989); Steven Shapin and Simon Schaffer, *Leviathan and the Air-Pump: Hobbes, Boyle, and the Experimental Life* (Princeton: Princeton University Press, 1985); Thomas S. Kuhn, *The Structure of Scientific Revolutions* (Chicago: University of Chicago Press, 1962); Imre Lakatos and Alan Musgrave, eds., *Criticism and the Growth of Knowledge* (Cambridge: Cambridge University Press, 1970); Bruno Latour, *Science In Action* (Cambridge, MA: Harvard University Press, 1987); Thomas J. Misa, "How Machines Make History, and How Historians (and Others) Help Them to Do So," *Science, Technology, and Human Values*, Vol. 13, Nos. 3 & 4 (1988), 308–331; and Karin D. Knorr-Cetina, "Introduction: The Micro-sociological Challenge of Macro-sociology," in Karin D. Knorr-Cetina and Aaron V. Cicourel, eds., *Advances in Social Theory and Methodology: Towards an Integration of Micro- and Macro-sociologies* (Boston: Routledge and Kegan Paul, 1981), 2–47.

2. Michael S. Mahoney, "The History of Computing in the History of Technology," *Annals of the History of Computing*, Vol. 10, No. 2 (1988), 113–125.

3. Computer *scientists* are, generally speaking, specialists in the logical design and programming of computers. Computer *engineers* design electronic hardware, that is, microchips and other devices. The boundary between these two disciplines is extremely fuzzy, but it is reflected in the two primary academic homes of "computer science" in its early years—mathematics and electrical engineering departments.

4. For example, Howard Gardner's history of cognitive science—the most comprehensive and intellectually rigorous to date—begins with an

introductory chapter entitled "What the Meno Wrought." He argues that the core of cognitive science is a theory of mental representation. That theory answers the questions about the nature of knowledge that Plato's doctrine of forms was created to resolve. See Howard Gardner, *The Mind's New Science* (New York: Basic Books, 1985), 4–5.

5. See Allen Newell, "Intellectual Issues in the History of Artificial Intelligence," in Fritz Machlup and Una Mansfield, eds., *The Study of Information: Interdisciplinary Messages* (New York: John Wiley, 1983), 187–228. (Newell, to his credit, makes explicit the limitations of such a history.) Other works in this genre include Pamela McCorduck, *Machines Who Think* (New York: W. H. Freeman, 1979); Jonathan Miller, *States of Mind* (New York: Pantheon, 1983); Howard Rheingold, *Tools for Thought: The People and Ideas Behind the Next Computer Revolution* (New York: Simon & Schuster, 1985); Frank Rose, *Into the Heart of the Mind* (New York: Harper & Row, 1984); Morton Hunt, *The Universe Within* (New York: Simon and Schuster, 1982); Jay David Bolter, *Turing's Man: Western Culture in the Computer Age* (Chapel Hill, NC: University of North Carolina Press, 1984); Bolter, *Writing Space: The Computer, Hypertext, and the History of Writing* (Hillsdale, NJ: Lawrence Erlbaum Associates, 1991); and John Haugeland, *Artificial Intelligence: The Very Idea* (Cambridge, MA: MIT Press, 1985).

6. The Difference Engine was a design for a powerful gear-based calculator that could produce mathematical tables and print them automatically. The Analytical Engine, also mechanical, would have been programmable by means of punched wooden cards (like those designed to control the Jacquard loom). Its ability to execute conditional loops made it strikingly similar to modern computers in concept. Neither machine was ever completed by Babbage, although a working model of the Difference Engine was successfully constructed at the London Science Museum in 1991. See Doron D. Swade, "Redeeming Charles Babbage's Mechanical Computer," *Scientific American*, Vol. 267, No. 8 (1993), 86–91.

7. A representative sample of books in this genre includes Stan Augarten, *Bit by Bit: An Illustrated History of Computers* (New York: Ticknor & Fields, 1984); Herman Goldstine, *The Computer from Pascal to von Neumann* (Princeton, NJ: Princeton University Press, 1972); and Michael R. Williams, *A History of Computing Technology* (Englewood Cliffs, NJ: Prentice-Hall, 1985). More narrowly focused volumes that follow the engineering/economic history pattern include Nancy Stern, *From ENIAC to UNIVAC: An Appraisal of the Eckert-Mauchly Computers* (Boston: Digital Press, 1981); James Chposky and Ted Leonsis, *Blue Magic: The People, Power, and Politics Behind the IBM Personal Computer* (New York: Facts on File, 1988); Franklin M. Fisher, James W. McKie, and Richard B. Mancke, *IBM and the US Data Processing Industry: An Economic History* (New York: Praeger, 1983); Emerson Pugh, *Memories That Shaped an Industry: Decisions Leading to IBM System/360* (Cambridge, MA: MIT Press, 1984); Glenn Rifkin and George Harrar, *The Ultimate Entrepreneur: The Story of Ken Olsen and Digital Equipment Corporation* (Chicago: Contemporary Books, 1988). Also

see the numerous special issues of *Annals of the History of Computing* devoted to particular computers and programming languages. At this writing the *Annals*, though it remains the field's sole journal, has published little social or cultural history of computers and virtually no political history. One noteworthy exception is William Aspray and Donald deB. Beaver, "Marketing the Monster: Advertising Computer Technology," *Annals of the History of Computing*, Vol. 8, No. 2 (1986), 127–143.

8. Mahoney, "The History of Computing," 114.

9. See Kenneth Flamm's definitive two-volume economic history of computers, *Targeting the Computer* (Washington, DC: Brookings Institution, 1987) and *Creating the Computer* (Washington, DC: Brookings Institution, 1988).

10. Walter A. McDougall argues forcefully for this viewpoint in . . . *the Heavens and the Earth: A Political History of the Space Age* (New York: Basic Books, 1985).

11. A lucid explanation of the historical significance of each term in the phrase "electronic stored-program digital computer" may be found in Augarten, *Bit by Bit*.

12. Paul N. Edwards, "The Army and the Microworld: Computers and the Militarized Politics of Gender," *Signs*, Vol. 16, No. 1 (1990), 102–127.

Chapter 1

1. Despite its contemporary association with science fiction, the term "cyborg" actually originated in science. Manfred Clynes and Nathan S. Cline coined the word in 1960 to describe human-machine systems for space exploration; in their definition, the cyborg "deliberately incorporates exogenous components extending the self-regulatory control function of the organism in order to adapt it to new environments." See Manfred Clynes and Nathan S. Kline, "Cyborgs and Space," *Astronautics* (September 1960), cited in Chris Hables Gray, "The Culture of War Cyborgs: Technoscience, Gender, and Postmodern War," *Research in Philosophy and Technology* (forthcoming).

2. Donna J. Haraway, "A Manifesto for Cyborgs: Science, Technology, and Socialist Feminism in the 1980s," *Socialist Review*, Vol. 15, No. 2 (1985), 65–107; Haraway, "The Promises of Monsters," in Lawrence Grossberg, Cary Nelson, and Paula A. Treichler, eds., *Cultural Studies* (New York: Routledge and Kegan Paul, 1992), 295–337.

3. The term "grand strategy," borrowed from political science and roughly equivalent in my usage to "geopolitics," refers to a nation's long-term, integrated political goals and the military means deployed to attain them.

4. Because of its source, this Air Force estimate must be regarded as conservative. This figure amounts to more than 5 percent of the estimated total population of the region.

5. The discussion of Igloo White is based on Paul Dickson, *The Electronic Battlefield* (Bloomington: Indiana University Press, 1976), 83–97; George L. Weiss, "Battle for Control of the Ho Chi Minh Trail," *Armed Forces Journal* (February 15, 1972), 19–22; "You Can't Run Wars with a Computer," *Business Week* (June 5, 1971), 122; and James Gibson, *The Perfect War: Technowar in Vietnam* (New York: Atlantic Monthly Press, 1987), 396–399.

6. Gregory Palmer, *The McNamara Strategy and the Vietnam War: Program Budgeting in the Pentagon, 1960–68* (Westport, CT: Greenwood Press, 1978), 7.

7. Martin Van Creveld, *Command in War* (Cambridge, MA: Harvard University Press, 1985), 244.

8. Ibid., 236.

9. H. Bruce Franklin, *War Stars: The Superweapon and the American Imagination* (New York: Oxford University Press, 1988), 7.

10. This was Senator Arthur Vandenberg's advice to Truman for how to push a costly foreign aid bill through the Republican-controlled Congress.

11. See Loren Baritz, *Backfire* (New York: Ballantine, 1985), chapter 1.

12. The Clifford Report, reprinted in Arthur Krock, *Memoirs: 60 Years on the Firing Line* (New York: Funk & Wagnall's, 1968), 422–482.

13. Major works of these historians include Fernand Braudel, *Civilization and Capitalism, Fifteenth to Eighteenth Centuries*, trans. Sian Reynolds (New York: Harper & Row, 1981–84); Immanuel Wallerstein, *The Modern World-System* (New York: Academic Press, 1974); Wallerstein, *The Politics of the World-Economy* (New York: Cambridge University Press, 1984).

14. Thomas J. McCormick, *America's Half-Century: United States Foreign Policy in the Cold War* (Baltimore: Johns Hopkins University Press, 1989), 5.

15. See Walter LaFeber, *The American Age: United States Foreign Policy at Home and Abroad Since 1750* (New York: W.W. Norton, 1989), 454 and passim.

16. Dean Acheson, "Authority of the President to Repel the Attack in Korea," *Department of State Bulletin* (July 31, 1950), 173–178.

17. General Douglas MacArthur, *Hearings Before the Committee on Armed Services and the Committee on Foreign Relations of the United States Senate, 82nd Congress, 1st session* to "Conduct an Inquiry into the Military Situation in the Far East and the Facts Surrounding the Relief of General of the Army Douglas MacArthur from his Assignments in the Area" (1951), 68, 81–83. Italics added.

18. James Chace and Caleb Carr, *America Invulnerable: The Quest for Absolute Security from 1812 to Star Wars* (New York: Summit Books, 1988), 251–252 and passim.

19. John Foster Dulles, *Hearings Before the Committee on Foreign Relations, United States Senate, 83rd Congress, 1st session* on "The Nomination of John Foster Dulles" (1953), 5–6.

20. Chace and Carr, *America Invulnerable*, 248.

21. NSC-68, as cited in Chace and Carr, *America Invulnerable*, 248.

22. Sherman Hawkins, personal communication.

23. Sherman Hawkins, "The Two Worlds of Shakespearean Comedy," in J. Leeds Barroll, ed., *Shakespeare Studies*, Vol. III (Cincinnati: The Center for Shakespeare Studies, 1968), 62–80.

24. See Northrop Frye, *The Anatomy of Criticism* (Princeton, NJ: Princeton University Press, 1957), pp. 182ff; Frye, *A Natural Perspective: The Development of Shakespearean Comedy and Romance* (New York: Columbia University Press, 1965).

25. We will return to this opposition in chapter 10. However, since I want to focus on the *structure* of the very particular closed world of the Cold War, the closed world/green world contrast will not play a central role in this book.

26. See Hayden White, *Metahistory* (Baltimore: Johns Hopkins University Press, 1973).

27. See Baritz, *Backfire*.

28. Alan Turing, "Computing Machinery and Intelligence," *Mind*, Vol. 59 (1950), 433–460. For a recent version of this debate, see Paul M. Churchland and Patricia Smith Churchland, "Could a Machine Think?," *Scientific American*, Vol. 256, No. 1 (1990), 32–37, and John R. Searle, "Is the Brain's Mind a Computer Program?," *Scientific American*, Vol. 256, No. 1 (1990), 26–31.

29. Alan Turing, "On Computable Numbers, with an application to the Entscheidungsproblem," *Proceedings of the London Mathematical Society*, Vol. 42 (1937), 251.

30. Ibid. The requirement of "sufficient power" refers essentially to the size of the computer's memory. It should be noted that this is only a principle or potential, since solving many problems would require time and computer memory on a scale completely beyond practical possibility. Interestingly, while Turing is often remembered for the problems his machine *can* solve, the mathematically significant point of his paper consists in a proof that there exist some mathematical problems that cannot even in principle be formulated as algorithms (fixed procedures) and therefore *cannot* be solved by a Turing machine.

31. Similarly, the first general-purpose, program-controlled digital calculator was constructed by Konrad Zuse in Germany in 1941 and used by the German war industry. Though they were discovered by the Allies after the war,

Zuse's machines were never followed up and had little influence on the main stream of computer development.

32. Among others, Herman Goldstine, one of the ENIAC's designers, makes this claim.

33. See Andrew Hodges, *Alan Turing: The Enigma* (New York: Simon and Schuster, 1983), Chapters 4 and 5.

34. I. J. Good, interviewed in Pamela McCorduck, *Machines Who Think* (New York: W. H. Freeman, 1979), 53.

35. John Markoff, "Can Machines Think? Humans Match Wits," *New York Times* (November 9, 1991), 1, 10.

36. Turing, "Computing Machinery and Intelligence," 442.

37. The ubiquity of phrases such as "electronic brain" and "giant brain" in 1950s press accounts of computing must be experienced to be fully appreciated.

38. Daniel C. Dennett, *Brainstorms: Philosophical Essays on Mind and Psychology* (Montgomery VT: Bradford Books, 1978); John Haugeland, *Artificial Intelligence: The Very Idea* (Cambridge, MA: MIT Press, 1985).

39. Sherry Turkle, *The Second Self: Computers and the Human Spirit* (New York: Simon and Schuster, 1984).

40. Ulric Neisser, *Cognitive Psychology* (New York: Appleton-Century-Crofts, 1967).

41. The term "cyberspace" was coined by William Gibson in his novel *Neuromancer* (New York: Ace Books, 1984).

42. See Michael Benedikt, ed., *Cyberspace: First Steps* (Cambridge, MA: MIT Press, 1991); Michael L. Dertouzos et al., "Communications, Computers, and Networks: How to Work, Play and Thrive in Cyberspace," *Scientific American*, Vol. 265, No. 3 (1991), special issue.

43. Donna J. Haraway, "The High Cost of Information in Post–World War II Evolutionary Biology," *Philosophical Forum*, Vol. XIII (1981–82), 244–278.

44. See Turkle on computers and "computational objects" as marginal objects in psychological discourse.

45. On the brittleness associated with highly computerized military forces, see Gene I. Rochlin, *The Computer Trap: Dependence and Vulnerability in an Automated Society* (Princeton, NJ: Princeton University Press, forthcoming).

46. See Haraway, "A Manifesto for Cyborgs" and "The Promises of Monsters."

47. See Paul N. Edwards, "Hyper Text and Hypertension: Hypertext,

Post-structuralist Critical Theory, and Social Studies of Science," *Social Studies of Science*, Vol. 24, No. 2 (1994), pp. 229–278.

48. See, for example, Fritz Machlup and Una Mansfield, eds., *The Study of Information: Interdisciplinary Messages* (New York: John Wiley, 1983), and Theodore Roszak, *The Cult of Information* (New York: Pantheon, 1986).

49. Turkle, *The Second Self*, 16.

50. Marvin Minsky, *The Society of Mind* (New York: Simon and Schuster, 1986).

51. Joseph Weizenbaum, *Computer Power and Human Reason: From Judgment to Calculation* (San Francisco: W. H. Freeman, 1976), 23–25.

52. Ibid., 18.

53. Raymond Williams, *Keywords: A Vocabulary of Culture and Society* (New York: Oxford University Press, 1976), 143–144.

54. Terry Eagleton, citing John B. Thompson, in *Ideology: An Introduction* (New York: Verso, 1991). Eagleton's rigorous examination of the term begins with a list of no fewer than sixteen common definitions.

55. Among the most sophisticated versions of this position is Gerald Cohen, *Karl Marx's Theory of History: A Defence* (Princeton: Princeton University Press, 1978).

56. Eagleton, *Ideology*, 8.

57. Thomas S. Kuhn, *The Structure of Scientific Revolutions* (Chicago: University of Chicago, 1962); Imre Lakatos and Alan Musgrave, eds., *Criticism and the Growth of Knowledge* (Cambridge: Cambridge University Press, 1970).

58. Imre Lakatos criticized Kuhn's theory for similar reasons, proposing as an alternative that science consists of possibly large numbers of competing "research programs" whose effects are less intellectually dominating. Imre Lakatos, "Falsification and the Methodology of Scientific Research Programmes," in Lakatos and Musgrave, eds., *Criticism and the Growth of Knowledge*, 91–196. For an analysis of discourse repertoires based in the sociological tradition, see G. Nigel Gilbert and Michael Mulkay, *Opening Pandora's Box : A Sociological Analysis of Scientists' Discourse* (New York: Cambridge University Press, 1984).

59. See Donna J. Haraway, *Primate Visions* (London: Routledge Kegan Paul, 1989), especially the introduction.

60. Karl Mannheim, *Ideology and Utopia: An Introduction to the Sociology of Knowledge*, trans. Louis Wirth and Edward Shils (New York: Harcourt Brace Jovanovich, 1936); Peter L. Berger and Thomas Luckmann, *The Social Construction of Reality* (New York: Anchor Books, 1966).

61. Trevor Pinch and Wiebe Bijker, "The Social Construction of Facts and Artifacts," in Wiebe Bijker, Thomas P. Hughes, and Trevor Pinch, eds., *The Social Construction of Technological Systems* (Cambridge, MA: MIT Press, 1987), 17–50.

62. John Law, "The Heterogeneity of Texts," in Michel Callon, John Law, and Arie Rip, eds., *Mapping the Dynamics of Science and Technology* (London: Macmillan, 1986), 67–83; Law, "Laboratories and Texts," in ibid., 35–50; Law, "Technology and Heterogeneous Engineering: The Case of Portuguese Expansion," in Bijker, Hughes, and Pinch, *The Social Construction of Technological Systems*, 111–134. See also the seminal work of Bruno Latour: Latour and Steve Woolgar, *Laboratory Life: The Social Construction of Scientific Facts* (London: Sage, 1979); Latour, "Give Me a Laboratory and I Will Raise the Earth," in Karin Knorr-Cetina and Michael Mulkay, eds., *Science Observed* (London: Sage, 1983), 141–170; and Latour, *Science In Action* (Cambridge, MA: Harvard University Press, 1987).

63. Michel Callon, "Some Elements of a Sociology of Translation: Domestication of the Scallops and the Fishermen of St. Brieuc Bay," in John Law, ed., *Power, Action, and Belief: A New Sociology of Knowledge?* (London: Routledge and Kegan Paul, 1986), 196–233; Callon and Bruno Latour, "Unscrewing the Big Leviathan: How Actors Macro-structure Reality and How Sociologists Help Them to Do So," in Karin D. Knorr-Cetina and Aaron V. Cicourel, eds., *Advances in Social Theory and Methodology* (Boston: Routledge and Kegan Paul, 1981), 277–303.

64. Wiebe Bijker, "The Social Construction of Bakelite: Toward a Theory of Invention," in Bijker, Hughes, and Pinch, eds., *The Social Construction of Technological Systems*; Bijker and John Law, eds., *Shaping Technology/Building Society: Studies in Sociotechnical Change* (Cambridge, MA: MIT Press, 1992).

65. Steven Shapin and Simon Schaffer, *Leviathan and the Air-Pump: Hobbes, Boyle, and the Experimental Life* (Princeton: Princeton University Press, 1985).

66. Thomas P. Hughes, *Networks of Power: Electrification in Western Society, 1880–1930* (Baltimore, MD: Johns Hopkins University Press, 1983); Hughes, "The Evolution of Large Technological Systems," in Bijker, Hughes, and Pinch, *The Social Construction of Technological Systems*, 51–82.

67. Peter Taylor, "Building on the Metaphor of Construction in Science Studies: Towards a Stronger Framework for Reconstructing the Heterogeneous Construction of Scientific Activity," manuscript, Department of Science and Technology Studies, Cornell University (1994).

68. Pinch and Bijker, "The Social Construction of Facts and Artifacts," 46.

69. Ludwig Wittgenstein, *Philosophical Investigations*, trans. G.E.M. Anscombe (New York: Macmillan, 1958), paragraph 7. All references to this work are to Wittgenstein's paragraph numbers rather than page numbers.

70. On this point see David Bloor, *Wittgenstein: A Social Theory of Knowledge* (New York: Columbia University Press, 1983), especially Chapter 3.

71. Wittgenstein, *Investigations*, paragraph 16.

72. Ibid., paragraph 199.

73. Ibid., paragraph 169.

74. Ludwig Wittgenstein, *On Certainty*, ed. and trans. G. E. M. Anscombe and G. H. von Wright (New York: Harper, 1969), paragraphs 109 and 110.

75. Shapin and Schaffer, *Leviathan and the Air-Pump*.

76. Michel Foucault, *Power/Knowledge: Selected Interviews and Other Writings 1972–1977*, ed. Colin Gordon (New York: Pantheon, 1980), 93.

77. Michel Foucault, *The History of Sexuality*, trans. Robert Hurley (New York: Vintage, 1980), 68–69.

78. "The history which bears and determines us has the form of a war rather than that of a language: relations of power, not relations of meaning. History has no 'meaning,' though this is not to say that it is absurd or incoherent. On the contrary, it is intelligible and should be susceptible of analysis down to the smallest detail—but this in accordance with the intelligibility of struggles, of strategies and tactics. Neither the dialectic, as logic of contradictions, nor semiotics, as the structure of communication, can account for the intrinsic intelligibility of conflicts. 'Dialectic' is a way of evading the always open and hazardous reality of conflict by reducing it to a Hegelian skeleton, and 'semiology' is a way of avoiding its violent, bloody and lethal character by reducing it to the calm Platonic form of language and dialogue." Foucault, *Power/Knowledge*, 114–115.

79. Foucault, *History of Sexuality*, 154.

80. Ibid., 100.

81. Ibid., 42.

82. See Part V, "Social Control and Privacy," in Charles Dunlop and Rob Kling, eds., *Computerization and Controversy* (New York: Academic Press, 1991), 410–522, for a selection of articles on this issue. Also see Shoshana Zuboff, *In the Age of the Smart Machine: The Future of Work and Power* (New York: Basic Books, 1988), especially Chapters 9 and 10.

83. See Gibson, *The Perfect War*. Gibson's analysis of official accounts of Vietnam also uses a Foucaultian framework.

84. See the chapter on "The Gentle Way in Punishment" in Michel Foucault, *Discipline and Punish: The Birth of the Prison*, trans. Alan Sheridan (New York: Vintage Books, 1977), for a description of such a struggle.

85. See Merritt Roe Smith's introduction to *Military Enterprise and Technological Change*, ed. Merritt Roe Smith (Cambridge: MIT, 1985), 1–37, and Thomas J. Misa, "How Machines Make History, and How Historians (and Others) Help Them to Do So," *Science, Technology, and Human Values*, Vol. 13, Nos. 3 & 4 (1988), 308–331.

86. See Claude Lévi-Strauss, *The Raw and the Cooked*, trans. John and Doreen Weightman (New York: Harper, 1969), especially the "Overture," and Paul Feyerabend, *Against Method* (London: New Left Books, 1975).

Chapter 2

1. General William Westmoreland, U.S. Army Chief of Staff, "Address to the Association of the U.S. Army." Reprinted in Paul Dickson, *The Electronic Battlefield* (Bloomington, IN: Indiana University Press, 1976), 215–223.

2. See Wiebe E. Bijker, "The Social Construction of Bakelite: Toward a Theory of Invention," in Bijker, Hughes, and Pinch, *The Social Construction of Technology*, 159–187.

3. *Mechanical* systems are classical Aristotelian machines (e.g., car engines) that perform work using the physical movement of levers, gears, wheels, etc. *Electromechanical* systems are machines powered by electric motors or electromagnets (e.g., vacuum cleaners), but part or all of whose function is still performed through the physical movement of parts. *Electronic* systems, by contrast, contain few or no moving parts. They consist entirely of electrical circuitry and perform their work through transformations of electric current (e.g., televisions or stereo systems).

The distinction between *digital* and *analog* methods corresponds closely to the more intuitive difference between *counting* and *measuring*. Digital calculation uses *discrete* states, such as the ratchet-like detents of clockwork gears (mechanical), the on-off states of relays (electromechanical switches), or the positive or negative electrical states of transistors (electronic), to represent discrete numerical values (1, 2, 3, etc.). These values can then be added, subtracted, and multiplied, essentially by a process of counting. Analog calculation, by contrast, employs *continuously variable* states, such as the ratio between the moving parts of a slide rule (mechanical), the speed of a motor's rotation (electromechanical), or the voltage of a circuit (electronic), to represent continuously variable numerical quantities (e.g., *any* value between 0 and 10). These quantities can then be physically combined to represent addition, subtraction, and multiplication, for example as someone might measure the perimeter of a room by cutting pieces of string to the length of each wall and then tying them together.

4. Michael R. Williams, *A History of Computing Technology* (Englewood Cliffs, NJ: Prentice-Hall, 1985), 78–83.

5. Pesi R. Masani, *Norbert Wiener, 1894–1964* (Boston: Birkhäuser, 1990), 68.

6. The transfer of the name "computer" to the machine was by no means immediate. The *Reader's Guide to Periodical Literature* does not list "computer" as a heading until 1957. Most news articles of the 1945–1955 period place the word, if they use it at all, in scare quotes. See Paul E. Ceruzzi, "When Computers Were Human," *Annals of the History of Computing*, Vol. 13, No. 3 (1991), 237–244; Henry S. Tropp et al., "A Perspective on SAGE: Discussion," *Annals of the History of Computing*, Vol. 5, No. 4 (1983), 375–398; Jean J. Bartik and Frances E. Holberton, interviewed by Richard R. Mertz, May 28, 1970, Smithsonian Computer Oral History Project, AC NMAH #196 (Archive Center, National Museum of American History, Washington, DC).

7. On the NDRC and OSRD, see Daniel Kevles, *The Physicists* (New York: Alfred A. Knopf, 1971); David Dickson, *The New Politics of Science* (New York: Pantheon, 1984); and Bruce L. R. Smith, *American Science Policy since World War II* (Washington DC: Brookings Institution, 1991).

8. Paul Forman, "Behind Quantum Electronics: National Security as Basis for Physical Research in the United States, 1940–1960," *Historical Studies in the Physical and Biological Sciences*, Vol. 18, No. 1 (1987), 152.

9. James Phinney Baxter, *Scientists Against Time* (Boston: Little, Brown, 1946), 21. On the Radiation Laboratory see also Karl L. Wildes and Nilo A. Lindgren, *A Century of Electrical Engineering and Computer Science at MIT, 1882–1982* (Cambridge, MA: MIT Press, 1985).

10. Forman, "Behind Quantum Electronics," 156–157.

11. See David Noble, *America By Design: Science, Technology, and the Rise of Corporate Capitalism* (Oxford: Oxford University Press, 1977), as well as Dickson, *New Politics of Science*, and Merritt Roe Smith, ed., *Military Enterprise and Technological Change* (Cambridge, MA: MIT Press, 1985). The phrase "iron triangle" is from Gordon Adams, *The Politics of Defense Contracting: The Iron Triangle* (New Brunswick, NJ: Transaction Books, 1982).

12. Wildes and Lindgren, *A Century of Electrical Engineering and Computer Science at MIT*, 289 and passim.

13. Ibid., 229. J.C.R. Licklider speculated that Bush's dislike of digital machines stemmed from an argument with Norbert Wiener over whether binary arithmetic was the best base to use. See John A. N. Lee and Robert Rosin, "The Project MAC Interviews," *IEEE Annals of the History of Computing*, Vol. 14, No. 2 (1992), 22.

14. See, for example, Bush's prescient article on the "memex," a kind of hypertext technology, in "As We May Think," *Atlantic Monthly* 176, 1945. Reprinted in Zenon Pylyshyn, ed., *Perspectives on the Computer Revolution* (Englewood Cliffs: Prentice-Hall, 1970), 47–59.

15. Herman Goldstine, *The Computer from Pascal to von Neumann* (Princeton, NJ: Princeton University Press, 1972), 135–136. As noted in chapter 1, the

British Colossus, not the ENIAC, was actually the first electronic digital computer. Because of British secrecy, Goldstine may not have known this when he wrote the cited lines in 1972.

16. Williams, *History of Computing Technology*; Stan Augarten, *Bit by Bit: An Illustrated History of Computers* (New York: Ticknor & Fields, 1984); 119–120.

17. Williams, *History of Computing Technology*, 275. The estimate of part failure probability is due to Herman Goldstine.

18. Ibid., 285.

19. In contemporary parlance the word "computer" refers to electronic digital machines with a memory and one or more central processing units. In addition, to qualify as a computer a device must be capable of (a) executing conditional branching (i.e., carrying out different sets of instructions depending upon the results of its own prior calculations) and (b) storing these instructions (programs) internally. Babbage's Analytical Engine had most of these features but would have stored its programs externally on punched wooden cards. Somewhat like the Analytical Engine, the EDVAC was only partially completed; thus it represents the first true computer *design* but not the first actually operating computer. Instead, the British Manchester University Mark I achieved that honor in 1948. See Williams, *History of Computing Technology*, 325; Augarten, *Bit by Bit*, 149.

20. Goldstine distributed this plan, under von Neumann's name but unbeknownst to him, as the famous and widely read "Draft Report on the EDVAC." Because von Neumann's name was on its cover, the misunderstanding arose that he was the report's sole author. But many of the Draft Report's key ideas actually originated with Eckert, Mauchly, and other members of the ENIAC design team. Because of this misunderstanding, which later escalated into a lawsuit, and the fame he acquired for other reasons, von Neumann has received more credit for originating computer design than he probably deserved. See Augarten, *Bit by Bit*, 136ff. The most essential feature of the so-called von Neumann architecture is serial (one-by-one) processing of the instruction stream.

21. Augarten, *Bit by Bit*, 130–131, citing Nancy Stern, *From ENIAC to UNIVAC* (Boston: Digital Press, 1981). Whether the ENIAC was actually able to solve this first problem is a matter of debate among historians. Kenneth Flamm, citing an interview with Stanley Frankel (Smithsonian Institution Computer Oral History Project, October 5, 1972, conducted by Robina Mapstone), holds that it failed and that the calculations were actually carried out later at Eckert and Mauchly's UNIVAC factory in Philadelphia. See Kenneth Flamm, *Targeting the Computer: Government Support and International Competition* (Washington, DC: Brookings Institution, 1987), 79.

22. Barnes, cited in Goldstine, *The Computer from Pascal to von Neumann*, 229.

23. Forman, "Behind Quantum Electronics," 152.

24. Harry S Truman, cited in Ambrose, *Rise to Globalism*, 86.

25. By the Cold War's end in 1989, peacetime military budgets routinely reached levels in excess of $300 billion. America had 1.5 million men and women in uniform, defense alliances with 50 nations, and military bases in 117 countries.

26. Ambrose, *Rise to Globalism*, xiii.

27. Ibid., 30, 53.

28. Rystad has named this interpretation the "Munich paradigm." See Göran Rystad, *Prisoners of the Past?* (Lund, Sweden: CWK Gleerup, 1982), 33 and passim.

29. Harry S Truman, *Memoirs: Years of Trial and Hope*, Vol. 2 (Garden City, NJ: Doubleday, 1956), 332–333.

30. Herbert York, Los Alamos physicist and Eisenhower advisor, noted that "because of Pearl Harbor you didn't have to discuss the notion of surprise attack. It was in your bones that the Russians are perfidious, that surprise attack is the way wars always start, and that appeasement doesn't work." Cited in Gregg Herken, *Counsels of War* (New York: Knopf, 1983), 125.

31. Atomic bombs, of course, were only one in a long line of weapons Americans (and others) believed would make war too terrible to fight. See H. Bruce Franklin, *War Stars*. (New York: Oxford University Press, 1988).

32. Baritz, *Backfire*, makes a compelling case for a similar argument.

33. On military social and political relations in the pre-revolutionary era, see Maury D. Feld, *The Structure of Violence: Armed Forces as Social Systems* (Beverly Hills: Sage, 1977). The tension between the authoritarian military ethic and democratic principles is the focus of a number of novels and films, such as Charles Nordhoff's *Mutiny on the Bounty* (Boston: Little, Brown, 1932) and Rob Reiner's 1992 film *A Few Good Men*, based on Aaron Sorkin's play of the same title.

34. I owe this point to Robert Meister.

35. See Samuel Huntington, *American Politics: The Promise of Disharmony* (Cambridge: Belknap Press, 1981).

36. Secretaries of War and Navy, joint letter to the National Academy of Sciences, cited in Vannevar Bush, *Science: The Endless Frontier* (Washington: U.S. Government Printing Office, 1945), 12.

37. Bowles, cited in Wildes and Lindgren, *A Century of Electrical Engineering and Computer Science at MIT*, 203.

38. Harry S Truman, cited in James L. Penick Jr. et al., *The Politics of*

American Science: 1939 to the Present, revised ed. (Cambridge, MA: MIT Press, 1972), 20.

39. The Vinson Bill creating the ONR, cited in ibid., 22.

40. For a somewhat different view of the ONR, see Harvey Sapolsky, *Science and the Navy* (Princeton, NJ: Princeton University Press, 1990). Among the best, and certainly to date the most detailed, discussions of the mixed motives of military support for postwar academic research is Stuart W. Leslie, *The Cold War and American Science: The Military-Industrial-Academic Complex at MIT and Stanford* (New York: Columbia University Press, 1993).

41. The Vinson Bill and the ONR Planning Division, cited in Penick et al., *The Politics of American Science*, 22–23.

42. Vinson, quoted in Kent C. Redmond and Thomas M. Smith, *Project Whirlwind* (Boston: Digital Press, 1980), 105.

43. Dickson, *New Politics of Science*, 118–119.

44. See Mina Rees, "The Computing Program of the Office of Naval Research, 1946–1953," *Annals of the History of Computing*, Vol. 4, No. 2 (1982), 103–113.

45. The figure of 20 percent for NSF support is generous, since the budget category used includes both mathematics and computer science research. On the DOE and NASA as auxiliary military agencies, see Flamm, *Targeting the Computer*, 46 and passim. My discussion in this chapter relies heavily on Flamm's published account; on some points I am indebted to him for personal communications as well. On NASA's role as a civilian cover for military research and the U.S. geostrategic aim of establishing international rights of satellite overflight (the "open skies" policy) in order to obtain intelligence about Soviet military activities, see Walter A. McDougall, *. . . the Heavens and the Earth: A Political History of the Space Age* (New York: Basic Books, 1985).

46. On Eckert, Mauchly, their machines, and their company, see Stern, *From ENIAC to UNIVAC*.

47. M. D. Fagen, ed., *A History of Engineering and Science in the Bell System* (Murray Hill, NJ: Bell Telephone Laboratories, 1978), 11.

48. Erwin Tomash and Arnold A. Cohen, "The Birth of an ERA: Engineering Research Associates, Inc., 1946–1955," *Annals of the History of Computing*, Vol. 1, No. 2 (1979), 83–97.

49. Julian Bigelow, "Computer Development at the Institute for Advanced Study," in N. Metropolis, J. Howlett, and Gian-Carlo Rota, eds., *A History of Computing in the Twentieth Century* (New York: Academic Press, 1980), 291–310.

50. Accounting for research and development costs is inherently problem-

atic, for example because of the sometimes fine line between procurement and development expenditures. Defense Department accounting practices do not always offer a clear-cut distinction between R&D and other budgets. See Flamm, *Targeting the Computer*, 94.

51. Flamm, *Targeting the Computer*, 94–96.

52. Nancy Stern, "The BINAC: A Case Study in the History of Technology," *Annals of the History of Computing*, Vol. 1, No. 1 (1979), 9–20; Stern, *From ENIAC to UNIVAC*. Shipped from Eckert and Mauchly's Philadelphia workshop to California, the BINAC failed to function after being reassembled. Northrop soon abandoned it in favor of analog computers.

53. Flamm, *Targeting the Computer*, 64, 76. The outbreak of the Korean War caused IBM chairman Thomas J. Watson, Sr., to reactivate IBM's military products division, providing an opportunity for Thomas J. Watson, Jr., to initiate electronic computer development there. (See James Birkenstock, interviewed by Erwin Tomash and Roger Stuewer, August 12, 1980. Charles Babbage Institute, University of Minnesota.)

54. Williams, *History of Computing Technology*, 334.

55. See Flamm, *Targeting the Computer*, 159 and passim and Flamm, *Creating the Computer: Government, Industry, and High Technology* (Washington, DC: Brookings Institution, 1988), 136–150.

56. Jay W. Forrester, "Computation Book," entry for 8/13/47. Magnetic Core Memory Records, 1932–1977, MC 140, Box 4, F 27. (Institute Archives and Special Collections, Massachusetts Institute of Technology, Cambridge, MA).

57. Redmond and Smith (*Project Whirlwind*, 75) quote a letter from Warren Weaver to Mina Rees to the effect that MIT professor Samuel Caldwell "would work only 'on research concerning electronic computing that will freely serve all science,' a view shared by many of his colleagues."

58. On Mauchly's security problems, see the Appendix to Augarten, *Bit by Bit*.

59. Forman, "Behind Quantum Electronics," 206.

60. Flamm, *Creating the Computer*, 17–18. Also see Harry Atwater, "Electronics and Computer Development: A Military History," *Technology and Responsibility*, Vol. 1, No. 2 (1982).

61. Ted Greenwood, *Making the MIRV* (Cambridge: Ballinger, 1975).

62. Frank Rose, *Into the Heart of the Mind* (New York: Harper and Row, 1984), 36.

63. Jay W. Forrester, "Managerial Decision Making," in Martin Greenberger, ed., *Computers and the World of the Future* (Cambridge, MA: MIT Press, 1962), 53.

64. Goldstine, *The Computer from Pascal to von Neumann*, 251.

65. Aiken's Mark I project was begun in 1937 and completed in 1943. Despite its large scale, neither this nor his subsequent computers had much influence on the main stream of computer development, largely due to his conservative technological approach (according to Augarten, *Bit by Bit*). Cannon reported Aiken's comment in his sworn testimony for *Honeywell v. Sperry Rand*, p. 17935, cited in Stern, *From ENIAC to UNIVAC*, 111.

66. *Calculation* is the mathematical and logical function: crunching numbers, analyzing data, working with Boolean (true-false) variables. *Communication* includes the transfer of text, data, and images among computers through networks, electronic mail, and fax. In *control* functions computers operate other machines: telephone switching networks, automobile engines, lathes. (The control of spacecraft from Earth provides an example of how the three functions are now integrated: computers calculate trajectories and determine which engines to burn, for how long, to achieve needed course corrections. They then control the spacecraft via digital communications, translated by the spacecraft's own computer into signals controlling the jets.) While pure calculation may be done at any speed and may be interrupted and resumed without loss, communication is often and control is almost always a *real-time* function. The person listening to a radio transmission or telephone call hears it at exactly the same pace and almost exactly the same time as the person speaking into the microphone; while the communication may be interrupted and resumed, this is usually inconvenient and undesirable. The servomechanism or mechanical linkage controlling a machine such as a car operates at exactly the same pace as the machine itself. If such control is interrupted, the machine stops or, worse, continues to operate "out of control."

67. Shannon, for example, presented an analog maze-solving machine to the eighth Macy Conference. Heinz von Foerster, ed., *Transactions of the Eighth Conference on Cybernetics* (New York: Josiah Macy Jr. Foundation, 1952), 173–180. Also see Shannon, "Computers and Automata," *Proceedings of the IRE*, Vol. 41 (1953), 1234–1241. Norbert Wiener, *Cybernetics* (Cambridge, MA: MIT Press, 1948), describes a wide variety of devices; after the war Wiener became deeply interested in prosthetics.

68. See Jan Rajchman, "Early Research on Computers at RCA," in Metropolis, Howlett, and Rota, *A History of Computing in the Twentieth Century*, 465–469.

69. David Mindell has recently argued that Bush's Differential Analyzer was not primarily a calculating engine, but a real-time control system. See David Mindell, "From Machinery to Information: Control Systems Research at MIT in the 1930s," paper presented at Society for the History of Technology Annual Meeting, Lowell, MA, 1994, 18.

70. The future role of digital equipment in communication itself was also far from clear, since almost all electronic communication technologies after the (digital) telegraph employed (analog) waveforms, converting sound waves

into radio or electrical waves and back again. However, digital switches—relays—were the primary *control* elements of the telephone network, and a source of some of the early work on digital computers. As we will see in chapter 6, Claude Shannon's wartime work on encryption of voice transmissions produced the first approaches to digitizing sound waves.

71. C. A. Warren, B. McMillan, and B. D. Holbrook, "Military Systems Engineering and Research," in Fagen, *A History of Engineering and Science in the Bell System*, 618.

72. 40,000 of the analog 40-mm gun directors designed by the Servomechanisms Lab were manufactured during the war. Its budget, by the war's end, was over $1 million a year. Wildes and Lindgren, *A Century of Electrical Engineering and Computer Science at MIT*, 211.

73. Mina Rees, "The Federal Computing Machine Program," *Science*, Vol. 112 (1950), 732. Reprinted in *Annals of the History of Computing*, Vol. 7, No. 2 (1985), 156–163.

74. George E. Valley Jr., "How the SAGE Development Began," *Annals of the History of Computing*, Vol. 7, No. 3 (1985), 218.

75. Fred J. Gruenberger, "The History of the JOHNNIAC," *Annals of the History of Computing*, Vol. 1, No. 1 (1979), 50.

76. Valley, "How the SAGE Development Began," 207. Italics added.

77. A simple example can illustrate this point. An ordinary thermostat is an analog control device with a digital (on/off, discrete-state) output. A coiled bimetal strip inside the thermostat varies in length continuously with the ambient temperature. To it is affixed a horizontal tube containing a drop of mercury. As the temperature drops, the coil shrinks and the tube tilts past the horizontal. This causes the mercury to flow to one end of the tube, closing an electrical connection that activates the heating system. As the temperature then rises, the coil slowly expands, the tube eventually tilts the other way, and the mercury flows to its other end. This deactivates the switch, and the heater turns off. This device might thus be said to compute a function: if temperature $t \leq$ temperature setting s, heater setting $h = 0$ (off). If $t < s$, $h = 1$ (on). This function *could* be computed using numerical inputs for t and s. But since the function is represented directly in the device, no conversion of physical quantities into numerical values is required.

78. Valley, "How the SAGE Development Began," 206.

79. See Stern, *From ENIAC to Univac*.

80. Albert S. Jackson, *Analog Computation* (New York: McGraw-Hill, 1960), 8.

81. Trevor Pinch and Wiebe Bijker, "The Social Construction of Facts and Artifacts," in Bijker, Hughes, and Pinch, *The Social Construction of Technology*, 17–50.

82. Defense Advanced Research Projects Agency, *Strategic Computing* (Washington, DC: Defense Advanced Research Projects Agency, 1983), 8.

83. Jonathan Jacky, "The Strategic Computing Program," in David Bellin and Gary Chapman, eds., *Computers in Battle* (New York: Harcourt Brace, 1987), 184.

84. For entrées to the large literature on this topic, see Gary Chapman, "The New Generation of High-Technology Weapons," in Bellin and Chapman, *Computers in Battle*, 61–100; Chris Hables Gray, *Computers as Weapons and Metaphors: The U.S. Military 1940–1990 and Postmodern War* (unpublished Ph.D. thesis, University of California, Santa Cruz, 1991); and Morris Janowitz, *The Professional Soldier* (Glencoe, IL: Free Press, 1960).

85. Colonel Francis X. Kane, USAF, "Security Is Too Important To Be Left to Computers," *Fortune*, Vol. 69, No. 4 (1964), 146–147.

86. Ferdinand Otto Miksche, "The Soldier and Technical Warfare," *Military Review*, Vol. 42, No. 8 (1962), 71–78; Major Keith C. Nusbaum, U.S. Army, "Electronic Despotism: A Serious Problem of Modern Command," *Military Review*, Vol. 42, No. 4 (1962), 31–39.

87. "Will 'Computers' Run Wars of the Future?," *U.S. News & World Report* (April 23, 1962), 44–48.

88. Westmoreland, "Address," 222.

Chapter 3

1. Mina Rees, "The Computing Program of the Office of Naval Research, 1946–1953," *Annals of the History of Computing*, Vol. 4, No. 2 (1982), 113.

2. In fact the Servo Lab, according to Gordon Brown, probably had as much to do with the idea for the simulator as de Florez: Gordon S. Brown, interviewed by Richard R. Mertz, May 28, 1970. Smithsonian Computer Oral History Project, AC NMAH #196 (Archive Center, National Museum of American History, Washington, DC). The Special Devices Division was later known as the Special Devices Center. The designation "analyzer" reflects the ASCA's origins in the Bush-MIT tradition of analog "differential analyzers."

3. Frank M. Verzuh, interviewed by William Aspray, February 20, 1984. Charles Babbage Institute, University of Minnesota.

4. Various accounts of Forrester's conversion to digital techniques are given in Kent C. Redmond and Thomas M. Smith, *Project Whirlwind: The History of a Pioneer Computer* (Boston: Digital Press, 1980); Mina Rees, "The Computing Program of the Office of Naval Research, 1946–1953," *Annals of the History of Computing*, Vol. 4, No. 2 (1982), 103–113; and George E. Valley Jr., "How the SAGE Development Began," *Annals of the History of Computing*, Vol. 7, No. 3

(1985), 196–226. A brief version of Forrester's own account may be found in Henry S. Tropp et al., "A Perspective on SAGE: Discussion," *Annals of the History of Computing*, Vol. 5, No. 4 (1983), 375–398.

5. Gordon Brown, director of the Servomechanisms Laboratory, recalled in an interview that from its inception, building and testing machines under conditions of actual use was a critical part of the Lab's ethos. "This issue of testing . . . was the great issue . . . on which the success of Whirlwind depended and also it was the issue on which it nearly foundered." Gordon S. Brown, interviewed by Richard R. Mertz, May 28, 1970. Smithsonian Computer Oral History Project, AC NMAH #196 (Archive Center, National Museum of American History, Washington, DC). Forrester and Everett walked a delicate line in moving away from the ASCA without compromising their applications-oriented approach. On the one hand they encouraged conclusions like that of Columbia's Francis Murray. Murray, in an independent report comparing Whirlwind to the IAS machine at Princeton, wrote that "the application of digital computation to simulation and control required the 'engineering development' of Whirlwind, a requirement not imposed upon the IAS computer, which was at liberty to follow a 'direction of interest to its own objective,' namely the consideration of 'purely scientific problems.'" (Murray, cited in Redmond and Smith, *Project Whirlwind*, 81.) Their machine, they argued, was a prototype, as opposed to a model, of a computer. On the other hand, once they had effectively abandoned the ASCA they could not commit to any particular application without a sponsor willing to fund the massive costs of development.

6. One exception was Eckert and Mauchly's BINAC computer, completed in 1949 under contract to Northrop Aircraft. See chapter 2, note 52.

7. On the analog research program, see P. A. Holst, "George A. Philbrick and Polyphemus," *Annals of the History of Computing*, Vol. 4, No. 2 (1982), 143–156; Mina Rees, "The Federal Computing Machine Program," *Science*, Vol. 112 (1950), 731–736; and Michael R. Williams, *A History of Computing Technology* (Englewood Cliffs, NJ: Prentice-Hall, 1985), chapter 5.

8. As one of Forrester's colleagues recalled, "until that time, the use of computers was, and for some time afterwards, basically scientific and engineering calculation. But the use of it for ship control, the use of it for fire control . . . was all Forrester. He was pushing very hard for real-time computer applications." David R. Israel, interviewed by Richard R. Mertz, April 22, 1970. Smithsonian Computer Oral History Project, AC NMAH #196 (Archive Center, National Museum of American History, Washington, DC).

9. George W. Brown, interviewed by Richard R. Mertz, March 15, 1973. Smithsonian Computer Oral History Project, AC NMAH #196. (Archive Center, National Museum of American History, Washington, DC).

10. The reason proffered to the ONR was that construction of the cockpit,

far simpler than its computer controller, should wait until the computer was ready.

11. Kenneth Flamm, *Creating the Computer* (Washington, DC: Brookings Institution, 1988), 54.

12. See the list of early computer projects in Flamm, *Creating the Computer*, 76–77.

13. This was an accident. Forrester had been installing some MIT-built servo equipment on the ship. While he was below decks, the ship sailed for a Pacific combat zone. He kept himself busy by maintaining the equipment until the ship was torpedoed and forced to return to port. (Gordon S. Brown interview, May 28, 1970.) Forrester's colleague Robert Wieser speculated in an interview that "with a fire control background, I think one of the first things that occurred to Jay . . . was the application of the computer to coordinated antisubmarine warfare and air defense, which has certain similarities with ASW that is a multiple tracking system with, if you like, a battle management kind of function built into it too." C. Robert Wieser, interviewed by Richard R. Mertz, March 20, 1970. Smithsonian Computer Oral History Project, AC NMAH #196 (Archive Center, National Museum of American History, Washington, DC).

14. Jay W. Forrester, "Computation Book," November 27, 1946, to December 10, 1948. Magnetic Core Memory Records, 1932–1977, MC 140, Box 4, F27 (Institute Archives and Special Collections, MIT Libraries, Cambridge, Massachusetts).

15. Forrester, in Henry S. Tropp et al., "A Perspective on SAGE: Discussion," *Annals of the History of Computing*, Vol. 5, No. 4 (1983), 376.

16. Forrester, interviewed by C. Evans, 1975. *Pioneers of Computing* Series, Science Museum of London. (Available from Charles Babbage Institute, University of Minnesota.)

17. Forrester, interviewed by Marc Miller, April 19, 1977. Oral History Collection, Institute Archives. Massachusetts Institute of Technology, Cambridge, MA.

18. Redmond and Smith, *Project Whirlwind*, 150.

19. Forrester, cited in ibid., 42. For more on Combat Information Centers, a World War II Navy innovation, see chapter 7.

20. Jay W. Forrester, "Discussion of the paper 'Application of Computing Machines to Guided Missile Problems' by Mr. Perry Crawford, Office of Naval Research" (undated, 1947?), Vannevar Bush Papers, Box 39, Library of Congress.

21. Project Whirlwind Report L-1, J. W. Forrester and R. R. Everett to Director, Special Devices Center, subj: "Digital Computation for Anti-submarine

Problem," October 1, 1947. Project Whirlwind Report L-2, J. W. Forrester and R. R. Everett to Director, Special Devices Center, subj: "Information System of Interconnected Digital Computers," October 15, 1947. Cited in Redmond and Smith, *Project Whirlwind*, 58, note 24.

22. Redmond and Smith, *Project Whirlwind*, 120.

23. Jay W. Forrester et al., "A Plan for Digital Information-Handling Equipment in the Military Establishment," Project DIC 6345, MIT Servomechanisms Laboratory, September 14, 1948. MIT Archives.

24. Forrester interview, April 19, 1977.

25. In a group retrospective organized by Henry Tropp in 1983, the principals involved, including Forrester, Everett, and military sponsors, emphasize this process of mutual influence (Tropp et al., "A Perspective on SAGE").

26. On MIT's institutional interests in ongoing military funding, see Henry Etzkowitz, "The Making of an Entrepreneurial University: The Traffic among MIT, Industry, and the Military, 1860–1960," in Everett Mendelsohn, Merritt Roe Smith, and Peter Weingart, eds., *Science, Technology, and the Military* (Boston: Kluwer Academic Publishers, 1988).

27. Gregg Herken, *Counsels of War* (New York: Knopf, 1983), 24. According to Herken, the surveys pointed out that despite strategic bombing, German arms production rose steadily until mid-1944; judged that effects on civilian morale had been relatively slight; and concluded that at least in the British case the cost of the bombing in lives and money had been higher than the damage caused to the other side.

28. Unification of the services under the OSD was not fully achieved until 1961; even then, interservice competition for roles, missions, and money continued.

29. Kenneth Schaffel, *The Emerging Shield: The Air Force and the Evolution of Continental Air Defense 1945–1960* (Washington, D.C.: Office of Air Force History, United States Air Force, 1991), 8 and passim.

30. Ibid., 36, citing official policy.

31. Ibid., 180.

32. H. Bruce Franklin, *War Stars* (New York: Oxford University Press, 1988), 96.

33. This film and a spate of others about heroic airmen also demonstrate how completely, by the mid-1950s, the Air Force and the notion of victory through air power had captured the popular imagination. See Franklin, *War Stars*.

34. Herken, *Counsels of War*, 97. The quotation from LeMay, reported by

Herken, originates with an unnamed Air Force intelligence officer present for the briefing.

35. British planes and antiaircraft guns successfully repulsed a major German air assault during the Battle of Britain in 1940 by shooting down only about 10 percent of the attacking planes.

36. Air Defense Policy Panel, report to Air Force Chief of Staff, August 14, 1947, cited in Schaffel, *The Emerging Shield*, 66.

37. Schaffel, *The Emerging Shield*, 67, citing *Preliminary Rand Report, subj.: Active Defense of the United States against Air Attack*, July 10, 1947, revised and reissued February 5, 1948.

38. Ibid., 75, citing Finletter Commission report dated January 1, 1948.

39. Ibid., 65–66.

40. See M. D. Fagen, ed., *A History of Engineering and Science in the Bell System* (Murray Hill, NJ: Bell Telephone Laboratories, 1978), chapter 7.

41. In the distorted mirror typical of superpower nuclear strategy, one reason for the Air Force belief in a Soviet kamikaze strategy may have been SAC's own plans, which involved something only a little better. SAC's medium-range bombers, even with aerial refueling and forward bases, could leave the USSR after a strike but could not return to the United States. SAC planned for pilots to ditch their aircraft in Afghanistan, Iran, Scandinavia, or northern Canada and attempt somehow to straggle home. B. Bruce-Briggs, *The Shield of Faith* (New York: Touchstone, 1988), 78.

42. Schaffel, *The Emerging Shield*, 77.

43. Ibid., 83 and passim.

44. The Korean War also provided an occasion for the application of Air Force strategic bombing doctrine. Less than three months after its entry into the war, the United States had already bombed every major city in North Korea. The use of nuclear weapons was seriously considered and publicly mentioned. But just as in World War II, the bombing failed to produce the automatic victory Mitchell had theorized. Nor did the nuclear threat.

45. Stephen Ambrose, *Rise to Globalism*, 4th ed. (New York: Penguin, 1985), 126.

46. Measured in constant 1981 dollars, spending on nuclear forces climbed from $9.6 billion in 1950 to $43.3 billion in 1953. Paul Bracken, *The Command and Control of Nuclear Forces* (New Haven: Yale University Press, 1984), 77.

47. Schaffel, *The Emerging Shield*, 116.

48. Cited in ibid., 158.

49. Valley, "How the SAGE Development Began," 205.

50. Redmond and Smith, *Project Whirlwind*, 154. In the end, Whirlwind cost nearly twice the panel's estimate.

51. Oddly, at the same time the panel "criticized the general conduct of computer research and development because it did 'not include sufficient emphasis on real-time computation'"—precisely Whirlwind's forte. Redmond and Smith, *Project Whirlwind*, 161.

52. Forrester's log-book entry for July 15, 1948, invites an even grander vision, invoking World War II icons: "Consideration should be given to establishing an independent set-up similar to that of the Radiation Laboratory, or perhaps even of the Manhattan District on a smaller scale." Jay W. Forrester, "Computation Book," November 27, 1946, through December 10, 1948. Magnetic Core Memory Records, 1932–1977, MC 140, Box 4, F 27 (Institute Archives and Special Collections, MIT Libraries, Cambridge, MA).

53. Killian, quoted in Karl L. Wildes and Nilo A. Lindgren, *A Century of Electrical Engineering and Computer Science at MIT, 1882–1982* (Cambridge, MA: MIT Press, 1985), 282–283.

54. Irving S. Reed, interviewed by Robina Mapstone, December 19, 1972. Smithsonian Computer Oral History Project, AC NMAH #196 (Archive Center, National Museum of American History, Washington, DC). According to Reed, the Valley committee granted Northrop $150 thousand to pursue radar and computer research for this project.

55. John V. Harrington, "Radar Data Transmission," *Annals of the History of Computing*, Vol. 5, No. 4 (1983), 370.

56. Schaffel, *The Emerging Shield*, 16.

57. Air Force Secretary Thomas K. Finletter called Lincoln Labs "the Manhattan Project of air defense." Cited in Samuel P. Huntington, *The Common Defense* (New York: Columbia University Press, 1961), 329.

58. F. W. Loomis, letter of transmittal, "Final Report of Project Charles," August 1, 1951, cited in Richard F. McMullen, *The Birth of SAGE, 1951–1958* (Air Defense Command Historical Study No. 33, 1965), 4.

59. McMullen, *The Birth of SAGE*, 8.

60. J. Robert Oppenheimer, "Atomic Weapons and Foreign Policy," *Foreign Affairs*, Vol. 31, No. 4 (1953), 531.

61. Zacharias, quoted in Herken, *Counsels of War*, 63.

62. Zacharias and an unnamed Project Charles scientist, quoted in ibid., 64.

63. Letter, Gen. Ennis C. Whitehead to Gen. Thomas D. White, December 14, 1953. Thomas D. White Papers, Box 1, Library of Congress.

64. Gen. Hoyt S. Vandenberg, cited in Kenneth Schaffel, "The US Air Force's Philosophy of Strategic Defense: A Historical Overview," in Stephen J. Cimbala, ed., *Strategic Air Defense* (Wilmington, DE: Scholarly Resources Inc., 1989), 15.

65. Vandenberg, "House Hearings on Air Force Appropriations for Fiscal Year 1954," March 6, 1953, 28–29, cited in McMullen, *The Birth of SAGE*, 13. Italics added.

66. Maj. Gen. Albert R. Shiely (ret.), in Tropp et al., "A Perspective on SAGE," 379.

67. Lt. Gen. S. R. Mickelsen, "ARAACOM to CONAD re: 'Integration of SAGE into CONAD Operations,'" February 15, 1956, cited in McMullen, *The Birth of SAGE*, 48.

68. Bruce-Briggs, *The Shield of Faith*, 100.

69. "2-Hour Warning Against Sneak Attack: Interview with Gen. Earle E. Partridge," *U.S News & World Report* (September 6, 1957), 77.

70. James Killian and A. G. Hill, "For a Continental Defense," *Atlantic Monthly* (November 1953), 37–38.

71. Bruce-Briggs, *The Shield of Faith*, 83.

72. Phil Gustafson, "Night Fighters Over New York," *Saturday Evening Post* (February 2, 1952), 33, 66.

73. Joseph Alsop and Stewart Alsop, "Matter of Fact: Air Defense Ignored in Political Shuffle," *Washington Post* (May 9, 1952); Alsop and Alsop, "We Accuse!," *Harper's* (October, 1954), 25–45.

74. Even core memory still sometimes finds a use, preferred over silicon RAM in high-radiation environments such as those encountered by satellites.

75. Forrester interview, April 19, 1977.

76. See Bracken, *The Command and Control of Nuclear Forces*.

77. Robert R. Everett, "WHIRLWIND," in N. Metropolis, J. Howlett, and Gian-Carlo Rota, eds., *A History of Computing in the Twentieth Century* (New York: Academic Press, 1980), 367–369 and passim, and Jay W. Forrester, "Reliability of Components," *Annals of the History of Computing*, Vol. 5, No. 4 (1983), 399–401.

78. This argument parallels Donald MacKenzie's discussion of notions of accuracy in ICBM development in *Inventing Accuracy: A Historical Sociology of Nuclear Missile Guidance* (Cambridge, MA: MIT Press, 1990).

79. "Modem" means "modulator/demodulator." A modem converts (or "modulates") digital information into analog form for telephone-line

transmission; the receiving modem "demodulates" the analog signal back into digital form. The system is necessary because telephone lines are optimized for voice (sound, an analog signal) transmission.

80. Thomas Parke Hughes's concept of technological system-building partly explains these relationships. Once the commitment to a nuclear strategy had been made, a whole host of technologies also had to be developed to support the nuclear forces: aircraft and missiles, obviously, but also command and early-warning systems (radar, communications networks, and high-flying reconnaissance planes). The commitment to nuclear forces rapidly ramified into extremely far-reaching commitments to aerospace, electronics, and communications technologies. From this perspective, computers resolved a key "reverse salient" in the system: slow and noisy human communications, information integration, and command decision-making. See Thomas Parke Hughes, "The Evolution of Large Technological Systems," in Wiebe Bijker, Thomas P. Hughes, and Trevor Pinch, eds., *The Social Construction of Technological Systems* (Cambridge, MA: MIT Press, 1987), 51–82.

81. Flamm, *Creating the Computer*, 87–89.

82. It was MIT that chose the SAGE computer supplier, and in the words of one SAGE engineer, "in those days, picking IBM was sort of a surprise, because Eckert-Mauchly and [Remington] Rand were the leaders commercially in high-speed computers of this type. . . . IBM had no name in computers." Israel interview, April 22, 1970. Smithsonian Computer Oral History Project, AC NMAH #196 (Archive Center, National Museum of American History, Washington, DC). Frank Verzuh confirms Israel's memory that Forrester and Everett initially looked to Remington Rand. (Rand merged with Sperry in 1955.) In 1952 Remington Rand owned Eckert and Mauchly's UNIVAC, and it had just acquired Engineering Research Associates, the company staffed by former Navy Communications Security Group members which produced the ATLAS (see chapter 4). Verzuh told Forrester he thought IBM would soon "gobble up the competitors again" (Verzuh interview, February 20, 1984).

83. Stan Augarten, *Bit by Bit: An Illustrated History of Computers* (New York: Ticknor & Fields, 1984), 208.

84. Robert P. Crago, in Tropp et al., "A Perspective on SAGE," 386.

85. Norman Taylor, personal communication to Karl L. Wildes and Nilo A. Lindgren, cited in Wildes and Lindgren, *A Century of Electrical Engineering and Computer Science at MIT*, 340.

86. Herbert D. Benington, "Production of Large Computer Programs," *Annals of the History of Computing*, Vol. 5, No. 4 (1983), 351.

87. Claude Baum, *The System Builders: The Story of SDC* (Santa Monica, CA: System Development Corporation, 1981).

88. Ibid., 32.

89. For extended descriptions of the SAGE system and how it operated, see Robert R. Everett, Charles A. Zraket, and Herbert D. Benington, "SAGE: A Data-Processing System for Air Defense," in *Proceedings of the Eastern Joint Computer Conference* (Washington, D.C.: Institute of Radio Engineers, 1957), reprinted in *Annals of the History of Computing*, Vol. 5, No. 4 (1983), 339–345; C. Robert Wieser, "The Cape Cod System," *Annals of the History of Computing*, Vol. 5, No. 4 (198), 362–369; John F. Jacobs, "SAGE Overview," *Annals of the History of Computing*, Vol. 5, No. 4 (1983), 323–329; and Schaffel, *The Emerging Shield*.

90. See chapter 1 on Operation Igloo White.

91. "Pushbutton Defense for Air War," *Life* 42:6 (1957), 62–67.

92. One of the most common genres of war story recounts a soldier's ingenious bypassing of formal structures to carry out orders more effectively.

93. Baum, *The System Builders*, 56 and passim. JOVIAL found wide application in other military command-control systems but never became a commercial software language. See Jean Sammet, *Programming Languages: History and Fundamentals* (Englewood Cliffs, NJ: Prentice-Hall, 1969).

94. Alan Borning, "Computer System Reliability and Nuclear War," *Communications of the ACM*, Vol. 30, No. 2 (1987), 112–131, presents a lengthy history of NORAD computer failures, some serious enough to lead to escalations in the alert status of nuclear forces. Problems of complexity and reliability in these systems became a cultural trope for nuclear fear, as reflected in films and novels from *Dr. Strangelove*, *Fail-Safe*, and *Colossus: The Forbin Project* in the 1960s to *War Games* and *The Terminator* in the 1980s, all of which involved some variation on the theme of computer-initiated nuclear holocaust.

95. Gen. Hoyt S. Vandenberg, "Suggested Remarks before the Joint Civilian Orientation Conference," March 26, 1953. Hoyt S. Vandenberg Papers, Box 91, Library of Congress.

96. IBM built an experimental transistorized machine, the AN/FSQ-32, for SAGE, but only one copy was made. Replacing the others would have been absurd: expensive, of course, but also strategically useless.

97. An enormous literature now exists on the interplay between computer technology and organizations. For entrées, see Charles Dunlop and Rob Kling, eds., *Computerization and Controversy* (New York: Academic Press, 1991); Kling and Walt Scacchi, "The Web of Computing: Computing Technology as Social Organization," *Advances in Computers*, Vol. 21 (1982), 3–85; Terry Winograd and Fernando Flores, *Understanding Computers and Cognition: A New Foundation for Design* (Norwood, NJ: Ablex, 1987); Lucy Suchman, *Plans and Situated Actions: The Problem of Human-Machine Communication* (New York: Cambridge University Press, 1987); and Shoshana Zuboff, *In the Age of the Smart Machine* (New York: Basic Books, 1988).

98. I borrow the phrase "unruly complexity" from Peter Taylor; see his "Building on the Metaphor of Construction in Science Studies," ms., Dept. of Science and Technology Studies, Cornell University, 1994.

99. Bracken, *The Command and Control of Nuclear Forces*, 12.

100. See Harry Collins, *Artificial Experts: Social Knowledge and Intelligent Machines* (Cambridge, MA: MIT Press, 1990); Hubert Dreyfus, *What Computers Can't Do: The Limits of Artificial Intelligence*, 2nd ed. (New York: Harper Colophon, 1979); Hubert Dreyfus and Stuart Dreyfus, *Mind Over Machine* (New York: Free Press, 1986); Joseph Weizenbaum, *Computer Power and Human Reason: From Judgment to Calculation* (San Francisco: W. H. Freeman, 1976); Winograd and Flores, *Understanding Computers and Cognition*.

101. See Suchman, *Plans and Situated Actions*.

102. Bracken, *The Command and Control of Nuclear Forces*, 55. In his chapter on Vietnam-era command systems, Martin Van Creveld (*Command in War* [Cambridge, MA: Harvard University Press, 1985]) concurs with Bracken about the provenance of computerized control.

103. "So many Air Defense Command centers were colocated with SAC bases that the Soviets could take out both with only 15 more bombs than it would take to eradicate SAC alone. Fewer than a hundred bombs not only would kill all of SAC that had not launched, but would decapitate the air defense apparatus as well" (Bruce-Briggs, *The Shield of Faith*, 129). As Bracken notes, this was part of a consistent policy. "The theory behind the 'soft' design for command and control was that the purpose of all of these systems was to get warning in order to launch a nuclear attack" (*The Command and Control of Nuclear Forces*, 188). A more mundane rationale was that SAGE commanders wanted access to the excellent officers' facilities at SAC bases.

104. Les Earnest, personal communication. See Bruce-Briggs, *The Shield of Faith*, 96, for an account of one such test.

105. I borrow the idea of ideological "work" from Vincent Mosco.

Chapter 4

1. I have stressed the military dimension of this discourse, but a similar analysis could be applied to economic policy in the postwar era. World-system theorists have emphasized the often very deliberate practice of policies designed to create, control, and manage a world free-market system centered around U.S. hegemony. See Thomas J. McCormick, *America's Half-Century: United States Foreign Policy in the Cold War* (Baltimore: Johns Hopkins University Press, 1989).

2. Herbert Simon, *Administrative Behavior* (New York: Macmillan, 1947), and *The New Science of Management Decision* (New York: Harper & Row, 1960).

3. See Norbert Wiener, *The Human Use of Human Beings* (Boston: Houghton Mifflin, 1950), and Karl Deutsch, *The Nerves of Government: Models of Communication and Control* (London: Free Press of Glencoe, 1963). An extended discussion of cybernetics and its origins is the subject of chapter 6.

4. On the wider uses of these theories see Steve Heims, *The Cybernetics Group* (Cambridge, MA: MIT Press, 1991), and Ida R. Hoos, *Systems Analysis in Public Policy* (Berkeley: University of California Press, 1972).

5. OR is largely responsible for the concept of "optimality" in logistics. Optimizing a system means finding the best possible *combination* of elements rather than attempting to maximize the performance of each one, focusing on improving the capacity of the "bottleneck" elements that drag down overall performance.

6. Fred Kaplan, *The Wizards of Armageddon* (New York: Simon & Schuster, 1983), 54. On operations research see James Phinney Baxter 3rd, *Scientists Against Time* (Boston: Little, Brown, 1948); P. M. S. Blackett, *Studies of War* (New York: Hill & Wang, 1962); and Alfred H. Hausrath, *Venture Simulation in War, Business, and Politics* (New York: McGraw-Hill, 1971).

7. At the time this book was being researched, those reports still remained classified.

8. Bruce L. R. Smith, *The RAND Corporation* (Cambridge, MA: Harvard University Press, 1966), 114.

9. The phrase is taken from Alain Enthoven and K. Wayne Smith, *How Much Is Enough? Shaping the Defense Program, 1961–69* (New York: Harper & Row, 1971).

10. Smith, *The RAND Corporation*, 9.

11. Rand R-184, "Fourth Annual Report" (Santa Monica: Rand Corporation, March 1950), 27. Cited in Kaplan, *The Wizards of Armageddon*, 91.

12. Rand Corporation, *The Rand Corporation: The First Fifteen Years* (Santa Monica, CA: Rand Corporation, 1963), 26–27. "In studies of policy analysis," this report continues, "it is not the theorems that are useful but rather the spirit of game theory and the way it focuses attention on conflict with a live dynamic, intelligent, and reacting opponent."

13. Roberta Wohlstetter, *Pearl Harbor: Warning and Decision* (Stanford, CA: Stanford University Press, 1962).

14. Kaplan, *The Wizards of Armageddon*, 90–110 and passim. The Wohlstetter study is A. J. Wohlstetter, F. S. Hoffman, R. J. Lutz, and H. S. Rowen, *Selection and Use of Strategic Air Bases* (Santa Monica: Rand R-266, 1954).

15. Kaplan, *The Wizards of Armageddon*, 89, 93.

16. In a satisfying twist on Thomas Hughes's military metaphor, the "reverse salients" here, aircraft and early warning systems, were not only technological but quite literally tactical. See Thomas Park Hughes, "The Evolution of Large Technological Systems," in Wiebe Bijker, Thomas P. Hughes, and Trevor Pinch, eds., *The Social Construction of Technological Systems* (Cambridge, MA: MIT Press, 1987), 51–82.

17. See, for example, Jonathan Schell, "Credibility and Limited War," in Jeffrey P. Kimball, ed., *To Reason Why: The Debate about the Causes of U.S. Involvement in the Vietnam War* (Philadelphia: Temple University Press, 1990), 121–140.

18. On limited war, counterforce, flexible response, and other Rand strategic theories of the late 1950s, see Herman Kahn, *On Thermonuclear War* (Princeton, NJ: Princeton University Press, 1960), and W. W. Kaufmann, ed., *Military Policy and National Security* (Princeton, NJ: Princeton University Press, 1956).

19. This phrase came into vogue in the early 1950s. A "weapon system" (or "weapons system") may include not only the weapon itself (such as a thermonuclear bomb), but the delivery vehicle (an intercontinental bomber), its logistical support (aerial refueling tanker plane, hangar), and its human elements (ground crew, pilots, etc.). The weapon-system concept allows direct cost comparisons based on the ability of various systems to perform a mission. In 1949 the Office of the Secretary of Defense organized a Weapons Systems Evaluation Group (WSEG) to apply OR techniques to evaluating new technologies; its first chair was Philip Morse of MIT, peripherally connected with the Whirlwind project and a major developer of wartime OR. Philip M. Morse, interviewed by Richard R. Mertz, December 16, 1970. Smithsonian Computer Oral History Project, AC NMAH #196. Archive Center, National Museum of American History, Washington, DC.

20. See Kaplan, *The Wizards of Armageddon*, 87.

21. See Carol Cohn, "Sex and Death in the Rational World of Defense Intellectuals," *Signs*, Vol. 12, No. 4 (1987), 687–718; Thomas B. Allen, *War Games* (New York: McGraw-Hill, 1987); Lynn Eden, "Sterilizing Destruction: The Imaginary Battlefield in Contemporary U.S. Nuclear Targeting," paper delivered at American Historical Association annual meeting, Chicago, December 28, 1991; Lynn Eden, *Constructing Destruction: An Historical Sociology of U.S. Nuclear War Planning* (book manuscript, Stanford University).

22. Jonathan Schell, *The Fate of the Earth* (New York: Avon Books, 1982), 140.

23. The 19th-century antirationalist Prussian strategist Carl von Clausewitz was responsible for the famous dictum "war is politics by other means." Unlike many of the Rand theorists, he had enormous respect for what he called "the fog of war"—the combination of powerful emotions, unreliable and often contradictory information, and a chaotic environment that renders rational

analysis impossible in battle—and for the inherent fragility of preprogrammed plans. See Carl von Clausewitz, *On War*, trans. Anatol Rapoport (New York: Penguin Books, 1982).

24. On the erotics of rationalism in general, and especially as applied to war, see Loren Baritz, *Backfire* (New York: Ballantine, 1985), chapter 1; Cohn, "Sex and Death"; Brian Easlea, *Fathering the Unthinkable: Masculinity, Scientists, and the Nuclear Arms Race* (London: Pluto Press, 1983); Paul N. Edwards, "The Army and the Microworld: Computers and the Militarized Politics of Gender," *Signs*, Vol. 16, No. 1 (1990), 102–127; Donna J. Haraway, *Primate Visions* (London: Routledge Kegan Paul, 1989); and Zoë Sofia, "Exterminating Fetuses," *Diacritics*, Vol. 14, No. 2 (1984), 47–59.

25. Joseph Weizenbaum, *Computer Power and Human Reason* (San Francisco: W. H. Freeman, 1976), 34.

26. U.S. Air Force Project RAND, "Air Force Advisory Group Agenda Material—Report of Operations," January 4–6, 1961. Box 1, Thomas D. White Papers, Library of Congress.

27. Fred J. Gruenberger, "The History of the JOHNNIAC," *Annals of the History of Computing*, Vol. 1, No. 1 (1979), 50. On the IBM 604 and the Card-Programmed Calculator, see Michael R. Williams, *A History of Computing Technology* (Englewood Cliffs, NJ: Prentice-Hall, 1985), 255–258 and passim.

28. By a general but not universal convention (which I follow in this book), names of computers, computer languages, and technical products related to computers are capitalized only when they are acronyms (e.g., ENIAC), not when they are proper names (e.g., Johnniac).

29. George W. Brown, interviewed by Richard R. Mertz, March 15, 1973. Smithsonian Computer Oral History Project, AC NMAH #196. (Archive Center, National Museum of American History, Washington, D.C.).

30. Paul Armer, interviewed by Robina Mapstone, April 17, 1973. Smithsonian Computer Oral History Project, AC NMAH #196. (Archive Center, National Museum of American History, Washington, D.C.).

31. Keith Uncapher, interviewed by Robina Mapstone, February 20, 1973. Smithsonian Computer Oral History Project, AC NMAH #196. (Archive Center, National Museum of American History, Washington, D.C.).

32. Robert L. Chapman, "Simulation in Rand's Systems Research Laboratory," in D. G. Malcolm, ed., *Report of System Simulation Symposium, 1957* (Baltimore: Waverly Press, 1958), 72.

33. Ibid.

34. Pamela McCorduck, *Machines Who Think* (New York: W. H. Freeman, 1979), 120 and passim.

35. Ibid., 129, 132 and passim; Herbert Simon, *Models of My Life* (New York: Basic Books, 1991).

36. Allen Newell, Herbert Simon, and J. C. Shaw, "The Logic Theory Machine," *IRE Transactions on Information Theory*, Vol. IT-2, No. 3 (1956). Simon's widely quoted statement to his class is the recollection of one of his students, Edward Feigenbaum, cited in McCorduck, *Machines Who Think*, 116.

37. T. C. Rowan, "Simulation in Air Force System Training," in Malcolm, *Report of System Simulation Symposium, 1957*, 84.

38. Claude Baum, *The System Builders* (Santa Monica, CA: System Development Corporation, 1981), 46.

39. Rowan, "Simulation in System Training," 87.

40. Simon noted that "the phrase 'information processing' was already part of [Allen Newell's] vocabulary by 1952 when [Simon] arrived" at Rand. Herbert Simon, quoted in McCorduck, *Machines Who Think*, 126.

41. See Allen, *War Games*. In 1979, a relatively serious nuclear false alert was caused by the accidental mounting of a training tape on an active warning system, sharply marking the equivalence of simulation and war-fighting equipment.

42. Ironically, Rand's limited-war doctrines gained a toehold as a result. To cut the defense budget, Eisenhower backed off from the goal of overwhelming nuclear superiority; limited, tactical nuclear war then replaced the empty threat of total central war as the linchpin of containment.

43. Eisenhower had reacted to the supposed "gap" with seemingly perplexing calm, because U-2 spy plane overflights of the USSR had revealed only *four* operational ICBMs by 1960. He could not divulge this knowledge, though, without revealing its source in the illegal and top-secret U-2 missions. See Walter A. McDougall, . . . *the Heavens and the Earth: A Political History of the Space Age* (New York: Basic Books, 1985).

44. Kaplan, *The Wizards of Armageddon*, 250.

45. B. Bruce-Briggs, *The Shield of Faith* (New York: Touchstone, 1988), 152.

46. Kaplan, *The Wizards of Armageddon*, 251. On McNamara's character and work, see also David Halberstam, *The Best and the Brightest* (New York: Random House, 1972), 263–323.

47. Kaplan, *The Wizards of Armageddon*, 253.

48. Alain C. Enthoven, "Cost-Effectiveness Analysis of Army Divisions: Remarks at the Fourth Annual U.S. Army Operations Research Symposium, March 1962" (revised April 1965), in Samuel A. Tucker, ed., *A Modern Design*

for *Defense Decision: A McNamara-Hitch-Enthoven Anthology* (Washington, D.C.: Industrial College of the Armed Forces, 1966), 185.

49. See Enthoven and Smith, *How Much Is Enough?*, 104–106.

50. McNamara, quoted in David Halberstam, "The Programming of Robert McNamara," *Harper's* (February 1971), 62.

51. Enthoven, "Cost-Effectiveness Analysis," 185.

52. Robert S. McNamara, "Managing the Department of Defense," *Civil Service Journal*, Vol. 4, No. 4 (1964), 3–4.

53. Ibid., 3.

54. McNamara, quoted in W. W. Kaufmann, *The McNamara Strategy* (New York: Harper & Row, 1964), 181.

55. McDougall, . . . *the Heavens and the Earth*, 332.

56. Morris Janowitz, *The Professional Soldier* (Glencoe, IL: Free Press, 1960).

57. Lieut. Col. David M. Ramsey Jr., U.S. Army, "Management or Command?," *Military Review*, Vol. 41, No. 10 (1961), 31.

58. Martin Van Creveld, *Command in War* (Cambridge, MA: Harvard University Press, 1985), 234–235.

59. On the evolution of management theories and their relation to social control and computing, see especially Harry Braverman, *Labor and Monopoly Capital* (New York: Monthly Review Press, 1974); David Noble, *Forces of Production* (New York: Oxford University Press, 1984); and Shoshana Zuboff, *In the Age of the Smart Machine* (New York: Basic Books, 1988).

60. "The term 'command system' is itself in one sense a misnomer. When the dichotomy between the 'directing' brain and the 'directed' body is set up, the fact that there is not a single member in any armed force in any time or place who has not performed some of the functions of command for at least part of the time is deliberately ignored" (Van Creveld, *Command in War*, 263). The phrase "command and control" actually sprang from a misinterpretation of the Air Force 465L computerized control system. Instead of understanding its name as "the Strategic Air Command's Control System," some officers apparently read it as "the Strategic Air Command-Control System," hence the phrase "command and control" (Les Earnest, personal communication).

61. See Paul Bracken, *The Command and Control of Nuclear Forces* (New Haven: Yale University Press, 1984); Ramsey, "Management or Command?"; and Van Creveld, *Command in War*. Despite the confusion it embodies, I will continue to use the phrase in its ordinary military sense.

62. Enthoven, quoted in Gregory Palmer, *The McNamara Strategy and the Vietnam War* (Westport, CT: Greenwood Press, 1978), 69.

63. Michael Getler, "Command and Control Overhaul Due," *missiles and rockets* (December 3, 1962), 16–18. Dalkey is quoted on pp. 17–18.

64. "Will 'Computers' Run Wars of the Future?," *U.S. News & World Report* (April 23, 1962), 44.

65. Baritz, *Backfire*, 240–241.

66. Gen. Thomas S. Power, SAC, cited in John Lewis Gaddis, *Strategies of Containment* (New York: Oxford University Press, 1981), 251n.

67. Alain C. Enthoven, "Systems Analysis and Decision Making," *Military Review*, Vol. 43, No. 1 (1963), 7–8.

68. NASA's blue-sky projects pushed miniaturization and computerized control very hard. Central computers at NASA ground control performed most of the missions' computational work, but the 1965 Gemini 3 carried its own computer controller. The first Apollo to orbit the moon relied entirely on its own guidance and control computer when out of reach of Earth communication on the far side of the moon. On NASA's contribution to computing see Donald B. Brick, James Stark Draper, and H. J. Caulfield, "Computers in the Military and Space Sciences," *IEEE Computer* (October 1984), 250–262.

69. James E. Webb, cited in McDougall, . . . *the Heavens and the Earth*, 281. The last sentence quoted is McDougall's, again citing Webb.

70. Memo, McNamara to Johnson, "South Vietnam," March 13, 1964, cited in Kaplan, *The Wizards of Armageddon*, 332.

71. James Gibson, *The Perfect War* (New York: Vintage Books, 1986).

72. Van Creveld, 239.

73. Cf. Gary Chapman, "The New Generation of High-Technology Weapons," in David Bellin and Gary Chapman, eds., *Computers in Battle* (New York: Harcourt, Brace, Jovanovich, 1987), 61–100.

74. Gibson, *The Perfect War*, 26 et passim.

75. Van Creveld, 252–253.

76. Gibson, *The Perfect War*, 124.

77. Van Creveld, 241.

78. McGeorge Bundy, in a memo to President Lyndon Johnson, May 22, 1964, cited in Kaplan, *The Wizards of Armageddon*, 333.

79. Cf. Kaplan, *The Wizards of Armageddon*, 329.

80. For example, NVA forces briefly seized and destroyed part of the American embassy compound in Saigon.

81. Gibson, *The Perfect War*, 14, 27.

82. "The big drive now is for standard, all-purpose computers that give top commanders maximum 'mission visibility' even though far from the scene. In the process, a new breed of military man is being created and the short-term command prerogatives of the field leaders may be considerably reduced" ("Baptism of fire for computers," *Business Week* [April 22, 1967], 136).

83. Kaplan, *The Wizards of Armageddon*, 336.

84. Robert S. McNamara, *The Essence of Security* (New York: Harper & Row, 1968), 142, 148.

85. Gen. William Westmoreland, U.S. Army Chief of Staff, "Address to the Association of the U.S. Army," October 14, 1969. Reprinted in Paul Dickson, *The Electronic Battlefield* (Bloomington, IN: Indiana University Press, 1976), 215–223.

86. Zuboff, *In the Age of the Smart Machine*, 349. On Jeremy Bentham's original Panopticon and the concept of panoptic power, see Michel Foucault, *Discipline and Punish*, trans. Alan Sheridan (New York: Vintage Books, 1977).

87. Van Creveld, *Command in War*, 264.

Chapter 5

1. Note that I use the subjective/objective opposition in the Hegelian sense of individual wants and desires vs. collective needs and goods, not the positivistic sense of imaginary vs. real or caprice vs. certainty.

2. Robert Meister, *Political Identity: Thinking Through Marx* (Cambridge, MA: Basil Blackwell, 1990), 43. Although I have adapted it heavily to my own ends, my discussion of political theory here largely reflects Meister's important work. Meister's book captures only a small part of a very large body of thought; I learned most of what I know about his ideas in classes, seminars, and private conversations at the University of California, Santa Cruz. Any faults in the concepts I discuss here are, of course, my own.

3. Ibid., 33.

4. Ibid., 42.

5. G. W. F. Hegel, *Hegel's Philosophy of Right*, trans. T. M. Knox (New York: Oxford University Press, 1952), 294 (addition to Paragraph 316).

6. In addition to the various works cited in chapter 1, see James Clifford and George E. Marcus, eds., *Writing Culture* (Berkeley: University of California Press, 1986); Terry Eagleton, *Ideology* (New York: Verso, 1991); Donna J. Haraway, *Primate Visions* (London: Routledge Kegan Paul, 1989); Julian Henriques et al., *Changing the Subject* (New York: Methuen & Co., 1984); Edward

Said, *Orientalism* (New York: Pantheon, 1978); Zoë Sofoulis, *Through the Lumen: Frankenstein and the Optics of Re-Origination* (unpublished Ph.D. thesis, University of California, Santa Cruz, 1988); and Hayden White, *The Content of the Form* (Baltimore: Johns Hopkins University Press, 1987).

7. Barry Barnes and David Bloor, "Relativism, Rationalism and the Sociology of Knowledge," in M. Hollis and S. Lukes, eds., *Rationality and Relativism* (Cambridge: MIT Press, 1982), 23, italics in original.

8. Ibid., 24.

9. Among the most sophisticated treatments of this theme in science studies may be found in Ian Hacking, *Representing and Intervening* (Cambridge: Cambridge University Press, 1983), and Peter Taylor, "Building on the Metaphor of Construction," manuscript, Dept. of Science and Technology Studies, Cornell University, Ithaca, NY (1994).

10. Haraway, *Primate Visions*, 8.

11. Mark Johnson, *The Body in the Mind* (Chicago: University of Chicago Press, 1987); George Lakoff, *Women, Fire, and Dangerous Things* (Chicago: University of Chicago Press, 1987); George Lakoff and Mark Johnson, *Metaphors We Live By* (Chicago: University of Chicago Press, 1980). For other, similar perspectives, as well as critique, see R. P. Honeck and Robert R. Hoffman, eds., *Cognition and Figurative Language* (Hillsdale, NJ: Lawrence J. Erlbaum, 1980), 393–423, and A. Ortony, ed., *Metaphor and Thought* (New York: Cambridge University Press, 1979).

12. Lakoff and Johnson, *Metaphors We Live By*, 15.

13. Ibid.

14. Ibid., 18.

15. Ibid., 13.

16. Ibid., 59.

17. Ibid., 90.

18. I owe this useful phrase to Hayden White.

19. Sherry Turkle, *The Second Self* (New York: Simon & Schuster, 1984), passim.

20. Richard Rorty, *Philosophy and the Mirror of Nature* (Princeton: Princeton University Press, 1979), chapter 1.

21. Turkle, *The Second Self*, 23–24. For a similar, rather frightening view of the power of this effect, see Frederick Crews, "The Myth of Repressed Memory," *New York Review of Books*, Vol. XLI, Nos. 19 & 20 (1994), 54–59, 51–57.

22. See chapter 7. Compare David E. Leary, ed., *Metaphors in the History of Psychology* (New York: Cambridge University Press, 1990), esp. Karl H. Pribram, "From metaphors to models: the use of analogy in neuropsychology," 79–103, and Robert R. Hoffman, Edward L. Cochran, and James M. Nead, "Cognitive metaphors in experimental psychology," 173–229.

23. I borrow these terms from Hayden White to describe how human thought and consciousness are often structured in the form of tropes, or figures of speech. See Hayden White, *Metahistory* (Baltimore: Johns Hopkins University Press, 1973), especially the Introduction, and *Tropics of Discourse* (Baltimore: Johns Hopkins University Press, 1978).

24. See Laurence D. Smith, "Metaphors of knowledge and behavior in the behaviorist tradition," in Leary, *Metaphors in the History of Psychology*, 239–266.

25. Sigmund Freud, *Civilization and its Discontents*, trans. James Strachey (New York: W. W. Norton, 1961).

26. Turkle, *The Second Self*, 286.

27. Quoted in ibid., 288.

28. In addition to Turkle, see (for example) J. David Bolter, *Turing's Man* (Chapel Hill, NC: University of North Carolina Press, 1984); Steven Levy, *Hackers* (New York: Anchor Press, 1984); and Joseph Weizenbaum, *Computer Power and Human Reason* (San Francisco: W. H. Freeman, 1976).

29. Turkle, *The Second Self*, 104–105.

30. On the professional practices of science, see, e.g., Bruno Latour and Steve Woolgar, *Laboratory Life* (London: Sage, 1979); Bruno Latour, *Science In Action* (Cambridge, MA: Harvard University Press, 1987); Evelyn Fox Keller, *Reflections on Gender and Science* (New Haven: Yale University Press, 1985); and Violet B. Haas and Carolyn Perrucci, eds., *Women in Scientific and Engineering Professions* (Ann Arbor: University of Michigan, 1984).

31. Programming, of course, is not the only way to interact with a computer; user applications provide simpler approaches. But their basic structure is similar. For the sake of simplicity I discuss only programming here.

32. See, e.g., Geoffrey K. Pullum, "Natural Language Interfaces and Strategic Computing," *AI & Society*, Vol. 1, No. 1 (1987), 47–58.

33. See Hubert Dreyfus, *What Computers Can't Do* (New York: Harper, 1979); Terry Winograd, *Language as a Cognitive Process* (Reading, MA: Addison-Wesley, 1983); and Terry Winograd and Fernando Flores, *Understanding Computers and Cognition* (Norwood, NJ: Ablex, 1987).

34. Jonathan Jacky, "Software Engineers and Hackers: Programming and Military Computing," in Paul N. Edwards and Richard Gordon, eds., *Strategic*

Computing: Defense Research and High Technology (unpublished ms., University of California, Santa Cruz, 1986).

35. See Sherry Turkle and Seymour Papert, "Epistemological Pluralism: Styles and Voices within the Computer Culture," *Signs*, Vol. 16, No. 1 (1990), 128-157, and Terry Winograd, "Thinking Machines: Can There Be? Are We?," in James J. Sheehan and Morton Sonsa, eds., *The Boundaries of Humanity* (Berkeley, CA: University of California Press, 1991).

36. This is one reason programming computers to play chess became a favorite puzzle of computer scientists in the early days of AI. Rule-based games remain a centerpiece of the culture of computer science—as well as one of its major products—from chess to video games to Dungeons and Dragons. Douglas Hofstadter's Pulitzer prize-winning *Gödel, Escher, Bach* (New York: Vintage, 1979) reflected the widespread fascination of computer professionals with art, literature, and music based on recursive and self-referential operations—art as a kind of mathematical game. *Gödel, Escher, Bach* remains to this day a kind of Bible for many young hackers.

37. For various articulations of the gender coding of modes of rationality, see Turkle, *The Second Self*; Brian Easlea, *Fathering the Unthinkable: Masculinity, Scientists, and the Nuclear Arms Race* (London: Pluto Press, 1983); Sally Hacker, "The Culture of Engineering: Woman, Workplace, and Machine," *Women's Studies International Quarterly*, Vol. 4, No. 3 (1981), 341-353; Haraway, *Primate Visions*; Sandra Harding, *The Science Question in Feminism* (Ithaca, NY: Cornell University, 1986).

38. On men's and women's moralities, see Carol Gilligan, *In a Different Voice* (Cambridge: Harvard University Press, 1982).

39. Terry Winograd, "Computers and Rationality: The Myths and Realities," in Richard Gordon, ed., *Microelectronics in Transition* (typescript, University of California, Santa Cruz, 1985), ms. p. 3.

40. See Nancy Hartsock, *Money, Sex, and Power* (New York: Longman, 1983); Keller, *Reflections on Gender and Science*; and Harding, *The Science Question in Feminism*.

41. Turkle quotes one programmer, the lone woman on a large team, whose experience of the machine tends toward "soft mastery": "I know that the guys I work with think I am crazy, but we will be working on a big program and I'll have a dream about what the program feels like inside and somehow the dream will help me through. When I work on the system I know that to everybody else it looks like I'm doing what everyone else is doing, but I'm doing that part with only a small part of my mind. The rest of me is imagining what the components feel like. It's like doing my pottery.... Keep this anonymous. I know this sounds stupid" (116–117). Turkle compares this to the kind of "fusion" experience described by the geneticist Barbara McClintock and offered by Evelyn Fox Keller as an alternative to the "masculine"

mode of subject/object separation in science [Evelyn Fox Keller, *A Feeling for the Organism* (San Francisco: W. H. Freeman, 1983)]. The interviewee's embarrassment about her own style reflects the rationalistic tendencies of computer culture.

42. Turkle, *The Second Self*, 14 and passim.

43. Winograd, *Understanding Natural Language*.

44. See Roger Schank, *The Cognitive Computer* (Reading, MA: Addison-Wesley, 1984); Weizenbaum, *Computer Power and Human Reason*; Dreyfus, *What Computers Can't Do*.

45. See Turkle, *The Second Self*, especially the chapter on "Hackers: Loving the Machine for Itself."

46. Cf. Turkle, *The Second Self*.

47. Levy, *Hackers*; Weizenbaum, *Computer Power and Human Reason*.

48. For a nuanced understanding of the place of isolation in male psychological identity, see the collective work of the Re-evaluation Counseling Communities, in particular the journal *Men*, ed. Chuck Esser (Seattle: Rational Island Publishers).

49. See the discussions of *The Terminator* and *Terminator 2* in chapters 1 and 10 of this volume. Also see Paul N. Edwards, "The Army and the Microworld: Computers and the Militarized Politics of Gender," *Signs*, Vol. 16, No. 1 (1990), 102–127, and Levy, *Hackers*. For the opening of alternatives to a masculine cyborg subject in the 1980s, see Donna J. Haraway, "A Manifesto for Cyborgs: Science, Technology, and Socialist Feminism in the 1980s," *Socialist Review*, Vol. 15, No. 2 (1985), 65–107, and Haraway, "The Promises of Monsters," in Lawrence Grossberg, Cary Nelson, and Paula A. Treichler, eds., *Cultural Studies* (New York: Routledge and Kegan Paul, 1992), 295–337.

Chapter 6

1. Warren McCulloch, "Why the Mind Is in the Head," in Lloyd Jeffress, ed., *Cerebral Mechanisms in Behavior* (New York: Wiley-Interscience, 1951), 42.

2. For this phrase and elements of my approach in this chapter, I am indebted to Donna J. Haraway, "The High Cost of Information in Post-World War II Evolutionary Biology," *The Philosophical Forum*, Vol. XIII, Nos. 2–3 (1982), 244–278.

3. Sigmund Freud, *Civilization and its Discontents*, trans. James Strachey (New York: Norton, 1961), 50.

4. Stephen Jay Gould, *The Mismeasure of Man* (New York: Norton, 1981), 223–232 and passim.

5. Cf. especially Michel Foucault, *The Birth of the Clinic*, trans. A.M. Sheridan Smith (New York: Vintage Books, 1975); *Madness and Civilization*, trans. Richard Howard (New York: Vintage Books, 1973); and *Discipline and Punish*, trans. Alan Sheridan (New York: Vintage Books, 1977).

6. Michel Foucault, "Truth and Power," in *Power/Knowledge: Selected Interviews and Other Writings 1972–1977*, trans. Colin Gordon (New York: Pantheon, 1980), 133.

7. Albert R. Gilgen, *American Psychology Since World War II* (Westport, CT: Greenwood Press, 1982), 39; Steve J. Heims, *The Cybernetics Group* (Cambridge, MA: MIT Press, 1991), 3.

8. James H. Capshew, *Psychology on the March: American Psychologists and World War II* (unpublished Ph.D. dissertation, University of Pennsylvania, 1986), 3.

9. Ibid., 289–290.

10. Seymour Sarason, *Psychology Misdirected* (Glencoe: Free Press, 1981), 1–2. Cited in Heims, *Cybernetics Group*, 3.

11. Cf. Haraway, "High Cost," and Donna Haraway, "Signs of Dominance: From a Physiology to a Cybernetics of Primate Society," in *Studies in History of Biology* (Baltimore: Johns Hopkins, 1983), Vol. 6, 131 and passim.

12. See the slightly different list in Howard Gardner, *The Mind's New Science* (New York: Basic Books, 1985), 38–43, as well as Allen Newell, *Unified Theories of Cognition* (Cambridge, MA: Harvard University Press, 1990).

13. On this paradigm shift see Walter B. Weimer and David S. Palermo, "Paradigms and Normal Science in Psychology," *Science Studies*, Vol. 3 (1973), 211–244; Erwin M. Segal and Roy Lachman, "Complex Behavior or Higher Mental Process: Is There a Paradigm Shift?" *American Psychologist*, Vol. 27 (1972), 46–55; David Palermo, "Is a Scientific Revolution Taking Place in Psychology?" *Science Studies*, Vol. 1 (1971), 135–155; and D. O. Hebb, "The American Revolution," *American Psychologist*, Vol. 15 (1960), 735–745.

14. B. F. Skinner, *Beyond Freedom and Dignity* (New York: Bantam, 1971) contains the most explicit account of this vision.

15. See the list of machine models in Edwin G. Boring, "Mind and Mechanism," *American Journal of Psychology*, Vol. LIX, No. 2 (1946), 184.

16. Cf. Edward C. Tolman, "Cognitive Maps in Rats and Men," *Psychological Review*, Vol. 55, No. 4 (1948), 189–208.

17. Norbert Wiener, *Cybernetics: Control and Communication in the Animal and the Machine* (Cambridge, MA: MIT Press, 1948); Claude Shannon, "The Mathematical Theory of Communication," *Bell System Technical Journal*, Vol. 27 (1948), 379–423, 623–656, reprinted in Shannon and Warren Weaver, *The*

Mathematical Theory of Communication (Urbana: University of Illinois Press, 1949).

18. The Hixon Symposium, September 20–25, 1948. Proceedings in Jeffress, *Cerebral Mechanisms in Behavior.*

19. Wiener, *Cybernetics.*

20. David Lipset, *Gregory Bateson: The Legacy of a Scientist* (New York: Prentice-Hall, 1980), 179.

21. Heims, *Cybernetics Group*, 14–15. Heims, the only historian to have studied the Macy Conferences extensively, communicates well the flavor of these extraordinary meetings and the sparks between the extraordinary scientists involved. Heims's work is the only comprehensive source on Macy meetings prior to the sixth conference, when transcripts began to be kept. Since Wiener did not coin the word "cybernetics" until 1947, I will refer to the group at this juncture as the "proto-cyberneticians" to avoid anachronism.

22. To illustrate using one of Wiener's examples, I reach for a pencil, using my eyes to guide my hand by continuously correcting any deviations from its proper course.

23. Behaviorist psychologists, as we have seen, excluded internal events from consideration on methodological and theoretical grounds. They abhorred talk of goals and purposes as causes occurring later in time than their supposed effects. See Brian D. Mackenzie, *Behaviorism and the Limits of Scientific Method* (London: Routledge and Kegan Paul, 1977). Mackenzie argues that S-R psychology's self-imposed methodological restrictions led to its eventual demise because its practitioners limited the domain of experimentation to a narrow segment of the scientifically interesting possibilities.

24. Kenneth J. W. Craik, "Theory of the Human Operator in Control Systems: I. The Operator as an Engineering System," written in 1945 and posthumously published in the *British Journal of Psychology*, Vol. 38 (1947), 56–61.

25. See the discussion, below, of John Stroud's presentation at the Macy Conferences. On origin stories in science, see Donna Haraway's introduction to her book *Primate Visions* (London: Routledge Kegan Paul, 1989).

26. The subtitle of Wiener's *Cybernetics.*

27. Arturo Rosenblueth, Norbert Wiener, and Julian Bigelow, "Behavior, Purpose and Teleology," *Philosophy of Science*, Vol. 10 (1943), 18–24.

28. The formula and definitions here follow William K. Estes, "Toward a Statistical Theory of Learning," *Psychological Review*, Vol. 57 (1950), 94–107, but could have been taken from almost any behaviorist publication.

29. Cf. MacKenzie, *Behaviorism.*

30. Rosenblueth, Wiener, and Bigelow, "Behavior, Purpose and Teleology," 18.

31. "Input" and "output," originally electrical engineering terms for currents entering and exiting circuits, had been adopted analogically by communications engineers to describe the entry and exit versions of *messages* traveling through a channel, already making apparent the similarities between energy and information that would emerge so strikingly from cybernetics.

32. Rosenblueth, Wiener, and Bigelow, "Behavior, Purpose and Teleology," 19.

33. Cybernetics purposely cast itself as a metatheory, an explanation of how everything is connected to everything else. To do so it made extensive use of a "literary device" Geoffrey Bowker calls the "Serres effect," namely the heterogeneous list. "Cyberneticians fill their texts with unexpected conjunctions (rust, an epidemic, and diffusion in a semi-conductor . . .). Thus a random page of Ashby's *Introduction [to Cybernetics]* discusses self-locking in oysters, irreversibly insoluble compounds, the dynamics of the cerebral cortex and absenteeism from unpleasant industries" (Geof Bowker, "The Age of Cybernetics, or How Cybernetics Aged," unpublished ms. [1991], University of Keele, 9). In its attempt to encompass heterogeneity, cybernetics' rhetoric resembles both my own concept of discourse and some current theories in science and technology studies (see chapter 1).

34. Rosenblueth, Wiener, and Bigelow, "Behavior, Purpose and Teleology," 23.

35. I owe this point to Donna Haraway.

36. Kenneth J. W. Craik, "Theory of the Human Operator in Control Systems, II: Man as an Example in a Control System," written in 1945 and posthumously published in *British Journal of Psychology*, Vol. 39 (1947), 142.

37. Boring, "Mind and Mechanism," 173–192.

38. Warren S. McCulloch and Walter Pitts, "A Logical Calculus of the Ideas Immanent in Nervous Activity," *Bulletin of Mathematical Biophysics*, Vol. 5 (1943), 115–133.

39. "In a discussion after a lecture of von Neumann, McCulloch mentioned that it was Turing's paper that had inspired their ideas" (Andrew Hodges, *Alan Turing: The Enigma* [New York: Simon & Schuster, 1983], 252n).

40. Heims, *Cybernetics Group*, 20.

41. Pamela McCorduck, *Machines Who Think* (New York: W. H. Freeman, 1979), 66, from interviews.

42. Wiener, in a letter to Rosenblueth, cited in Steve Heims, *John von Neumann and Norbert Wiener* (Cambridge, MA: MIT Press, 1980), 185.

43. Heims, *Von Neumann and Wiener*, 203–204.

44. Heims, *Cybernetics Group*, 19–20.

45. The phrase "cybernetics group" is taken from Heims's book.

46. McCulloch and Pitts, "Logical Calculus."

47. See Heims, *Von Neumann and Wiener*, 209 and passim, and John von Neumann, "The General and Logical Theory of Automata," in Jeffress, *Cerebral Mechanisms in Behavior*, 1–41.

48. Proceedings of the last five Macy conferences were published as *Transactions of the Conference on Cybernetics*, 5 vols., ed. Heinz von Foerster (New York: Josiah Macy Jr. Foundation, 1950–1955). This quotation is from *Transactions of the Seventh Conference*, 11.

49. Alan Turing, "Computing Machinery and Intelligence," *Mind*, Vol. 59 (1950), 442.

50. Von Foerster, *Sixth Conference*, 112–145.

51. Steve Heims, "Encounter of Behavioral Sciences with New Machine-Organism Analogies in the 1940s," *Journal of the History of the Behavioral Sciences*, Vol. 11 (1975), 372. In this article Heims gives an account of the unpublished fourth Macy conference, based on interviews with some participants.

52. Ibid., 372.

53. Von Foerster, *Seventh Conference*, 44.

54. Jeffress, *Cerebral Mechanisms in Behavior*, 97.

55. Heims, *Cybernetics Group*, 257.

56. Von Foerster, *Sixth Conference*, 27–28. This refers to a question dating to William James, namely how long what is perceived as the "present moment" endures.

57. Ibid.

58. Ibid., 32.

59. Ibid., 33.

60. Heims, *Cybernetics Group*, 261. Stroud's sole article is John Stroud, "The Fine Structure of Psychological Time," in H. Quastler, ed., *Information Theory and Psychology* (Glencoe, IL: Free Press, 1955), 174–207. Note that the guided missile remained a central metaphor, fifteen years after the Macy Conferences.

61. For a textbook summary of the field, oriented toward engineering applications, see E. C. Poulton, *Tracking Skill and Manual Control* (New York: Academic Press, 1974).

62. Edwin G. Boring and M. Van de Water, eds., *Psychology for the Fighting Man* (Washington, DC: Infantry Journal Press, 1943), 24.

63. R. L. Gregory, *Eye and Brain* (New York: McGraw-Hill, 1966), 91–92.

64. M. D. Fagen, ed., *A History of Engineering and Science in the Bell System* (Murray Hill, NJ: Bell Telephone Laboratories, 1978), 165.

65. Ibid., 317.

66. Shannon and Weaver, *The Mathematical Theory of Communication*, 26.

67. Ibid., 26.

68. David Palermo, *Psychology of Language* (Glenview, IL: Scott, Foresman, 1978), 20.

69. Shannon and Weaver, *The Mathematical Theory of Communication*, 95, italics added. Note the military analogy.

70. Claude E. Shannon, "Programming a Computer for Playing Chess," *Philosophical Magazine*, Vol. 41 (1950), 256–275; Claude E. Shannon, "Computers and Automata," *Proceedings of the IRE*, Vol. 41 (1953), 1234–1241.

71. Shannon and Weaver, *The Mathematical Theory of Communication*, 96.

72. Ibid., 98.

73. Von Foerster, *Seventh Conference*, 150.

74. Ibid., 154.

75. J. C. R. Licklider, presentation on "The manner in which and extent to which speech can be distorted and remain intelligible," in Von Foerster, *Seventh Conference*, 99.

76. Ibid.

77. Charles W. Bray, ed., *Human Factors in Military Efficiency: Summary Technical Report of the Applied Psychology Panel, NDRC*, Vol. 1 (Washington, DC: U.S. Government Printing Office, 1946), xi.

78. Cf. Foucault, *Discipline and Punish*.

79. Jay W. Forrester, "Managerial Decision Making," in Martin Greenberger, ed., *Computers and the World of the Future* (Cambridge: MIT Press, 1962), 53, partially cited above (chapter 2).

Chapter 7

1. The word "cyborg" was coined in 1960 to describe human-machine systems for space exploration. See chapter 1, note 1.

2. S. S. Stevens, "Machines Cannot Fight Alone," *American Scientist,* Vol. 34 (1946), 397.

3. OSRD Report No. 901, "The Performance of Communication Equipment in Noise," Psycho-Acoustic Laboratory, Harvard University (1942). Harvard University Archives.

4. Letter, Psycho-Acoustic Laboratory to Admiral J. A. Furer, Navy Coordinator of Research and Development, August 16, 1943. Harvard University Archives.

5. Reported in *Business Week* (January 19, 1946), 54.

6. Morse went on, later in the war, to join and then head the Operations Research Group (see chapter 4). In 1949 he chaired the Pentagon's Weapon Systems Evaluation Group (WSEG), and in the mid-1950s he became the first director of MIT's Computation Center. Philip M. Morse, interviewed by Richard R. Mertz, December 16, 1970. Smithsonian Computer Oral History Project, AC NMAH #196. (Archive Center, National Museum of American History, Washington, DC).

7. George A. Miller, F. M. Wiener, and S. S. Stevens, *Transmission and Reception of Sounds under Combat Conditions*, Summary Technical Report of Division 17, Section 3, NDRC (Washington, DC: NDRC, 1946), 2.

8. James H. Capshew, *Psychology on the March* (unpublished Ph.D. dissertation, University of Pennsylvania, 1986), 127, 131.

9. Cumulative list of ONR staff, PAL. Harvard University Archives.

10. Just as with today's military establishment, some of this activity covered ground already thoroughly explored by the private sector. For example, Psycho-Acoustic Lab files contain letters from several sleeping aid manufacturers demanding to know why their (existing) products were not considered for testing. In one case, the Flents Co. provided documentation—ignored by the lab and the Navy—proving its ear stopples at least as effective against noise as the V-51R Ear Wardens designed at public expense by PAL. Harvard University Archives.

11. Press release, University News Office, Harvard University, January 6, 1946. Harvard University Archives.

12. Flint, Michigan, *Journal,* September 17, 1944. Harvard University Archives.

13. Miller, Wiener, and Stevens, *Transmission of Sounds*, 4.

14. See Bruno Latour, "Give Me a Laboratory and I Will Raise the Earth," in Karin Knorr-Cetina and Michael Mulkay, eds., *Science Observed* (London: SAGE, 1983), 141–170. Latour describes the process by which laboratories create fresh sources of power by transforming the world in their own image.

World War II was, indeed, displaced into scientific laboratories; the laboratories, in turn, reemerged in the battlefield in the form of forward-based research centers and experiments, technological devices, and scientifically designed means of destruction.

15. Miller, Wiener, and Stevens, *Transmission of Sounds*, 4.

16. The problem of hearing loss was turned over to the Committee for Medical Research, though PAL did continue to work on hearing aids for the over 40,000 deafness "casualties" of the war. More recent research has shown that repeated exposure to noise of the intensity of the PAL experiments can cause serious physical damage, resulting in tinnitus (ringing in the ears) and permanent hearing impairment.

17. Internal PAL memo, December 3, 1943. Harvard University Archives.

18. George A. Miller, *Language and Communication* (New York: McGraw-Hill, 1951), v.

19. OSRD, "Performance in Noise," 1.

20. Miller, Wiener, and Stevens, *Transmission of Sounds*, 2.

21. "Proposed Program for Systems Research on Information Systems," Bureau of Ships, Navy Department, memorandum (undated), about 1945–46. Harvard University Archives.

22. Capshew, *Psychology on the March*, 134–135. Also see S. S. Stevens, "Autobiography," in Gardner Lindzey, ed., *A History of Psychology in Autobiography*, Vol. 6 (Englewood Cliffs, NJ: Prentice-Hall, 1974), 393–419.

23. See William J. McGill, "Origins of Mathematical Psychology," in William Hirst, ed., *The Making of Cognitive Science: Essays in honor of George A. Miller* (New York: Cambridge University Press, 1988), 3–18.

24. S. S. Stevens, ed., *Handbook of Experimental Psychology* (New York: Wiley, 1951).

25. Letter from John E. Karlin (Bell Labs) to S. S. Stevens, December 30, 1952. Harvard University Archives. Note the cybernetic metaphor.

26. On the concept of scientific laboratories as obligatory passage points, see Michel Callon and Bruno Latour, "Unscrewing the Big Leviathan: How Actors Macro-structure Reality and How Sociologists Help Them to Do So," in Karin D. Knorr-Cetina and Aaron V. Cicourel, eds., *Advances in Social Theory and Methodology* (Boston: Routledge and Kegan Paul, 1981), 277–303; Latour, "Give Me a Laboratory"; and Latour, *Science in Action* (Cambridge, MA: Harvard University Press, 1987).

27. See Marvin Denicoff, "AI Development and the Office of Naval

Research," in Thomas C. Bartee, ed., *Expert Systems and Artificial Intelligence* (Indianapolis: Howard W. Sams & Co., 1988), 271–287.

28. Terry Winograd, "Strategic Computing Research and the Universities," in Charles Dunlop and Rob Kling, eds., *Computerization and Controversy* (New York: Academic Press, 1991), 704–716.

29. S. S. Stevens, "A Proposal for Psycho-Acoustic Research in the Post-war Period," submitted to members of the NDRC, May 21, 1945. Harvard University Archives.

30. M. D. Fagen, ed., *A History of Engineering and Science in the Bell System* (Murray Hill, NJ: Bell Telephone Laboratories, 1978), 227.

31. G. A. Miller and P. Viek, "An Analysis of the Rat's Responses to Unfamiliar Aspects of the Hoarding Situation," *Journal of Comparative Psychology*, Vol. 37 (1944), 221–231; P. Viek and G. A. Miller, "The Cage as a Factor in Hoarding," *Journal of Comparative Psychology*, Vol. 37 (1944), 203–210; G. A. Miller, "Concerning the Goal of Hoarding Behavior in the Rat," *Journal of Comparative Psychology*, Vol. 38 (1945), 209–212; G. A. Miller and L. Postman, "Individual and Group Hoarding in Rats," *American Journal of Psychology*, Vol. 59 (1946), 652–668.

32. Miller, interviewed in Bernard J. Baars, *The Cognitive Revolution in Psychology* (New York: Guilford Press, 1986), 203.

33. George A. Miller, *The Design of Jamming Signals for Voice Communications*, unpublished Ph.D. dissertation, Harvard University, 1946; Miller, Wiener, and Stevens, *Transmission of Sounds*.

34. G. A. Miller, G. A. Heise, and W. Lichten, "The Intelligibility of Speech as a Function of the Context of the Test Materials," *Journal of Experimental Psychology*, Vol. 41 (1951), 329–335; George A. Miller and Jennifer A. Selfridge, "Verbal Context and the Recall of Meaningful Material," *American Journal of Psychology*, Vol. 63 (1950), 176–185.

35. Miller, *Language and Communication*, v–vi.

36. Ibid., 5.

37. Ibid., 6. Miller and Chomsky collaborated on some of Chomsky's most important early papers in the late 1950s. See below.

38. See "George A. Miller," in Gardner Lindzey, ed., *A History of Psychology in Autobiography*, Vol. VIII (Stanford, CA: Stanford University Press, 1989), 391–420.

39. Noam Chomsky, "Review of B. F. Skinner's *Verbal Behavior*," *Language*, Vol. 35 (1959), 26–58.

40. G. A. Miller and F. C. Frick, "Statistical Behavioristics and Sequences of

Responses," *Psychological Review*, Vol. 56 (1949), 311–325; F. C. Frick and G. A. Miller, "A Statistical Description of Operant Conditioning," *American Journal of Psychology*, Vol. 64 (1951), 68–77; G. A. Miller, "Finite Markov Processes in Psychology," *Psychometrika*, Vol. 17 (1952), 149–167; G. A. Miller and W. J. McGill, "A Statistical Description of Verbal Learning," *Psychometrika*, Vol. 17 (1952), 369–396. See Colin Cherry, *On Human Communication* (Cambridge: MIT Press, 1957), 39, for a short history of the Markov technique.

41. W. R. Garner and H. W. Hake, "The Amount of Information in Absolute Judgments," *Psychological Review*, Vol. 58 (1951), 446–459.

42. Wendell R. Garner, "The Contribution of Information Theory," in Hirst, *Making Cognitive Science*, 32.

43. Miller, interviewed in Baars, *Cognitive Revolution*, 218.

44. "Norbert was around, and we read his books, and we all were impressed with things like the Turing test and the Rosenblueth-Wiener-Bigelow paper on feedback systems that set their own goals.... [T]o say that the brain is a Turing Machine... seemed to mean that there could be some theory: you could create some Universal Turing Machine that would work just like the brain!... At the time these things seemed very powerful." Miller, interviewed in Baars, *Cognitive Revolution*, 213.

45. George A. Miller, "What Is Information Measurement?," *American Psychologist*, Vol. 8 (1953), 3–11.

46. Letter, Leonard C. Mead (Head, Human Engineering Section, Special Devices Center) to S. S. Stevens (PAL), September 10, 1946. Harvard University Archives.

47. John G. Darley, "Psychology and the Office of Naval Research: A Decade of Development," *American Psychologist*, Vol. 12, No. 6 (1957), 314. These two branches accounted for between one-quarter and one-half of ONR Psychology Division contracts in the first postwar decade. Toward the decade's end they were increasingly overshadowed by group and personnel psychology research.

48. Allen Newell and Herbert Simon, *Human Problem Solving* (Englewood Cliffs, NJ: Prentice-Hall, 1972), 881.

49. George A. Miller, "The Magical Number Seven, Plus or Minus Two: Some Limits on our Capacity for Processing Information," *Psychological Review*, Vol. 63, No. 2 (1956), 81–97.

50. Ibid., 95.

51. George A. Miller, "Information and Memory," *Scientific American*, Vol. 195 (1956), 43.

52. Ibid., 96.

53. Chomsky's ideas about formal models of language greatly excited Miller. With his knowledge of information theory, psychology of language, a growing interest in computer models, and his high visibility in the field, Miller was an ideal collaborator for Chomsky. (Miller appears to have christened the field of psycholinguistics with his article "Psycholinguistics" in G. Lindzey, ed., *Handbook of Social Psychology* [Cambridge, MA: Addison-Wesley, 1954], 693–708.) In 1958, the two published a study of finite state languages: Noam Chomsky and George A. Miller, "Finite State Languages," *Information and Control*, Vol. 1 (1958), 91–112.

54. George A. Miller, "A Very Personal History," unpublished presentation to Cognitive Science Workshop, MIT, June 1, 1979, ms. 6, 9.

55. Marvin Minsky, "Steps Toward Artificial Intelligence," *Proceedings of the IRE*, Vol. 49 (1961), 8–29.

56. George A. Miller, Eugene Galanter, and Karl H. Pribram, *Plans and the Structure of Behavior* (New York: Holt and Co., 1960), 3.

57. Karl Lashley, "The Problem of Serial Order in Behavior," in Lloyd Jeffress, ed., *Cerebral Mechanisms in Behavior* (New York: Wiley-Interscience, 1951), 112–136.

58. Stevens, "Autobiography," 408; Norbert Wiener, *Cybernetics* (New York: Wiley, 1948).

59. Miller, Galanter, and Pribram, *Plans and the Structure of Behavior*, 3.

60. Ibid., 28.

61. Ibid., 29.

62. Simon later argued that, despite many citations, Miller failed to give him and Newell enough credit in *PSB*. See Herbert A. Simon, *Models of My Life* (New York: Basic Books, 1991), 224.

63. Miller, Galanter, and Pribram, *Plans and the Structure of Behavior*, 57.

64. Ibid., 38.

65. Ibid., 56.

66. Ibid., 30.

67. *Social Sciences Citation Index*. Statistics for the earlier period 1960–1965, when the book was having its most profound and immediate effects, are not available.

68. Miller, "A Very Personal History."

69. Donald A. Norman and Willem J. M. Levelt, "Life at the Center," in Hirst, *Making Cognitive Science*, 102.

70. George Miller and Jerome Bruner, "Center for Cognitive Studies: General Description," April 8, 1960. Harvard University Archives.

71. Ibid.

72. Ibid.

73. Ibid.

74. Miller, in an interview with the *Baltimore Sun* (October 23, 1982). Neisser, who consolidated the field with his *Cognitive Psychology* (New York: Appleton-Century-Crofts, 1967), which explicitly touted the computer metaphor, repudiated it almost bitterly nine years later in *Cognition and Reality* (San Francisco: Freeman, 1976).

Chapter 8

1. The quasi-official versions of AI history, based primarily on interviews with the founders, are Daniel Crevier, *AI: The Tumultuous History of the Search for Artificial Intelligence* (New York: Basic Books, 1993), and Pamela McCorduck, *Machines Who Think* (New York: W. H. Freeman, 1979). Other semipopular, interview-based works include Howard Rheingold, *Tools for Thought* (New York: Simon & Schuster, 1985), and Frank Rose, *Into the Heart of the Mind* (New York: Harper & Row, 1984). The founders' own accounts include Allen Newell, "Intellectual Issues in the History of Artificial Intelligence," in Fritz Machlup and Una Mansfield, eds., *The Study of Information: Interdisciplinary Messages* (New York: Wiley, 1983), 187–228; the Appendix to Allen Newell and Herbert Simon, *Human Problem Solving* (Englewood Cliffs, NJ: Prentice-Hall, 1972); and Herbert A. Simon, *Models of My Life* (New York: Basic Books, 1991). A serious intellectual history of cognitive science, including AI, is Howard Gardner, *The Mind's New Science* (New York: Basic Books, 1985).

2. See John von Neumann, *The Computer and the Brain* (New Haven: Yale University Press, 1958), posthumously published.

3. One of the most significant offshoots of the cybernetic approach was early work on neural network modeling. On this approach and why it languished for two decades after the early 1960s, despite important early successes, see Mikel Olazaran Rodriguez, *A Historical Sociology of Neural Network Research* (unpublished Ph.D. thesis, Department of Sociology, University of Edinburgh, 1991). The best-known of the 1950s neural-model schemes was the "perceptron." See Frank Rosenblatt, *Principles of Neurodynamics* (New York: Spartan, 1962), and Marvin Minsky and Seymour Papert, *Perceptrons* (Cambridge, MA: MIT Press, 1969).

4. Interestingly, while Turing is usually remembered for the extremely wide range of problems his hypothetical machine can solve, the mathematically significant point of his paper consists in a proof that there exist some problems that *cannot*, even in principle, be solved by a Turing machine.

5. See Michael S. Mahoney, "The History of Computing in the History of Technology," *Annals of the History of Computing*, Vol. 10, No. 2 (1988), 113–125, and William F. Aspray, "The Scientific Conceptualization of Information: A Survey," *Annals of the History of Computing*, Vol. 7, No. 2 (1985), 117–140.

6. Andrew Hodges, *Alan Turing: The Enigma* (New York: Simon & Schuster, 1983), 106.

7. Stan Augarten, *Bit by Bit* (New York: Ticknor & Fields, 1984), 145. Similarly, of the early computer pioneers only Howard Aiken had studied the work of Charles Babbage, the nineteenth-century inventor of a mechanical protocomputer.

8. As Winograd and Flores point out, this notion of opacity is part of the ideology of cognitivism and may in fact be violated in situations of breakdown. See Terry Winograd and Fernando Flores, *Understanding Computers and Cognition: A New Foundation for Design* (Norwood, NJ: Ablex, 1987).

9. Example borrowed from Augarten, *Bit by Bit*, 213.

10. Cf. Steven Levy, *Hackers* (New York: Anchor Press, 1984), Part I.

11. Morse, cited in Karl L. Wildes and Nilo A. Lindgren, *A Century of Electrical Engineering and Computer Science at MIT, 1882–1982* (Cambridge, MA: MIT Press, 1985), 335.

12. Mahoney, 120, 117. On programming as a craft skill and its work organization, see Philip Kraft, *Programmers and Managers* (New York: Springer-Verlag, 1977).

13. See Henry S. Tropp et al., "A Perspective on SAGE: Discussion," *Annals of the History of Computing*, Vol. 5, No. 4 (1983), 386.

14. On the history of programming languages see Jean Sammet, *Programming Languages: History and Fundamentals* (Englewood Cliffs, NJ: Prentice-Hall, 1969).

15. Jerry A. Fodor, *The Language of Thought* (New York: Thomas Y. Crowell, 1975).

16. Oliver G. Selfridge, "Pattern Recognition and Modern Computers," *IRE Proceedings of the 1955 Western Joint Computer Conference* (1955); Oliver G. Selfridge, "Pandemonium, a Paradigm for Learning," in D. V. Blake and A. M. Utley, eds., *Proceedings of the Symposium on Mechanisation of Thought Processes* (London: H. M. Stationery Office, 1959).

17. McCorduck, *Machines Who Think*, 134.

18. Ibid., 116.

19. A. Newell, J. C. Shaw, and H. A. Simon, "Elements of a Theory of

Human Problem Solving," *Psychological Review*, Vol. 65, No. 3 (1958), 151–166, italics added.

20. Allen Newell, "Physical Symbol Systems," *Cognitive Science*, Vol. 4 (1980), 135–183.

21. For a more extended discussion of differences between AI and cybernetics, see Newell, "Intellectual Issues."

22. McCorduck, *Machines Who Think*, 67.

23. Claude Shannon and John McCarthy, *Automata Studies* (Princeton, NJ: Princeton University Press, 1956).

24. My account of the conference follows McCorduck, *Machines Who Think*, chapter 5, and Gardner, *The Mind's New Science*, chapter 6.

25. From the grant proposal for the Dartmouth conference, cited in McCorduck, *Machines Who Think*, 99.

26. Donald O. Hebb, *Organization of Behavior* (New York: Wiley, 1949).

27. Marvin Minsky, *Neural Nets and the Brain-Model Problem* (Ph.D. thesis, Princeton University, 1954).

28. McCorduck, *Machines Who Think*, 86.

29. On the LISP language, see John Haugeland, *Artificial Intelligence: The Very Idea* (Cambridge, MA: MIT Press, 1985), 147–157.

30. Cited in Hubert Dreyfus, *What Computers Can't Do*, 2nd ed. (New York: Harper Colophon, 1979), 81–82.

31. McCorduck, *Machines Who Think*, 42.

32. Explicit references to the Enlightenment project for an encyclopedia of all human knowledge are common in the AI literature. See Edward Feigenbaum and Pamela McCorduck, *The Fifth Generation: Japan's Computer Challenge to the World* (Reading, MA: Addison-Wesley, 1983), and Douglas Lenat et al., "Cyc: Toward Programs with Common Sense," *Communications of the ACM*, Vol. 33, No. 8 (1990), 31–49.

33. Jay W. Forrester, "Whirlwind and High-Speed Computing" (1948), reprinted in Kent C. Redmond and Thomas M. Smith, *Project Whirlwind* (Boston: Digital Press, 1980), 225–236.

34. D. N. Arden et al., *Report of the Long Range Computation Study Group* (Massachusetts Institute of Technology, 1961), 20. MIT Archives.

35. Ibid., 25, 31.

36. McCarthy, interviewed in McCorduck, *Machines Who Think*, 217.

37. Marvin Denicoff, "AI Development and the Office of Naval Research," in Thomas C. Bartee, ed., *Expert Systems and Artificial Intelligence* (Indianapolis: Howard W. Sams & Co., 1988), 272.

38. Ibid.

39. Sidney G. Reed, Richard H. Van Atta, and Seymour J. Deitchman, *DARPA Technical Accomplishments*, Vol. 1 (Alexandria, VA: Institute for Defense Analyses, 1990), 1.

40. Richard J. Barber Associates Inc., *The Advanced Research Projects Agency 1958–1974* (Washington, DC: National Technical Information Service, 1975), I-26.

41. Arthur L. Norberg and Judy E. O'Neill, with Kerry J. Freedman, *A History of the Information Processing Techniques Office of the Defense Advanced Research Projects Agency*. The Charles Babbage Institute, University of Minnesota (October, 1992), 33.

42. See Terry Winograd, "Strategic Computing Research and the Universities," in Charles Dunlop and Rob Kling, eds., *Computerization and Controversy* (New York: Academic Press, 1991), 704–716.

43. "Message from the President of the United States Relative to Recommendations Relating to Our Defense Budget," 87th Congress, 1st Session, 28 March 1961, House Documents, Document No. 123, 8, cited in Norberg and O'Neill, *A History of the Information Processing Techniques Office*, 35.

44. Robert S. Englemore, "AI Development: DARPA and ONR Viewpoints," in Bartee, *Expert Systems and Artificial Intelligence*, 215.

45. Barber Associates, *The Advanced Research Projects Agency 1958–1974*, V-49.

46. Richard H. Van Atta, Sidney G. Reed, and Seymour J. Deitchman, *DARPA Technical Accomplishments*, Vol. 2 (Alexandria, VA: Institute for Defense Analyses, 1991), 13–14.

47. J. C. R. Licklider, interviewed in John A. N. Lee and Robert Rosin, "The Project MAC Interviews," *IEEE Annals of the History of Computing*, Vol. 14, No. 2 (1992), 15.

48. Wildes and Lindgren, *A Century of Electrical Engineering and Computer Science at MIT*, 247.

49. Jerome Wiesner, cited in ibid., 265.

50. Ibid., 257.

51. Lee and Rosin, "The Project MAC Interviews," 17.

52. Norberg and O'Neill, *A History of the Information Processing Techniques Office*, 37–38.

53. J. C. R. Licklider, quoted in Rheingold, *Tools for Thought*, 140.

54. J. C. R. Licklider, quoted in Norberg and O'Neill, *A History of the Information Processing Techniques Office*, 39.

55. This and all subsequent quotations in this section are from J. C. R. Licklider, "Man-Computer Symbiosis," *IRE Transactions on Human Factors in Electronics*, Vol. HFE-1 (1960), 4–11.

56. Italics added.

57. Arden et al., *Report of the Long Range Computation Study Group*, 43.

58. See Sherry Turkle, *The Second Self* (New York: Simon & Schuster, 1984).

59. Licklider, "Man-Computer Symbiosis," 6.

60. Lee and Rosin, "The Project MAC Interviews," 18.

61. Barber Associates, *The Advanced Research Projects Agency 1958–1974*, V-50.

62. Van Atta, Reed, and Deitchman, *DARPA Technical Accomplishments*, Vol. 2, 13–14.

63. Ibid.

64. Honeywell had purchased the GE computer operation (ibid., 19–14).

65. J. C. R. Licklider, "The Early Years: Founding IPTO," in Bartee, *Expert Systems and Artificial Intelligence*, 220.

66. Robert Sproull, quoted in *IEEE Spectrum*, Vol. 19, No. 8 (1982), 72–73.

67. Kenneth Flamm, *Targeting the Computer* (Washington, DC: Brookings Institution, 1987), 53.

68. In industry, DARPA's funding share has been considerably lower. But until the 1980s, when expert systems began to find commercial applications, industry research in AI took a back seat to academic work.

69. Vannevar Bush, *Science: The Endless Frontier* (Washington, DC: U.S. Government Printing Office, 1945), 12.

Chapter 9

1. Fred Halliday, *The Making of the Second Cold War* (London: Verso, 1986). I will use a modified form of Halliday's periodization, marking the first Cold War as the period between 1947 (containment doctrine) and 1969 (the Nixon administration). Cold War II ran from 1979 (Carter's rightward shift) to 1989 (the collapse of East Germany). When referring to the Cold War *tout court*, I intend the entire period 1947–1989.

2. Stephen Ambrose, *Rise to Globalism*, 4th ed. (New York: Penguin, 1985), 235–236.

3. Ibid., 238.

4. Cyrus Vance, cited in ibid., 304.

5. See chapter 4.

6. President Jimmy Carter, cited in Ambrose, *Rise to Globalism*, 307.

7. Halliday, *The Making of the Second Cold War*, 17–18.

8. President Ronald Reagan, 1985 State of the Union address, cited in Walter LaFeber, *The American Age: United States Foreign Policy at Home and Abroad Since 1750* (New York: Norton, 1989), 677.

9. Halliday, *The Making of the Second Cold War*, 235.

10. Vannevar Bush, *Science: The Endless Frontier* (Washington, DC: U.S. Government Printing Office, 1945), 12.

11. Kenneth Flamm, *Creating the Computer* (Washington, DC: Brookings Institution, 1988), 94 et passim.

12. Cited in Kenneth Flamm, *Targeting the Computer* (Washington, DC: Brookings Institution, 1987), 88.

13. If Department of Energy and NASA funds were included in the totals, the military-related share of research funds remained well above 50 percent. (The DOE supports the nuclear weapons laboratories and production of nuclear weapons materials; a large percentage of NASA's work, especially after the Apollo missions ended, was devoted to the space shuttle, whose primary mission was the delivery of military satellites.)

14. U.S. House of Representatives, *Failures of the NORAD Attack Warning System* (Hearings before a Subcommittee of the Committee on Government Operations, United States House of Representatives, 97th Congress, First Session, May 19–20, 1981).

15. Robert C. Aldridge, *First Strike!* (Boston: South End Press, 1983), 245.

16. Paul Bracken, *The Command and Control of Nuclear Forces* (New Haven: Yale University Press, 1984), 54–56.

17. See Charles Perrow, *Normal Accidents* (New York: Basic Books, 1984).

18. A cogent overview of computer failures and the associated problems of command may be found in Alan Borning, "Computer System Reliability and Nuclear War," *Communications of the ACM*, Vol. 30, No. 2 (1987), 112–131.

19. Bracken, *The Command and Control of Nuclear Forces*, 59, 2, and passim.

20. Jonathan Jacky, "Software Engineers and Hackers: Programming and Military Computing," in Paul N. Edwards and Richard Gordon, eds., *Strategic Computing: Defense Research and High Technology* (unpublished ms.).

21. James Fallows, *National Defense* (New York: Random House, 1981), 44.

22. See Jonathan Jacky, "Ada's Troubled Debut," *The Sciences*, Vol. 27, No. 2 (1987), 20–29.

23. See James Fallows and Franklin C. Spinney, *Defense Facts of Life* (Boulder: Westview Press, 1985).

24. President Ronald Reagan, address to the nation on military affairs, March 23, 1983.

25. William Broad, *Star Warriors* (New York: Simon & Schuster, 1985), and *Teller's War* (New York: Simon & Schuster, 1992).

26. T. K. Jones, cited in LaFeber, *The American Age*, 672. See also Robert Scheer, *With Enough Shovels* (New York: Random House, 1982).

27. Danny Cohen, *A Report to the Director, Strategic Defense Initiative Organization* (Eastport Study Group, December 1985), 5.

28. See David Parnas, "Computers in Weapons: The Limits of Confidence," in David Bellin and Gary Chapman, eds., *Computers in Battle* (New York: Harcourt Brace, 1987), 123–154; Jonathan Jacky, "Safety-Critical Computing: Hazards, Practices, Standards, and Regulation," in Charles Dunlop and Rob Kling, eds., *Computerization and Controversy* (New York: Academic Press, 1991), 612–631; and Brian Cantwell Smith, "Limits of Correctness in Computers," in ibid., 632–646.

29. Parnas, it should be noted, was no radical, having served for years as a consultant to the Navy on software issues.

30. James C. Fletcher, "The Technologies for Ballistic Missile Defense," *Issues in Science and Technology* (Fall, 1984), 25.

31. Col. David Audley, cited in Eric Roberts and Steve Berlin, "Computers and the Strategic Defense Initiative," in Bellin and Chapman, *Computers in Battle*, 166.

32. The controversy about the SDI's technical and strategic feasibility had effectively already taken place twice before, first with the SAGE bomber defense system and then again with the Nixon-era anti–ballistic missile (ABM) proposals. All of the arguments against the SDI—including the impossibility of adequate software, the simplicity and cost-effectiveness of enemy countermeasures, and the massive surge in arms racing such systems were likely to spawn—had already been raised in the ABM debate. Teller's vision notwithstanding, nothing that would affect the validity of these arguments had changed since the 1970s.

33. See John Bosma, "Arms Control, SDI, and the Geneva Conventions," in Zbigniew Brzezinski, ed., *Promise or Peril: The Strategic Defense Initiative* (Washington, DC: Ethics and Public Policy Center, 1986), 339–363.

34. John Bosma, "The Selling of Star Wars," *Harper's* (June 1985), 22–24.

35. One such consideration was the ongoing indirect strategy of bleeding the Soviet economy to death through arms racing, to which the Freeze would have put an immediate halt. SDI, by contrast, massively amplified the arms race by extending it into defensive technology and countermeasures.

36. DARPA officials have insisted on referring to Strategic Computing as a "program" rather than an "initiative," largely in order to differentiate it from the SDI. However, the program's scale and the major advances it planned clearly justify the use of the grander term "initiative," persistently employed by both popular and academic writers.

37. See Hubert Dreyfus, *What Computers Can't Do*, 2nd ed. (New York: Harper Colophon, 1979).

38. SHRDLU and the microworlds approach are fully described in Terry Winograd, *Understanding Natural Language* (New York: Academic Press, 1972). For analysis and critiques, see John Haugeland, ed., *Mind Design* (Cambridge: MIT Press, 1981).

39. The most sympathetic view of commercial expert systems is Edward Feigenbaum, Pamela McCorduck, and H. Penny Nii, *The Rise of the Expert Company* (New York: Times Books, 1988). For critiques of the expert-systems approach, see Harry Collins, *Artificial Experts* (Cambridge, MA: MIT Press, 1990), and Hubert Dreyfus and Stuart Dreyfus, *Mind Over Machine* (New York: Free Press, 1986). In a talk at Cornell University in 1992, Dreyfus reported that in a DARPA-sponsored study, he and his brother had located only *two* expert systems that performed all of their described functions and reliably achieved better results than human beings. For a brief history of ideas in AI, see Allen Newell, "Intellectual Issues in the History of Artificial Intelligence," in Fritz Machlup and Una Mansfield, eds., *The Study of Information* (New York: Wiley, 1983), 187–228.

40. See Collins, *Artificial Experts*, and Dreyfus and Dreyfus, *Mind Over Machine*.

41. Douglas Lenat et al., "Cyc: Toward Programs with Common Sense," *Communications of the ACM*, Vol. 33, No. 8 (1990), 31–49.

42. Flamm, *Targeting the Computer*, 52–53.

43. Ibid., 54.

44. Defense Advanced Research Projects Agency, *Strategic Computing—New Generation Computing Technology: A Strategic Plan for its Development and Application to Critical Problems in Defense* (Washington, DC, 1983).

45. See Defense Advanced Research Projects Agency, "Annual Reports" on Strategic Computing (Washington, DC, 1985, 1986). Following the second annual report in 1986, DARPA ceased providing public information on Strategic Computing.

46. In 1986 reporter Susan Faludi uncovered evidence of just such misrepresentation of system capabilities at the Palo Alto research corporation Advanced Decision Systems (ADS). Susan Faludi, "The Billion Dollar Toy Box," *West Magazine, San Jose Mercury News* (November 23, 1986).

47. See Terry Winograd, "Strategic Computing Research and the Universities," in Dunlop and Kling, *Computerization and Controversy*, 704–716.

48. See Sidney G. Reed, Richard H. Van Atta, and Seymour J. Deitchman, *DARPA Technical Accomplishments*, Vol. 1. (Alexandria, VA: Institute for Defense Analyses, 1990), chapter 21.

49. See the assessments of Jacky, "The Strategic Computing Program," and Mark Stefik, "Strategic Computing at DARPA: Overview and Assessment," *Communications of the ACM*, Vol. 28, No. 7 (1985), 690–704.

50. On the moral and legal issues surrounding autonomous weapons, see Gary Chapman, "The New Generation of High-Technology Weapons," and John Ladd, "Computers and War: Philosophical Reflection on Ends and Means," both in Bellin and Chapman, *Computers in Battle*, 61–100, 297–314.

51. In a generally accepted genealogy of hardware and software, the first four "generations" of digital computers were constituted, in essence, by successive hardware innovations leading to order-of-magnitude reductions of scale. Vacuum tube, transistor, integrated circuit (IC), and very-large-scale integrated circuit (VLSI) technologies superseded each other at roughly ten-year intervals. Machine code, assembly languages, symbolic languages (such as FORTRAN or COBOL), and structured programming languages (Pascal) represented software innovations roughly corresponding to these stages. The fifth generation, as envisioned in the early 1980s, would involve machines built around decentralized "parallel" architectures, which execute numerous instructions simultaneously. The equivalent software innovation would be "intelligent knowledge-based systems" descended from the expert-systems branch of AI. The new generation was expected to understand natural language, "reason" like human beings within limited domains of knowledge, "see" with excellent acuity and comprehension, and command other "intelligent" processes.

52. See Tohru Moto-oka, "The Fifth Generation: A Quantum Jump in Friendliness," *IEEE Spectrum*, Vol. 20, No. 11 (1983), 46–47, and Edward Feigenbaum and Pamela McCorduck, *The Fifth Generation* (Reading, MA: Addison-Wesley, 1983).

53. Janet Raloff, "DOD Targets Fifth Generation Too," *Science News*, Vol. 125, No. 21 (1984), 333; Robert Cooper and Robert Kahn, "SCS: Toward

Supersmart Computers for the Military," *IEEE Spectrum*, Vol. 20, No. 11 (1983), 53–55; Robert Kahn, "A New Generation in Computing," *IEEE Spectrum*, Vol. 20, No. 11 (1983), 36–41.

54. Kahn and Cooper's *IEEE Spectrum* articles (see previous note) were part of a special issue on fifth generation computing, printed back-to-back with articles about the Japanese initiatives.

55. DARPA, *Strategic Computing*, 4.

56. Quotes from Cooper and Keyworth in this and the preceding paragraph, and discussion of relation between SDI and SCI, are taken from Jacky, "The Strategic Computing Program."

57. DARPA, *Strategic Computing*, 1.

58. Ibid.

59. David Mizell, in a talk at the Conference on Strategic Computing: History, Politics, Epistemology, organized by the Silicon Valley Research Group at the University of California, Santa Cruz, March 2–3, 1985.

Chapter 10

1. See, for example, Donna J. Haraway, *Primate Visions* (London: Routledge Kegan Paul, 1989); Wiebe Bijker, Thomas P. Hughes, and Trevor Pinch, eds., *The Social Construction of Technological Systems* (Cambridge, MA: MIT Press, 1987); Wiebe E. Bijker and John Law, eds., *Shaping Technology/Building Society: Studies in Sociotechnical Change* (Cambridge, MA: MIT Press, 1992); Karin D. Knorr-Cetina, "Introduction: The Micro-sociological Challenge of Macro-sociology," in Knorr-Cetina and Aaron V. Cicourel, eds., *Advances in Social Theory and Methodology* (Boston: Routledge and Kegan Paul, 1981), 2–47; Bruno Latour, "Give Me a Laboratory and I Will Raise the Earth," in Karin Knorr-Cetina and Michael Mulkay, eds., *Science Observed* (London: SAGE, 1983), 141–170; Latour, *Science In Action* (Cambridge, MA: Harvard University Press, 1987); Michel Callon and Bruno Latour, "Unscrewing the Big Leviathan: How Actors Macro-structure Reality and How Sociologists Help Them to Do So," in Knorr-Cetina and Cicourel, *Advances in Social Theory and Methodology*, 277–303.

2. Among the best and most comprehensive studies of this multidimensional character of scientific discourse is Steven Shapin and Simon Shaffer, *Leviathan and the Air-Pump: Hobbes, Boyle, and the Experimental Life* (Princeton: Princeton University Press, 1985).

3. Louis Althusser, "Ideology and Ideological State Apparatuses," in Louis Althusser, *Lenin and Philosophy*, ed. and trans. Ben Brewster (New York: Monthly Review Press, 1971), 127–188.

4. In telling and retelling our stories, both to others and to ourselves, we constitute a semipermanent identity. This point becomes most obvious when our self-constituting narratives are challenged or altered. The archetypal modern experience of the narrative construction of subjectivity is psychotherapy, where one works to recover lost stories, to retell other stories, and to link them all together along some new kind of plot line.

5. See chapter 1. The closed-world and green-world paradigms I use here descend from Sherman Hawkins and Northrop Frye respectively. However, I have fit them to my own purposes; responsibility for any defects in my discussion is entirely my own. I am profoundly indebted to Sherman Hawkins (personal communication) for my understanding of these themes.

6. In this book I usually use the word "cyborg" as a generic term for sentient entities anywhere along a continuum from humans with artificial parts to fully artificial beings. This usage has become conventional in cultural studies, largely due to Haraway's influence (e.g., Donna J. Haraway, "A Manifesto for Cyborgs," *Socialist Review*, Vol. 15, No. 2 [1985], 65–107). In science fiction (SF), however, conventions are normally rather different. In SF, *robots* are electromechanical; they may, but need not, superficially resemble human beings, and they generally have computerized minds. Isaac Asimov's 1950s robots had "positronic" brains. *Androids* are robots made to look as much as possible like human beings, but generally without biological components; their minds, too, are usually computers. Biologically engineered human-like entities like *Blade Runner*'s "replicants" are also sometimes called "androids." *Cyborgs* are blends of organism (often but not always human) and machine, ranging from biological brains with mechanical bodies to robots covered with biological flesh. *AIs* are artificial intelligences, usually self-aware but disembodied computers of great intellectual powers. Boundaries between these categories are fuzzy and by no means universal. Here I shall continue to use "cyborg" generically, sometimes substituting Turkle's suggestive term "second self"; see Sherry Turkle, *The Second Self: Computers and the Human Spirit* (New York: Simon & Schuster, 1984).

7. Karel Capek, *R.U.R. (Rossum's Universal Robots), a Fantastic Melodrama*, trans. Paul Selver (Garden City, NJ: Doubleday, 1923).

8. The rest of this chapter interprets a variety of science fiction and film. I am painfully aware that there is a large body of critical literature that covers these texts and films and explores, I am sure, many of the themes I address here. For various reasons, though, I have chosen not to consult this literature.

9. The "fail-safe" idea refers to a system that requires an explicit command to proceed with an attack. The system's default, absent such a command, is for the bombers to return home. In theory such a system would be immune to catastrophic failure.

10. Michael Smith has analyzed exhibitions from the 1964 World's Fair in New York City, revealing a remarkable range of images representing closed,

artificial systems in hostile environments, such as space capsules, undersea communities inhabiting pressurized domes, underground houses, and space stations. Many of the images Smith collected depict smiling, prosperous white nuclear families going about their ordinary daily activities inside the walls of these bubble-like shells. As Smith observes, on a symbolic level all of these amount to fallout shelters: tiny artificial containers in which the white nuclear family would avoid the perils of nuclear holocaust. Michael Smith, lecture at Stanford University, 1993.

11. Compare Anne McCaffrey, *The Ship Who Sang* (New York: Ballantine Books, 1970), about a spaceship controlled by a disembodied but definitely female brain.

12. Steven Levy, *Hackers* (New York: Anchor Press, 1984), 205.

13. See Zoë Sofoulis, *Through the Lumen* (unpublished Ph.D. thesis, University of California at Santa Cruz, 1988).

14. Stan Augarten, *Bit by Bit* (New York: Ticknor & Fields, 1984), 280.

15. A direct reference to Asimov's 1950s-era robot stories. For a collection of these, see Isaac Asimov, *Robot Visions* (New York: RoC Press, 1990).

16. Haraway, "Manifesto for Cyborgs," 99.

17. Ibid., 65.

18. Although my analysis generally approaches the film as a separate artistic whole, I need to use occasional references to the book to justify its inclusion here. The film takes its central problematic—the issues of sentience and slavery for human-created subjectivities—and much of its iconography from the book's Turing-test issues about the differences between humans and the electronic androids.

19. Turkle, *The Second Self*, 62 and passim.

20. See Jacques Derrida, *Of Grammatology*, trans. Gayatri C. Spivak (Baltimore: Johns Hopkins University Press, 1976), especially Spivak's introduction and the section on Freud's "Mystic Writing Pad."

21. For a statement of cyberpunk's aesthetic goals, see Bruce Sterling's introduction to Sterling, ed., *Mirrorshades: The Cyberpunk Anthology* (New York: Ace Books, 1986). Another excellent introduction is Larry McCaffery, ed., *Storming the Reality Studio* (Durham, NC: Duke University Press, 1991).

22. William Gibson, *Neuromancer* (New York: Ace Books, 1984), 52.

23. Ibid., 3.

24. Haraway, "Manifesto for Cyborgs," 82, citing Zoë Sofia.

Epilogue

1. See the discussion in chapter 3.

2. Statistics compiled by Win Treese, *The Internet Index*, available at WWW. openmarket.com/diversions/internet-index/ December 16, 1994. At this writing, the most perceptive treatment of the evolution, social implications, and vulnerabilities of massively interlinked computer networks is Gene I. Rochlin, *The Computer Trap: Dependence and Vulnerability in an Automated Society* (Princeton, NJ: Princeton University Press, forthcoming).

3. Michael L. Dertouzos et al., "Communications, Computers, and Networks: How to Work, Play and Thrive in Cyberspace," special issue of *Scientific American*, Vol. 265, No. 3 (1991).

4. On the National Information Infrastructure, see Clinton Administration, *Agenda for Action* (Washington, D.C.: The White House, 1994), and Computer Professionals for Social Responsibility, *"White Paper" on the National Information Infrastructure* (Washington, DC: CPSR, 1994).

5. This point emerged in conversation with Donna Haraway.

6. The accuracy of computer-guided weapons, though far greater than that of any previous technology, was not nearly as high as the Pentagon led the public to believe. The six-month build-up to the war made it possible to emplace and test an enormous logistical infrastructure (including computer links); had this infrastructure been absent, or had it been stressed by immediate combat, the war might have—indeed, probably would have—followed quite a different course. See the chapter on the Gulf War in Rochlin, *The Computer Trap*.

7. See J. L. McClelland, D. E. Rumelhart, and G. E. Hinton, "The Appeal of Parallel Distributed Processing," in James L. McClelland and David E. Rumelhart, eds., *Parallel Distributed Processing*, Vol. 1 (Cambridge, MA: MIT Press, 1986), 3–44; Paul Churchland, *A Neurocomputational Perspective* (Cambridge, MA: MIT Press, 1989); Paul M. Churchland and Patricia Smith Churchland, "Could a Machine Think?" *Scientific American*, Vol. 256, No. 1 (1990), 32–37; Andy Clark, *Associative Engines: Connectionism, Concepts, and Representational Change* (Cambridge, MA: MIT Press, 1993); and Mikel Olazaran Rodriguez, *A Historical Sociology of Neural Network Research* (unpublished Ph.D. thesis, Department of Sociology, University of Edinburgh, 1991).

8. See chapter 6.

9. See Philip E. Agre, *The Dynamic Structure of Everyday Life* (Cambridge: Cambridge University Press, forthcoming); Philip E. Agre and David Chapman, "What Are Plans For?" (AI Memo 1050, MIT Artificial Intelligence Laboratory, 1988); and Rodney A. Brooks, "Intelligence Without Representation," *Artificial Intelligence*, Vol. 47 (1991), 139–159.

10. A computer-graphics term for seamless three-dimensional transformation.

11. The phrase is due to Langdon Winner, *Autonomous Technology* (Cambridge, MA: MIT Press, 1977).

Index

2001: A Space Odyssey, 2, 305, 308, 311, 321–325, 339
HAL, 314, 321–326, 329, 340, 357
AAF (Army Air Force). *See* Air Force
Aberdeen Proving Ground. *See* Ballistics Research Laboratory
Acheson, Dean, 9, 11–12, 277, 279
ADIS. *See* Air Defense Integrated System
AEC. *See* Atomic Energy Commission
AF. *See* Air Force
Aiken, Howard, 17, 66, 81, 188
Air Defense Integrated System (ADIS), 96–97
Air Force (AF), 3, 25, 91, 104, 107, 128–129, 138, 206, 261–265, 268, 287, 296, 319
  Big L systems, 107, 272, 319
  Cambridge Research Center (AFCRC), 92, 267
  culture of, 82, 84–85, 94, 108, 110–111, 118–119
  Human Factors Operations Research Group, 227, 267
  nuclear weapons and, 83–90, 109–110
  offensive vs. defensive strategy in, 82–90, 93–96, 98–99, 108
  and Rand Corporation, 113, 115–126
  resistance to computers, 71, 96–97, 106
  role in promoting high-technology weaponry, 57, 64, 82, 106, 108, 111
  support for computer development, 60, 64, 90, 106, 108, 122, 132
Airplane Stability and Control Analyzer (ASCA), 76–78
Analog computers, 60, 75, 76, 87, 102, 175, 254, 263. *See also* Servomechanisms; Differential analyzer
  advantages of, 67–68, 70
  vs. digital computers, 43–44, 45n3, 49, 66, 67–70, 77–78, 80, 96–97, 190, 246
  disadvantages of, 77
  electronic, 45n3, 67, 70, 77, 78
  gun directors, 45, 67, 87
  as metaphors in cybernetics, 195
  WWII-era tradition of, 44, 46, 49, 66, 70
Analog-digital conversion, 69, 92
  in Project X speech coding system, 200
  SAGE and, 75, 100
Antiaircraft guns, 65, 115, 206
  computers and, 49, 67, 77
  as metaphors in cybernetics, 186, 235
  and origins of cybernetic theory, 1, 147, 180, 196
Armed forces (U.S.)
  academic and industrial collaboration with, 46, 47–48, 51–53, 58–65, 67–68, 110, 115, 175, 211, 219–222, 241, 272
  administrative centralization of, 5–6, 71, 96, 129–131, 134

Armed forces (U.S.) (cont.)
  command tradition vs. managerial approach, 5–6, 7, 70–72, 106–109, 126–134, 139–145, 206
  support for R&D, 43–53, 61–70, 221, 281–283 (*see also* Air Force; DARPA; ONR; SAGE)
  women in, 25, 167
Arms race, 88, 136, 284, 293, 318
Army, U.S., 45, 52, 57, 60, 61, 83, 89, 127, 129, 138
  and antiaircraft defenses, 67, 87, 96, 104
  and psychology, 176, 199
  and SCI, 296–297
Army Air Force (AAF). *See* Air Force
ARPA. *See* Defense Advanced Research Projects Agency
Artificial intelligence (AI)
  air defense simulations and origins of, 124
  and cognitivism, 2, 19, 20, 147, 179, 235, 252, 255–256, 267–268, 271–273
  and cybernetics, 239–241, 255, 265
  and cyborg discourse, 20–22, 26, 173, 255–256, 260, 266–268, 271–273, 275, 300
  DARPA support for, 64, 259, 260–264, 267–271, 293–301
  Dartmouth Conference, 124, 229, 239, 252–254, 259, 260
  expert systems, 18, 71, 251, 273, 295, 340, 356
  founders of (*see* entries for McCarthy, Minsky, Newell, Simon)
  microworlds, 171, 172, 295, 301, 330
  as military technology, 27, 173, 264–273, 275, 293–301
  in science fiction, 2, 22–27, 276, 303–351, 357–365
    embodied vs. disembodied, 306, 314–315, 326, 329–330, 332, 336–337, 340, 348, 349
  SCI and, 64, 293–301
  SDI and, 298, 299, 301

time-sharing and, 240, 256–259, 266, 268–270
Turing test and, 16, 243–244
Atanasoff, John, 17, 50
Atomic Energy Commission (AEC), 47, 61, 115
Autonomous machines, 24, 29, 301, 316, 362
Autonomous weapons
  vs. automatic weapons, 20
  SCI and, 25, 275, 294, 296–297, 300

Babbage, Charles, 17, 242
Ballistic Missile Early Warning System (BMEWS), 107, 284–285
Ballistics Research Laboratory (BRL), 45–47, 49–50
Barnes, Major General Gladeon, 51–52
Bateson, Gregory, 181, 189, 195
BBN. *See* Bolt Beranek and Newman
Behaviorism
  vs. cognitivism, 19, 179, 209, 223–225, 232–234, 236
  and cybernetics, 180–184, 187, 193, 202, 204, 209, 240
  reflex machine metaphors, 162–164
Bell Laboratories, 17, 102, 221, 223, 253, 323
  and Nike antiaircraft defense system, 60, 67, 87
  Project X speech encryption system, 200–202 (*see also* Shannon)
  and Psycho-Acoustic Laboratory, 216, 220, 225
Beranek, Leo, 212, 214, 263, 264
Bigelow, Julian, 180–187, 189, 193, 207, 226, 230, 264. *See also* Wiener
Bijker, Wiebe, 33–34, 44, 70
Bilateralism. *See* Closed world; Containment; Truman Doctrine
BINAC, 60–61, 69, 92, 247
Blade Runner, 2, 303, 309, 314, 315, 342–346, 349
BMEWS. *See* Ballistic Missile Early Warning System

Bolt Beranek and Newman (BBN), 264, 270
Boolean logic, 188, 200
Boring, Edwin G., 187–188, 219
Bracken, Paul, 109, 131, 285–286
BRL. *See* Ballistics Research Laboratory
Bruner, Jerome, 233–234
Bush, George, 354
Bush, Vannevar, 77, 281
 and analog computer tradition at MIT, 48
 attitude toward digital computers, 50
 differential analyzer, 46
 and OSRD, 46–47, 271
 Rapid Arithmetical Machine, 48
 Rockefeller Differential Analyzer (RDA), 48, 67
 Science: The Endless Frontier, 58–59, 271

C^3I. *See* Command and control
Cameron, James, 22, 357
Carter, Jimmy, 298
 and renewed Cold War, 278–281, 283, 301
Census Bureau, 60–61
Center for Cognitive Studies, 209, 222, 233–235
Central Intelligence Agency (CIA), 87, 277
Chomsky, Noam, 164, 202, 224–225, 229, 234, 263
CIA. *See* Central Intelligence Agency
Clausewitz, Karl von, 121, 144
Closed world. *See also* Cold War; Containment; Cyborg
 AI as, 256, 272–273, 295, 312
 capitalist societies as, 10–12, 107–108
 Communist societies as, 10–12, 13, 53, 107–108
 computerized command-control systems as infrastructure of, 1, 7, 25, 41, 44, 72–73, 75, 90, 103, 107–111, 113, 114, 145, 260, 272–273, 275, 313 (*see also* SAGE)
 computers and, 14–15, 99, 125, 133
 cyberspace as, 308, 312, 331–332
 definitions of, 1–3, 12, 307–309
 in fiction and theater, 12–13, 15, 22–27, 276, 303–351, 357–364
 games and simulations as, 172
 gender in, 326, 330, 342, 346, 350, 357–364
 inverted worlds, 309, 334, 336, 349
 vs. green world, 13, 306, 309–312
 metaphors for
  Berlin Wall, 13
  fallout shelters, 320–321
  Igloo White, 6
  Iron Curtain, 13
  SAGE, 104–106, 109–111, 308
  SCI, 275, 299–301
  SDI, 275, 292, 293, 299–301, 303, 308
  spaceships and space stations, 307, 308, 321, 334
  War Room, 39, 112, 307, 316–320
 nested enclosures, 103, 308, 320–321, 323, 326, 340–341
 nuclear war and, 14, 315–321, 328–331, 357–359, 362, 363
 politics, 1, 8–15, 19, 107–111, 137–138, 142–145, 272, 276–281, 298
 Rand Corporation research and, 120, 123, 312
 simulations as, 28, 172, 312
 subjectivity and experience within, 15, 28, 99, 143, 172–173, 178, 276, 303–306, 329, 331–332, 340–351, 355
 transformed by Cold War's end, 354, 362–365
Closed-world discourse, 40, 83, 279
 containment as central metaphor of, 8, 113, 272
 and cyborg discourse, 21–22, 27, 147–148, 173, 268, 272–273, 275–276, 303
 definitions of, 1–3, 7–12, 15
 supports for, 90
  computers, 15, 38–39, 41, 99, 108, 113, 169

Cognitive psychology, 19, 313, 340.
   See also Cognitivism
   vs. behaviorism, 179, 224–225
   and cyborg discourse, 21, 147
   definitions, 178–180
   historiography of, 22, 147
   Licklider, J.C.R., and, 260, 267–268
   Miller, George, and, 222, 237
   MIT Symposium on Information
      Theory, 228–229
   Psycho-Acoustic Laboratory and,
      209, 222–237
Cognitive science, 2, 19, 229, 235.
   See also Cognitive psychology
Cognitivism, 2. *See also* Cognitive
      psychology
   AI and, 239–241, 254–256
   computers
      as metaphors in, 147–148,
         160–162, 179, 234–237, 252
      as tools in, 179, 234–237
   cybernetics and, 45, 179–180,
      239–241
   and cyborg discourse, 172–173,
      272–273
   and human-machine integration,
      20
Cold War, 6–7, 54. *See also* Closed
      world; Containment
   antimilitarism and U.S. response to,
      56–58, 145
   Carter-Reagan renewal of (Cold
      War II), 278–284
   containment doctrine, 8–12
   coupling of U.S. and USSR in, 286
   globalism, 8–12, 107–111
   grand strategy in, 2, 44, 55, 113
   high technology as U.S. response
      to, 1, 7–8, 57–60, 65–66, 103,
      110–111, 118–121, 130–133, 142,
      275, 288–290, 293, 301, 317–318
   linkage, 276–277
   Reagan era, 279–281
   Vietnam War and, 134–138 (*see also*
      Vietnam War)
   WWII events as icons in
      atomic bomb, 55–56

Maginot Line, 55–56, 86, 90, 95
Manhattan Project, 55–56, 89,
   134–135
Munich accords, 55–56, 136, 286
Pearl Harbor, 55–56, 118–119
radar, 55
Colossus computer. *See* Turing, Alan
Colossus: The Forbin Project, 2, 323,
   325–327
Combat Information Centers (CICs),
   81, 218
Command and control systems, 266
   Big L systems, 107
   centralizing effects of, 6, 132–134,
      138
   and closed world, 15, 44
   concept of, 79–83, 131–132
   DARPA and, 261–262, 267–270
   human-machine integration in, 20,
      205–207, 214, 215, 226, 240, 260,
      271–273
   as ideology, 15, 72, 110–111, 145
   nuclear early warning and nature
      of, 100–101, 131, 285–286
   SAGE as first computerized,
      93–101, 104–111
   SDI and AI in, 290, 299–301
Computer languages
   Ada, 245, 287
   assembly languages, 103, 124,
      244–245
   Short Code, 247
   COBOL, 245, 249–250
   compilers, 103, 245, 247–250, 256,
      269, 287
   debates over value of, 248
   FORTRAN, 245, 249, 251, 265
   interpreters, 103, 247
   IPL, 124, 251–253
   JOVIAL, 107
   LISP, 170, 245, 254, 258, 267
   machine language, 244–248, 252,
      255
   Pascal, 169–170, 245
Computer networks, 19, 25, 64, 268,
   286, 290, 293, 308, 315, 339–340,
   353–354

ARPANET, 270, 353
cyberspace, 347–351, 354
SAGE and, 100–101, 102
World Wide Web, 354, 364–365
Computers. *See also* Analog computers; Digital computers
definitions of, 27–28, 45n3, 51n19
Containment, 1, 10, 109, 113, 119, 320. *See also* Closed world; Cold War
as Cold War strategic doctrine, 6, 8–12, 138, 278–281
Control
digital computers and, 43, 66–67, 67n66, 70
SAGE and, 104–109
Whirlwind and real-time, 75–81, 97, 101
Cooper, Robert, 71, 299
Craik, Kenneth, 182, 185–186, 196, 216
Crawford, Perry, 77, 79–80
Cybernetics, 1–2, 19, 45, 114, 187–209, 225–226, 263. *See also* Wiener, Norbert; Macy Conferences
and AI, 239–241, 243, 250–256, 356
cybernetic psychology, 179–186, 207, 230, 234
human-machine integration and, 147, 210, 213, 218
servo/organism analogies in, 175, 180–187, 191, 193–199, 203
and *Plans and the Structure of Behavior*, 230–233
weapon system metaphors in, 182, 184, 186, 196–199, 205–207
Cyberspace, 19–20, 303, 308, 312, 331, 332, 342, 347–351, 354, 365. *See also* Computer networks
Cyborg, 197, 199, 203, 211, 264, 300
definitions of, 19–21, 210
in fiction and theater, 22–27, 303–351, 357–365
1980s rehabilitation of, 338–340, 363

in green world, 332–340
as inhabitant of closed world, 2–3, 22, 27, 147–148, 172–173, 304, 313–315
subjectivity of, 23, 24, 165–169, 172–173, 237, 273, 303, 304–306, 312–315
mind as recombinant theater, 340–351
Cyborg discourse, 233, 236–237, 246
and closed-world politics, 2
definitions of, 2, 21–22, 147–148
politics of, 22, 178, 266–267, 273
supports for, 39, 186–187, 255–256

DARPA. *See* Defense Advanced Research Projects Agency
DDR&E. *See* Director of Defense Research and Engineering
Defense Advanced Research Projects Agency (DARPA), 20, 64, 71, 234, 353. *See also* Artificial Intelligence
Information Processing Techniques Office (IPTO), 240, 262, 267–271, 296
origin of, 260–262
Department of Defense. *See* Armed forces, Army, Navy, Air Force, OSD
Differential analyzers, 46, 48, 49, 69, 77
Digital computer, 16–18, 51n19, 62, 64, 190, 242, 244–246. *See also* Analog computers; Computers; SAGE
advantages of, 49, 70
electronic, 45n3, 69
problems of, 43–44, 50, 66–70
SAGE and, 75–82, 90–111
seen as calculators only, 43, 66–67, 67n66, 78
Director of Defense Research and Engineering (DDR&E), 129, 261–262
Discourse. *See also* Closed-world discourse; Cyborg discourse; Metaphor

Discourse (cont.)
  concept of, 30–41
  language-games and, 34–37
  support, concept of, 38–41 (*see also* Foucault)
  technology and, 36–37, 40, 41
Dr. Strangelove, 303, 318–321
Dulles, John Foster, 12, 119, 277

EAL. *See* Electro-Acoustic Laboratory
Eckert, J. Presper, 50, 60, 61, 69, 247. *See also* ENIAC, UNIVAC, BINAC
EDVAC, 50–51, 62, 77, 187, 246
Eisenhower, Dwight, 12, 47, 58, 119, 126, 129, 136
  space policy, 134, 260
  support for SAGE, 95
Electro-Acoustic Laboratory (EAL), 209, 212–215, 218, 263
Electronic battlefield, 1, 43, 65, 71, 106, 114, 134, 144, 145, 172, 237, 273, 282, 300. *See also* Vietnam; Igloo White; Westmoreland
Empire Strikes Back, The. *See Star Wars* trilogy
ENIAC, 17, 46, 48–52, 60, 61, 65, 67, 77, 187, 189, 246
Enthoven, Alain, 127–128, 132–134. *See also* McNamara; Office of Systems Analysis
Everett, Robert, 78, 80, 92
Expert systems. *See* Artificial intelligence

Fail-Safe, 303, 316–318
Feigenbaum, Edward, 251, 298
Flamm, Kenneth, 53, 61, 102, 283
Florez, AF Capt. Luis de, 76
Formal machines, 184, 191, 193, 198, 204, 240–241. *See also* Formal/mechanical modeling
Formal/mechanical modeling, 224, 225, 229, 232, 233, 251, 273, 357
Forrester, Jay, 63, 65, 71, 76–82, 90, 91–95, 144, 207, 256, 262

Foucault, Michel, 32, 34, 37–41, 103, 151, 169
  discourse, concept of, 38–40, 120
  power and knowledge, 39–40, 152, 176
Frank, Lawrence K., 181, 189
Fremont-Smith, Frank, 181, 189
Freud, Sigmund, 176
  hydraulic system metaphor, 163–164
Frick, Frederick, 225, 227, 267

Galanter, Eugene, 212, 230–233, 255
Game theory 190, 225. *See also* Rand Corporation
Garner, Wendell, 212, 216, 225
Gender, 176. *See also* Women
  and computers, 45–46, 167–172, 321, 326, 330
  and war, 24–25, 167, 172, 326
Gerard, Ralph, 192–194, 204
Gibson, James William, 138–140, 142
Gibson, William, 303, 342, 346–350
Goldstine, Herman, 49, 50, 66. *See also* ENIAC
Green world, 26, 306, 330, 346, 349, 350
  cyborg subjectivity in, 336–338
  definitions of, 13, 309–312
  in *Star Wars* trilogy, 332–338
  in *2001: A Space Odyssey,* 323–325

Hegel, G.W.F., 150–151, 166
Heims, Steve, 190, 194, 199
Hill, Albert, 94, 98
Hitch, Charles, 127–128. *See also* Office of Systems Analysis
Hixon Symposium, 175, 198, 230, 253
Hughes, Thomas, 33
Human-machine integration, 1–2, 123, 147, 173, 210, 220, 226, 235, 236, 240, 259, 262, 268. *See also* Cybernetics; Psycho-Acoustic Laboratory

IAS (Institute for Advanced Study) machine (computer), 61, 63, 81

IBM, 3, 61, 63–64, 122, 249, 257, 258, 270, 275, 282, 327
  604 calculator, 121
  704 computer, 124, 257–258
  AN/FSQ-7 computer, 97, 101–102, 104
  Card-Programmed Calculator, 121, 123
  Defense Calculator (701 computer), 61, 122, 124, 251
  involvement in SAGE, 101–102
Ideology, 30–31. *See also* Discourse
Igloo White, 3–5, 6, 40, 106, 140, 142–143, 272, 282. *See also* Vietnam War
Information theory, 66, 114, 180, 184, 189, 193, 234, 253, 272
  Miller, George, and, 225, 229, 235
  PAL and spread of, 209–210, 212, 221–222, 236
Integrated circuits (ICs), 64, 108, 270, 285
Intercontinental ballistic missiles (ICBMs), 64, 107, 108, 110, 277, 279, 319
  and Cold War, 7, 57–58, 136
  and SDI, 289–290

Johnniac. *See* Rand Corporation
Johnson, Lyndon, 5, 134, 136, 137, 141, 279

Kahn, Herman, 116, 316, 318, 320
Kaplan, Fred, 118–119, 126, 141, 143
Killian, James R. Jr., 91, 98
Kissinger, Henry, 276–279, 298
Köhler, Wolfgang, 194–195
Korean War, 11, 82, 89, 263

Language. *See* Discourse; Metaphor; Wittgenstein; Lakoff
Language-game. *See* Wittgenstein
Lashley, Karl, 225, 230
Latour, Bruno, 33, 220
LeMay, AF General Curtis, 85, 119, 126, 293. *See also* Strategic Air Command

Lévi-Strauss, Claude, 41
Licklider, J. C. R.
  as head of ARPA IPTO, 240, 260, 262, 263, 267–272
  Macy Conference presentation, 205–206
  "Man-Computer Symbiosis," 264–267, 299
  at MIT, 226, 262–263
  at Psycho-Acoustic Laboratory, 212, 216
  work with SAGE, 226
Lincoln Laboratory, 63, 93–94, 96–98, 102, 103, 124, 132, 222, 236, 250, 264, 267, 268. *See also* MIT; SAGE; Whirlwind
Lincoln Transition System, 94, 97
  presentation group, 226, 235, 262–263, 265, 268
  TX-0 computer, 257, 265
Lorente de Nó, Rafael, 189–190

MacArthur, General Douglas, 11–12
Macy Conferences, 187–205, 240. *See also* Cybernetics
  Cerebral Inhibition Meeting, 180–181, 189
  organization of, 187–189
  presentations
    Gerard, Ralph, 192
    Köhler, Wolfgang, 194–195
    Licklider, J.C.R., 205–206
    Lorente de Nó, Rafael, 190
    Rosenblueth, Arturo, 190
    Shannon, Claude, 199–204
    Stroud, John, 196–198
    Von Foerster, Heinz, 193, 195
    Von Neumann, John, 190
    Wiener, Norbert, 190
Maginot Line, 55–56, 86, 90, 95, 99. *See also* Cold War
Manhattan Project, 48, 52, 55–56, 89, 95, 288. *See also* Cold War
Masculinity. *See* Gender
Mauchly, John, 50, 60, 61, 63, 69, 247. *See also* ENIAC; UNIVAC; BINAC

McCarthy, John, 124, 202, 229. See also Artificial intelligence
   and AI, 239–271
   and J. C. R. Licklider, 264, 270
   LISP, 254
   time-sharing, 257–259
McCulloch, Warren, 175, 181, 185, 240, 243, 263, 356
   and Macy Conferences, 187–191, 198
McCulloch-Pitts neural network theory, 188, 190, 191, 196, 200, 250, 356
   George Miller and, 226
   Minsky, Marvin and, 253
   Turing machines and, 243
McNamara Line, 142, 145, 272. See also Igloo White
McNamara, Robert, 71, 126, 135
   centralization of Defense Department, 5, 129–134
   and nuclear strategy, 132, 133, 136
   and Office of Systems Analysis, 113, 127–129
   and Planning Programming Budgeting System (PPBS), 5–6
   as Secretary of Defense, 5, 127–130, 268–269
   and Vietnam War, 136–138, 141, 144
Mead, Margaret, 181, 189
Meister, Robert, 149–150
Metaphor, 147–173
   in psychological theories, 147–148, 158–173 (see also Cognitivism)
   Lakoff-Johnson theory of, 148, 153–158, 160–161
   politics of, 147–153, 157–158, 165–173
   relation to tools, 14, 27–30, 37
   subjectivity and 21, 147–160, 165–173, 304–306, 341–342 (see also Closed world; Cyborg discourse; Subject positions)
Miller, George, 209, 212, 216, 217, 220, 253, 255, 262, 263, 267, 268. See also Psycho-Acoustic Laboratory

career of, 222–237
Plans and the Structure of Behavior (PSB), 210, 230–233, 235, 255, 266
Minsky, Marvin, 28, 229, 253–255, 258, 263–264, 270
MIT, 28, 45–49, 67, 102, 165, 172, 189, 222, 226, 247–248, 257–258, 263–264, 266, 267, 269–270, 296
   and air defense, 75–82, 90–99 (see also SAGE)
   Radiation Laboratory, 47–48, 55, 58, 180, 216, 263
   Research Laboratory of Electronics (RLE), 48, 255, 257, 263
   Servomechanisms Laboratory, 76, 78
Mitchell, AAF Commander Billy, 84–85
The Court-Martial of Billy Mitchell (film), 85, 315
Moore School of Engineering, 46, 48, 77, 187. See also ENIAC
Morse, Philip, 212, 248, 263
Mutual orientation, 81–82, 134, 222, 237, 289

NASA (National Aeronautics and Space Administration), 60, 135, 260
National Defense Act, 84, 129
National Defense Research Committee (NDRC), 46, 180, 200, 206, 223, 248
   Division 17.3, 212, 216 (see also PAL)
National Science Foundation (NSF), 53, 58–59, 60, 212, 219, 221, 258, 261, 282–283
Navy, U.S., 52, 59, 60, 78–80, 83, 116, 128, 199, 218. See also ONR; Special Devices Center
NDRC. See National Defense Research Committee
Neisser, Ulric, 19, 212, 222, 235
Neural networks, 356–357, 360. See also McCulloch-Pitts neural network theory

Neuromancer, 303, 309, 346–350
Newell, Allen, 209, 227, 229, 230, 231–232, 250–255, 260, 265
Rand air defense simulation project, 122–124
Nixon, Richard, 276–277, 282–283, 298
NORAD (North Atlantic Air Defense Command), 22, 25, 97, 107, 206, 285, 290, 326, 328–330
Northrop Aircraft Corporation, 60, 61, 69, 92
NSF. *See* National Science Foundation
Nuclear strategy, 1, 58, 88, 89, 138, 290. *See also* Rand Corporation; Air Force
and closed-world discourse, 15, 125, 312
counterforce, 119, 141, 279
flexible response, 119, 132–133
as game, 14, 113, 117–120, 328–331
massive retaliation, 12, 119, 132
mutual assured destruction, 293, 319
prompt use, 84–86, 110
simulation and, 14, 118–121, 125, 312, 329
systems analysis and, 113, 116–121
NVA (North Vietnamese Army). *See* Vietnam War

Office of Naval Research (ONR), 43, 48, 116, 301
computer development and, 67–68, 75, 78–79, 81–82, 91, 249, 259–260, 268–269 (*see also* SAGE)
and post-WWII science, 59–60
psychology and, 212, 219–220, 221, 227
pure vs. applied research, 59, 63, 79, 221, 259–260
Office of Scientific Research and Development (OSRD), 46–47, 52, 215, 261, 271
Office of Systems Analysis (OSA), 113, 127–130, 132, 133–134, 137.

See also McNamara; Enthoven; Hitch
role in Vietnam War, 140–141
Office of the Secretary of Defense (OSD), 84, 93, 127–130, 133, 260, 269
and Vietnam War, 5–6
ONR. *See* Office of Naval Research
Operation Igloo White. *See* Igloo White
Operations research (OR), 126, 127, 143, 219, 221
and AI, 259–260, 271
Rand Corporation and, 113–121
vs. systems analysis, 116–117
Oppenheimer, J. Robert, 94
OSA. *See* Office of Systems Analysis
OSD. *See* Office of the Secretary of Defense
OSRD. *See* Office of Scientific Research and Development

PAL. *See* Psycho-Acoustic Laboratory
Paradigm, 33, 34, 37, 39, 153
AI, 294–295
behaviorist, 162–163, 209
cognitive, 178–179, 209, 212, 216, 221, 222, 228–229, 233
and concept of discourse, 30–32, 40, 41
cybernetic, 181, 185, 193–196
digital computation as, 68, 82
in nuclear strategy, 115, 117
Partridge, AF General Earle, 87, 96, 97
Persian Gulf War, 353–356, 357, 360
Pinch, Trevor, 33–34, 70
Pitts, Walter, 188–189, 191, 198, 203, 240, 263. *See also* McCulloch-Pitts neural network theory
Planning Programming Budgeting System (PPBS), 5–6, 128, 130, 132, 134, 143. *See also* OSA
Pribram, Karl, 212, 222, 230–233, 255
Project Charles, 93–95, 262. *See also* SAGE

Project MAC, 269–270
Project Whirlwind, 63, 75, 82, 90–92, 97, 99–101, 103, 222, 227. *See also* SAGE; Whirlwind computer
Psycho-Acoustic Laboratory (PAL), 180, 205, 209, 227, 230, 235–237, 240, 248, 254, 262, 263, 267, 268.
See also Cognitive psychology
chain of communication, 215–216
combat noise control, 210–215
language engineering, 216–218
Miller, George and, 222–227
post-WWII era, 219–222

Radar, 66, 117, 123–124, 218, 250, 285. *See also* SAGE, Air Force
and analog technology, 67–69, 87, 101
and Cold War strategy, 55, 57, 85–86, 87, 88–89, 107, 319
role in SAGE system, 91–94, 96–97, 101, 104, 108
tracking and cybernetics, 182, 185, 193, 216
WWII and, 47, 51, 67, 90
Radiation Laboratory. *See* MIT
Rand Corporation, 7, 113, 115–125, 127, 130, 193, 222, 229, 235, 250, 255, 316
air defense simulation and training, 122–125
and air defense strategy, 86, 96
game theory and, 114, 116–119
Johnniac computer, 122, 124
and nuclear strategy, 1, 85–86, 116–121, 132, 312
role in computer development, 61, 68, 103, 121–125, 251, 270
and U.S. strategy in Vietnam, 141, 143
Reagan, Ronald, 275, 283, 299, 339
closed-world discourse and, 275, 298, 354
and renewed Cold War, 278, 279–281, 290, 301
and Strategic Defense Initiative, 1, 7, 288–293, 298

Rees, Mina, 67–68
Research Laboratory of Electronics (RLE). *See* MIT
Return of the Jedi. See Star Wars trilogy
Rosenblueth, Arturo, 180–187, 189–191, 193, 207, 226, 230, 240, 264
Rusk, Dean, 136, 277

SAC. *See* Strategic Air Command
SAGE, 62, 69, 75–111, 114, 118, 122, 127, 226, 261–262, 266, 319, 320.
See also Closed world; Air Force; Radar; Cold War
Air Force ambivalence toward, 94–99, 106–111
analog alternative to, 96–97
as computerized command-control system, 79–82, 90, 93, 106, 107–111, 142, 143, 205
influence on computer development, 99, 101, 124–125, 132, 249, 264, 265
influence on computer industry, 101–103
political symbolism of, 104–111
Project Whirlwind and, 76–82, 90–92, 97, 99–101, 103
role of civilian scientists in, 81–83, 90–95, 98–99, 110–111
and SDI, 288–290, 293, 308
Schaffer, Simon, 33, 36–37
SCI. *See* Strategic Computing Initiative, DARPA
SDC. *See* Systems Development Corporation
SDI. *See* Strategic Defense Initative
Second self, 19, 21, 172, 187, 259, 267, 312, 314, 323. *See also* Cyborg
Servomechanisms, 66, 76. *See also* Antiaircraft weapons
in antiaircraft guns, 45, 67
as metaphor in cybernetics, 175, 180–182, 184, 187, 191, 193–199, 203, 231
Servomechanisms Laboratory. *See* MIT

Shakespeare, William, 12–13, 346
Shannon, Claude, 67, 188, 207, 224, 229, 240, 263, 265
 and Dartmouth AI conference, 253–254
 at Macy Conferences, 191, 199–205
 communication theory of, 66, 114, 180, 200–205, 214, 225
 Weaver, Warren and, 202–203
Shapin, Steven, 33, 36–37
Simon, Herbert, 114, 124, 209, 227, 229, 230, 231–232, 250–255, 260, 265
Skinner, B.F., 162–164, 179, 219, 225
Special Devices Center, U.S. Navy, 63, 76, 78, 80, 226–227. *See also* ONR; Project Whirlwind
Sputnik, 129, 134, 260
SRL. *See* System Research Laboratory
Star Wars (film trilogy), 2, 297, 303, 311–312, 315, 332–338
Stevens, Stanley Smith (S.S.), 212, 219–220, 223, 225
Strategic Air Command (SAC), 6, 22, 57, 85–86, 87, 95, 107, 110, 118–119, 126, 127, 133, 315, 318
Strategic Computing Initiative (SCI), 20, 25, 64, 71, 275, 293–301, 337
Strategic Defense Initiative (SDI), 1, 275, 288–293, 298–299, 301, 303, 308
Stroud, John, 196–199
Subject positions, 2, 27, 148–150, 165, 170, 173, 340. *See also* Closed world; Cyborg discourse
 definition, 304–306
 discourses and, 40
Subjectivity. *See* Closed world; Closed world discourse; Cyborg discourse
System Research Laboratory (SRL, Rand Corp.), 122–123
Systems analysis, 5, 15, 113–114, 133, 143. *See also* Rand Corporation; Office of Systems Analysis
 definition of, 116–117
Systems Development Corporation (SDC), 103, 107, 125, 249, 261–262, 269, 270
Systems Research Laboratory (SRL), 218–219, 227

Terminator 2: Judgment Day, 315, 357–364
Terminator, The, 22–27, 309
Tools. *See also* Computers
 computers as, 7, 21, 27–30, 37, 70, 71, 113, 133, 143, 147, 169, 241, 337
 and discourse, 27–41
 language as tool, 36–37
 metaphors, 14, 28–30, 37
Tron, 331–332, 347
Truman Doctrine. *See* Containment
Truman, Harry, 9, 59, 86, 136, 271
 defense policies of, 5, 8, 10–12, 52, 54–55, 88–89, 95, 129
 and nuclear strategy, 58, 88–89
Turing machines, 15–18, 78, 114, 184, 187, 200, 241–244. *See also* McCulloch-Pitts theory; Artificial intelligence
Turing test, 16, 18, 159–160, 243–244, 315
 Blade Runner and, 343
 and multiculturalism in science fiction, 339
 as support for cyborg discourse, 18, 21–22, 39, 159–160, 161, 166, 186, 303
Turing, Alan, 19, 188, 192, 199, 246, 250, 251, 322
 Colossus computer, 17, 62
 WWII code-breaking, 17–18
Turkle, Sherry, 19, 21, 28, 160, 161, 165–167, 169, 171–172, 267, 343. *See also* Second self

UNIVAC, 60–62, 79

Valley, George, 69
Valley Committee, 90–94

Van Creveld, Martin, 5–7, 130–131, 141, 145
Vandenberg, AF General Hoyt, 84, 89, 95
Vietnam War, 3–7, 15, 40, 72, 107, 114, 127, 136–145, 275, 277, 279, 301, 321, 355. *See also* Igloo White
  computers and, 15, 138–145, 282
  information pathologies of, 6, 141
  North Vietnamese Army (NVA), 139–143, 145
Von Foerster, Heinz, 193–196
Von Neumann, John, 124, 195, 199, 200, 225, 226, 240, 262
  and Alan Turing, 243
  computer/brain analogy, 190, 191–192, 194
  EDVAC design, 50–51
  ENIAC project, 50–51
  and IAS computer, 61, 122
  and John McCarthy, 253
  and Macy Conferences, 187–194
  and Rand Corporation, 116, 117, 122

War Games, 303, 328–331
Weaver, Warren, 202–203, 204. *See also* Shannon, Claude
Weizenbaum, Joseph, 28–30, 121, 172, 234
Westmoreland, General William, 43, 65, 71, 72, 114, 138, 144–145. *See also* Vietnam; Electronic battlefield
Whirlwind computer, 60, 62, 77–79, 91–92, 97, 99–103, 248–249, 256–257. *See also* Project Whirlwind; SAGE
Wiener, Norbert, 45, 225, 233, 240, 250, 263, 272, 357. *See also* Cybernetics; Information theory
  and antiaircraft weapons, 45
  "Behavior, Purpose, and Teleology" (1943), 181–187, 193, 207, 226, 230–231, 264
  and cybernetics, 114, 180, 225, 226

and information theory, 66–67, 114, 184
and Macy Conferences, 180, 181, 187–207
Wiesner, Jerome, 91, 263
Willow Run Research Center. *See* Air Defense Integrated System
Winograd, Terry, 170, 221, 294–295
Wittgenstein, Ludwig, 33, 34–37, 38, 40, 156, 157
Wohlstetter, Albert, 116, 118–119, 125, 127
Women, 249, 342. *See also* Gender
  as human computers, 45–46
  in military forces, 25, 139
World War II, 83, 90, 94, 126, 130
  and computer development, 44–51
  and computers, 62, 68, 243–244
  and cybernetic theory, 45, 175, 180–188
  effect on science, 47–48, 58–59, 176–178, 271
  operations research in, 115, 119
  and origins of cognitivism, 2, 20, 147, 175, 177, 188, 205–219, 271–273
  transition into Cold War, 10, 52, 53–57, 85
Worldwide Military Command and Control System (WWMCCS), 6, 107, 108, 206, 270, 286, 290

Zaccharias, Jerrold, 94–95